P9-CLS-762

HAWKING
HAWKING

ALSO BY CHARLES SEIFE

*Virtual Unreality: Just Because the Internet
Told You So, How Do You Know It's True?*

Proofiness: The Dark Arts of Mathematical Deception

*Sun in a Bottle: The Strange History of Fusion
and the Science of Wishful Thinking*

*Decoding the Universe: How the New Science
of Information Is Explaining Everything in the Cosmos,
from Our Brains to Black Holes*

*Alpha and Omega:
The Search for the Beginning and End of the Universe*

Zero: The Biography of a Dangerous Idea

HAWKING HAWKING

THE SELLING OF A SCIENTIFIC CELEBRITY

CHARLES SEIFE

BASIC BOOKS
New York

Copyright © 2021 by Charles Seife

Cover design by Chin-Yee Lai
Cover image copyright © David Montgomery via Getty Images
Cover copyright © 2021 by Hachette Book Group, Inc.

Hachette Book Group supports the right to free expression and the value of copyright. The purpose of copyright is to encourage writers and artists to produce the creative works that enrich our culture.

The scanning, uploading, and distribution of this book without permission is a theft of the author's intellectual property. If you would like permission to use material from the book (other than for review purposes), please contact permissions@hbgusa.com. Thank you for your support of the author's rights.

Basic Books
Hachette Book Group
1290 Avenue of the Americas, New York, NY 10104
www.basicbooks.com

Printed in the United States of America

First Edition: April 2021

Published by Basic Books, an imprint of Perseus Books, LLC, a subsidiary of Hachette Book Group, Inc. The Basic Books name and logo is a trademark of the Hachette Book Group.

The Hachette Speakers Bureau provides a wide range of authors for speaking events. To find out more, go to www.hachettespeakersbureau.com or call (866) 376-6591.

The publisher is not responsible for websites (or their content) that are not owned by the publisher.

Adapted and redrawn figures are thanks to Shanthi Chandrasekar.

Print book interior design by Jeff Williams.

Library of Congress Cataloging-in-Publication Data

Names: Seife, Charles, author.
Title: Hawking Hawking : the selling of a scientific celebrity / Charles
    Seife.
Description: First edition. | New York : Basic Books, 2021. | Includes
    bibliographical references and index.
Identifiers: LCCN 2020038838 | ISBN 9781541618374 (hardcover) | ISBN
    9781541618381 (ebook)
Subjects: LCSH: Hawking, Stephen, 1942–2018. | Physicists—Great
    Britain—Biography.
Classification: LCC QC16.H33 S45 2021 | DDC 530.092 [B]—dc23
LC record available at https://lccn.loc.gov/2020038838

ISBNs: 978-1-5416-1837-4 (hardcover); 978-1-5416-1838-1 (ebook)

LSC-C

Printing 1, 2021

CONTENTS

PART III INSPIRAL

PROLOGUE

The *Daily Mail* loved Stephen Hawking far more than Stephen Hawking loved the *Daily Mail*. Even by UK tabloid standards, the *Mail*'s science coverage was either laughable or infuriating, depending on your point of view. The paper's pages were always chock-full of headlines about scientific research—research often hyped by the *Mail*'s writers almost beyond the point of recognition. All the better to grab the attention of an audience.

Nobody could grab an audience like Stephen Hawking, so his name regularly graced the tabloid pages. Usually not in a flattering light. The professor was typically either a harbinger of doom—warning of imminent death due to global warming, robot rebellion, alien invasion, or other catastrophes—or he was at the center of some sort of scandal about his sex life or his marriages or abuse allegations. But in early 2018, just before Hawking's death, the *Daily Mail* broke new ground.

"Has Stephen Hawking Been Replaced with a 'Puppet'?" the headline read. "Conspiracy Theorists Claim the REAL Professor Is DEAD and a 'Puppet' Has Taken His Place—and Reveal the SIX Clues That Support the Idea."[1]

In a surprisingly long and detailed article, the tabloid set out evidence that sometime in the mid-1980s, the esteemed physicist had been replaced with an impostor. As outrageous as this theory sounds, the *Mail* explained how supposed anomalies in his appearance as he aged (particularly the look of his teeth), his unexpectedly long survival with a disease that typically kills in a couple of years, and a number of other clues suggested that the original Stephen Hawking had died and that a facsimile had been foisted on the public. "The voice we hear," said the article, "is the result of NASA astrophysicists typing information into a computer—information they want . . . to push on a gullible and unsuspecting public, fans of Hawking who hang on to—what they believe to be his—every word."

Even for the bizarre parallel reality conjured by tabloid writers, this was way out there. They had ventured into this sort of territory only once before, almost exactly fifty years prior. In 1969, the tabloid circuit lit up with rumors that the Beatles' Paul McCartney had been killed in an auto accident and replaced with a doppelgänger.

However, comparing Stephen Hawking to Paul McCartney doesn't quite capture the nature of Hawking's singular celebrity. Throughout all of history, there might be perhaps three or four scientists whose fame and renown among the public could be compared to Hawking's: Einstein, Newton, Galileo—maybe Darwin. For the media and for the public at large, Hawking had become the ultimate symbol of the triumph of the mind. He was the world's smartest man, an unmatched brain who spent his time unraveling the deepest mysteries of the universe.

The *Mail*'s suggestion that Hawking had been replaced by a simulacrum was just the most extreme and absurd version of how the press and the public had portrayed Hawking for decades. The professor's image had been built into a towering contradiction: On one hand, Hawking appeared to the world to be something more than human, his mind so transcendent that he was in a class by himself. He inhabited an intellectual plane above the realm of normal humanity. Yet, on the other hand, he could be treated as a nearly inanimate object. Hawking suffered from a neurological disease that slowly robbed him of the ability to move of his own volition, to speak except through a computer-generated voice. It was a short leap for a thoughtless person to imagine Hawking to be artificial, to be some sort of technology-assisted homunculus rather than a real human. As the *Daily Mail* so rudely put it, it was not even always possible to tell whether the voice emanating from his computer was truly under the control of the being sitting in the wheelchair.*

---

* The public treasured any encounter with Hawking that seemed to express his real, unfiltered personality. For example, he was notorious for his bullheaded indifference to other vehicles when driving his wheelchair on the streets around Cambridge University—and people found this utterly charming. When Hawking died in 2018, one of the viral discussions on Twitter was a long thread filled with stories of people who nearly ran him over with their cars.

By the time he died in 2018, it was almost impossible to discern Hawking the human underneath the layers of accumulated symbolism; from the public's point of view, he had become a caricature rather than a living person. Even though everyone who knew Hawking has described him as one of the most stubborn, willful people they've ever met, it was incredibly hard to distinguish his true will, to sense the authentic being underneath the public persona.

To understand Stephen Hawking, one has to turn back the clock. During the last third of his life, Hawking was firmly entrenched as the world's most celebrated living scientist, yet his actual scientific contributions were more or less irrelevant to his fame. Though he was a regular fixture in the media, the press attention wasn't usually related to his science. Hawking's research during the years of his greatest popularity would be largely discounted and have little lasting impact on the world of physics. He was like a collapsed star: space around him glowed brightly with his energy, but at core he was but a faint reflection of what he once had been.

Not long before, Hawking had been a supernova. The middle third of his life was a spectacular and brilliant transformation. Over the course of two decades, he transmogrified himself from a fairly obscure physicist, laboring with his colleagues (and his rivals) to understand the conditions at the very beginning of the universe, into an international celebrity. Into the world's smartest man. Into the scientific equivalent of The Beatles. It was a metamorphosis at once immensely satisfying and intensely painful. By the time it was complete, Hawking had broken with much of his past and constructed a myth to replace it.

Only in the first third of Hawking's life—before he achieved his status and fame—does the real human being behind the legend begin to appear. The backward flow of narrative time slowly restores Hawking to his primeval brilliance. By traveling back to Hawking's youth, one can understand how he came to the key scientific insights upon which he built his name. One can discern the roots of his need to become a famous communicator of science. And one can understand the mortal fears of a young man racing to establish a legacy—and a family—as a deadly disease appeared poised to claim him at any moment.

Unlike a scientific discovery, which becomes easier to understand as time moves forward and as more and more researchers contribute their knowledge, the life of Stephen Hawking becomes clearer as time moves backward, as the accumulated layers of celebrity and legend are stripped away. In the end, Stephen Hawking the human being becomes something very different from the Stephen Hawking so beloved by the public.

Public Hawking was famous for being the world's smartest man, the apex of scientific intellectualism. Human Hawking was brilliant, but he knew he was surrounded by equally brilliant peers who labored in semi-obscurity. Public Hawking was the world's greatest science communicator. Hawking the human had more difficulty communicating than just about any other person on the planet; by the time he was famous, he could only express a few words per minute, if that. Public Hawking stoically shrugged off his physical ailments as a mere inconvenience. Human Hawking's disability had, quite naturally, shaped every part of his existence: his outlook, his science, his family life, and eventually his fame. To the public, everything Hawking did was extraordinary and different and courageous—it was a spectacle when he spoke, when he ate, when he danced, when he worked, when he loved. To Hawking, there was no courage in merely being himself.

Even Hawking's colleagues and rivals had to take pains to distinguish the human from the legend. "I'm not speaking of him as a pure intellectual who rides through the universe on his magic wheelchair," says Leonard Susskind, a physicist at Stanford University who battled Hawking about the properties of black holes. "I'm speaking of him as a human being. You know, none of us were ever really able to know him."[2]

Turn back the clock, and what emerges is a real human: petulant, arrogant, and callous as well as warm, witty, and brilliant. Complex. Fascinating. Singular.

What emerges is Stephen Hawking.

PART I

RINGDOWN

O sages standing in God's holy fire
As in the gold mosaic of a wall,
Come from the holy fire, perne in a gyre,
And be the singing-masters of my soul.
Consume my heart away; sick with desire
And fastened to a dying animal
It knows not what it is; and gather me
Into the artifice of eternity.

—WILLIAM BUTLER YEATS,
"Sailing to Byzantium"

CHAPTER 1

NEXT TO NEWTON (2018)

———————

Only thrice in the past hundred years has a scientist been buried in Westminster Abbey. There was Ernest Rutherford, who figured out the structure of the atom; there was J. J. Thomson, who discovered the electron; and then there was Stephen Hawking.

On June 15, 2018, Hawking's ashes were interred in the floor of the cathedral, laid to rest underneath a slate-black stone just a few feet away from the graves of Isaac Newton and Charles Darwin.

Hawking publicly disavowed any comparison to Newton—whenever anyone made such a suggestion, he would rubbish it as "media hype." Yet the public loved to make the connection: like Newton, Hawking was the most famous physicist of his day; at Cambridge, Hawking occupied the Lucasian chair, the very office that Newton had held three hundred years before him; Hawking and Newton had both devoted much of their lives to understanding the mysteries of the gravitational force. Even in death, Hawking could not escape the association with Newton. Not only are the two scientists buried within a few strides of one another, but they have the same epitaph. Newton's black tombstone is featureless but for an inscription in Latin: "Hic depositum est quod mortale fuit Isaaci Newtoni." Hawking's bears the same words in English, with his name substituted for Newton's: "Here lies what was mortal of Stephen Hawking."[1]

Though smaller than Newton's gravestone, Hawking's is more elaborate. The epitaph curves gently around a set of swirls engraved into the slate, swirls that seem to move toward an elliptical void: clouds of gas falling into the maw of a black hole. To the left, there is an equation whose letters seemingly defy gravity:

$$T = \hbar c^3/8\pi GMk$$

Almost nobody who visits the gravestone understands what those symbols mean. But to Stephen Hawking, that equation was the key to transcending the mortal.

· · ·

Until he died in 2018, Hawking was one of the most recognized human beings on the planet—and probably the easiest to spot. Almost unable to move his body, seated awkwardly in his motorized wheelchair and accompanied by an entourage of nurses, he was unable to go anywhere incognito. Not that he cared to.

The public adored Hawking without knowing precisely why. Einstein had his theory of relativity and Newton his universal gravitation, but the vast majority of people who admired Hawking knew little about what he had done to deserve his reputation. Nor did they understand why, in the press, he was always compared to Einstein or Newton, a comparison he modestly rejected but at the same time worked very hard to cultivate. And even those who had a glimmer of Stephen Hawking's science saw only a tiny fraction of what made Hawking Hawking. For he was not just Hawking the physicist, Hawking the celebrity; he was Hawking the showman, Hawking the husband and father, Hawking the symbol.

These facets warred among themselves: the very moment Hawking achieved celebrity, his marriage collapsed and his family shattered. Hawking the human depended on his students to be caretakers and nursemaids even as Hawking the physicist wanted to groom them to be his intellectual offspring. He was perhaps the most celebrated communicator of physics in the world, yet he had extraordinary difficulty making himself understood. Even the most straightforward-seeming element of his persona, his ability as a first-class physicist, is much more complicated than it seems at first. Scientists viewed Hawking as a mind of a very high order—but at the same time, many rolled their eyes at some of his later work, trashing it as all but worthless. The real Hawking lies underneath this complex skein of tangled and contradictory narratives.

As with the black holes that he studied, there are incredible forces that prevent outsiders from glimpsing Hawking's inner self. But there is a real person who lies beyond the event horizon of his celebrity.

That singularity contained legion: an important scientist whose importance is almost universally misunderstood; a person who suffered deeply and also caused deep suffering; a celebrity scientist who broke the mold of his forebears and fundamentally changed the concept of a scientific celebrity.

Most people who know anything of Hawking are blinded by a flash from his life, an image of a tumultuous decade—from 1980 through 1990—when he transformed himself from a well-respected but obscure scientist in a neglected corner of physics into one of the most recognized names on the planet. But like a supernova that briefly outshines the rest of its host galaxy, Hawking's celebrity both attracted attention and distracted it—simultaneously inviting the gaze of countless millions and hiding the star itself, a quivering and naked object that shed everything that once clung to it.

• • •

At the abbey's Great Lectern, Benedict Cumberbatch—an actor who had portrayed Hawking in one of the many movies, plays, and television shows about the physicist's life—read a passage from the Wisdom of Solomon:

> For it is he who gave me unerring knowledge of what exists,
> to know the structure of the world and the activity of the elements;
> the beginning and end and middle of times,
> the alternations of the solstices and the changes of the seasons,
> the cycles of the year and the constellations of the stars.

Unerring he was not, but Stephen Hawking had made it his profession to try to understand the beginning and the end of the universe. When he began his research in the early 1960s, his field, cosmology, was a sleepy backwater, an area of study that hadn't seen a substantial advance for decades. By the time he died, it was arguably the most exciting field in physics, an area that was (and still is) generating Nobel Prize after Nobel Prize for transforming our understanding of how the universe came to be.

Hawking's first substantial piece of research was an important discovery about the beginning of the universe. At the time, in 1965, there were two competing models of how the cosmos was born: whether it was eternally renewing itself, or whether it had been born in a gigantic explosion now known as the Big Bang. For his PhD thesis, Hawking proved that if the universe began with a Big Bang, then it had to have started as a singularity: a point where the laws of physics no longer make sense, an infinitesimal but infinite blemish on the fabric of space and time. A place where mathematics itself breaks down. This was a stunning insight; if one believed in the Big Bang, one had to accept that the laws of physics as we know them are inadequate to describe the birth of our cosmos. This idea—now known as the singularity theorem—ignited Hawking's career.

As he grew in confidence and stature, Hawking became a key figure in solidifying what is now the dominant theory of how the very early universe expanded, a theory known as *inflation*. But Hawking himself thought his most important contribution to cosmology was his work on an ambitious, radical, and controversial theory in which he attempted to calculate the quantum-mechanical "wavefunction of the universe." Not only did Hawking believe that his theory described the very starting point of space and time in our cosmos, but he was convinced that it did so in a way that made God unnecessary. "What place, then, for a creator?" he asked, much to the chagrin of many theologians (and some scientists) around the world.[2]

However, Hawking's most important scientific work wasn't about the birth of our universe or its wavefunction, but about a different kind of singularity: the singularity at the heart of a black hole. Hawking devoted much of his life to understanding how these mysterious objects behaved, and in the most important moment in his scientific lifetime, he realized they had a bizarre property that nobody else had imagined they might possess.

Black holes are astronomical objects whose gravitational attraction is so powerful that nothing venturing too close can escape—not even light. Black holes are born when a large star dies; when the fusion engine at its heart shuts down, it collapses under the force of its own gravity. In a fraction of a second, the entire weight of a star bears down on itself, first crushing the matter into an undifferentiated glob of atoms, then crushing the atoms themselves—and then, finally, it becomes a singularity. Because the gravitational pull around the collapsed star is so great, nothing can venture close to that singularity and escape to tell the tale. It's as if the former star is now surrounded by an invisible shroud marking the point of no return: cross this so-called event horizon, and you are doomed, unable to return home, destined to fall into the black hole no matter how hard you struggle.

Because black holes swallow light, they are as black as black can be; they are the ultimate absorbers, gobbling any illumination rather than reflecting it. But in the 1970s, Hawking had a surprising realization: black holes aren't really perfectly black after all. They radiate particles, including light particles, in all directions. Under most circumstances, this radiation—now known as *Hawking radiation*—is incredibly weak, far too weak to be detected at any reasonable distance. However, just the fact that the radiation exists had some profound implications. Because if a black hole radiates energy, this means it must eventually evaporate—explode—in a burst of radiation. This, in turn, implies that the matter and energy swallowed by the black hole must eventually be released. And, as Hawking was the first to realize, the release of that matter and energy leads to a seemingly irreconcilable clash between the two mainstays of modern physics: Einstein's theory of relativity and quantum theory. The discovery of Hawking radiation not only upended the conventional wisdom about black holes, but seemed like a major milestone in the quest to resolve the conflict between the two theories. Perhaps he could even replace them with an overarching "theory of everything."

"I would say, in retrospect, [Hawking] has three great contributions to science. One is the singularity theorems," says John Preskill, a physicist and friend of Hawking's. "And one is the idea about the wavefunction of the universe. But the most important, by far, is the discovery of Hawking radiation and its implications."[3]

The equation inscribed on Hawking's tombstone is the main formula for Hawking radiation—the temperature of a black hole as a function of its mass, which, in turn, dictates the amount and type of radiation that it emits—all superimposed on the black hole that it describes.

• • •

As the grand organ in Westminster Abbey swelled, hundreds of voices sang in unison to an old English melody:

> Father, hear the prayer we offer:
> not for ease that prayer shall be,
> but for strength that we may ever
> live our lives courageously.

At the age of twenty-one, Stephen Hawking was diagnosed with a deadly neurological disease, amyotrophic lateral sclerosis (ALS), and given two or three years to live. For the next fifty-five years of his life, he lived under a death sentence, uncertain of whether he would see his next birthday. Everything that Hawking did in his life—discovering new physics, writing best-selling nonfiction books, jet-setting around the world, raising three children—was done against the backdrop of a relentless ailment that robbed him of his ability to walk, to write, to speak, to feed himself, to move almost any of the muscles in his body. Yet Hawking bristled when anyone conflated his perseverance in the face of his disease—his almost pigheaded stubbornness—with bravery. "I find it a bit embarrassing in that people think I have great courage," he told a reporter in 1990. "But it's not as if I had a choice and deliberately chose a difficult path. I have just done the only thing open to me in the situation."[4]

By the time he published *A Brief History of Time*, the runaway best seller that made him a household name in the late 1980s, Hawking used a motorized wheelchair; he was unable to talk and almost entirely unable to move. His only means of communication was through a computer that he controlled with a rocker switch in his hand; he would laboriously assemble a sentence with that switch and send it to a speech synthesizer that would pronounce his words.

At events, people would gather around Hawking and not quite know how to interact with him, treating him with a curious mix of awe and condescension. Intellectually, they knew he was one of the world's most prominent physicists. Yet his disability was so profound that they almost reflexively treated him like a toddler, cooing with delight and amazement at whatever he said or did. In 2011, Jane Fonda wrote about her visit with Hawking:

> I dropped to my knees next to Stephen's chair, reminding him what Beethoven had said and asked him if, like Beethoven, his disease had enabled him to go further, deeper in his understanding of his research—of the origins of the universe. . . . As I waited, I rested my head on his shoulder, looking closely at him, at the subtle movements in his face as he concentrated on what he was "writing." And all I could think about was that this man, imprisoned in a wasted body, was able to comprehend matters that are presumed to lie far outside the realm of human understanding.
>
> After about 5 minutes, letters and then words began to slowly appear on the screen: "It . . . freed . . . me . . ." Ah haaa!! [Director] Moises [Kaufman] and I looked at each other in delight, certain that our hypothesis was about to be proven—sure that Stephen was about to say something like, "it freed me to grasp the origins of the universe . . ." We waited for the sentence to be finished, another few minutes . . . and then, there it was: "It freed me to stop teaching!!!" and a computerized voice said it aloud so everyone heard. I looked at Stephen and noticed what appeared to be a sly grin. I'd been told he had a playful sense of humor. He had just demonstrated it! And we all had a good laugh. He didn't have to teach anymore!!! That's what ALS had done for him. Of course!!![5]

Almost automatically, visitors would imbue Hawking not just with deep wisdom, but also with childlike simplicity. He was becoming a guru, a symbol. A metaphor—and one that was almost too perfect. Immobile in his wheelchair, Hawking was a being of pure intellect, a man whose powerful mind allowed him to travel to domains where nothing else in the universe could venture.

Hawking was well aware that the myth, the archetype, was powerful enough to swallow all traces of the real human being underneath. "That Stephen is some sort of pure mind because of his disability, I think that hurt him a lot," says Christophe Galfard, one of the many PhD students Hawking advised over the years. "The man as a scientist, not just the image of the scientist, was somehow diluted." So Hawking struggled to prevent that from happening. In his home life, Hawking refused to discuss, much less make concessions to, his disability, almost to a pathological extent; this obstinacy became a sore spot in his first marriage. As a physicist, he tried to produce ideas of such depth and importance that his physical impediments would be seen as irrelevant. "I would like to be thought of as a scientist who just happens to be disabled, rather than as a disabled scientist," he would say. Yet throughout his life, Hawking suspected that people saw his disability as a mitigating factor, a reason to judge him differently from other physicists. Or worse, as something that would come to define him. His fears were well founded.[6]

The disability was central to Hawking the Metaphor, even if it was in many ways peripheral to Hawking the Human. And as much as he wanted people to see beyond his condition, Hawking realized, to his chagrin, that his disability was right at the core of his public persona.

· · ·

In addition to the family, friends, and notable personages at Hawking's interment ceremony, 1,000 members of the public—drawn by lottery from a pool of more than 25,000—packed into the abbey to pay their respects. Those in attendance were surprised to hear Hawking's electronic voice reverberate through the church one last time, backed by a slow, meandering synth-jazz piece written for the occasion by the composer Vangelis. "I am very aware of the preciousness of time. Seize the moment. Act now. I have spent my life travelling across the universe inside my mind."[7]

Simultaneously, a 100-foot-wide radio telescope in Spain beamed the very same words, accompanied by the very same music, into space—aimed at a nearby black hole dubbed 1A 0620-00. For the next 3,500 years, Hawking's words will be winging their way at the speed of light toward their destination, and toward their destruction.

Hawking would have appreciated the spectacle.

Long before *A Brief History of Time* made him an international celebrity, Hawking loved showmanship and had a natural talent for grabbing the spotlight. Even his fellow scientists could fall under his spell. Physicist Lenny Susskind, who battled against some of Hawking's ideas about the workings of black holes over the years, would tell of how Hawking could capture the attention in a room, how all conversation would suddenly go quiet as the gathered physicists realized that Hawking was going to say something. And then there would be this long pause as he crafted his answer on the computer, and the tension would build, often to be deliberately deflated several minutes later with a simple yes or no answer. Or it could be a devastating bon mot: "He was a very, very witty man. With one word, he could puncture . . . one funny comment could just explode in your face and you'd wind up laughing." With the public, he was even more effective: people left a Hawking lecture inspired and awed.[8]

Hawking was not just a master of grabbing the spotlight; he knew precisely how to use it once it was on him. Even as a graduate student he created a splash—and started building his image as a towering intellect—by engineering a public confrontation with the world's foremost astrophysicist of the time. And when he was at center stage, he used his wicked, self-deprecating sense of humor to win goodwill and to give the impression of deep humility: no mean feat given that even his closest friends and colleagues have described him as arrogant and superlatively stubborn.

Hawking's interviews, his travels, his talks, his writings—even much of his advocacy—all helped reinforce his role as the foremost scientific celebrity of our time. The photographs of Hawking on his various adventures—wheeling his wheelchair in the Antarctic wastelands, or grinning as he floats free of the bonds of gravity in a zero-g airplane flight—were iconic, reproduced in newspapers and websites across the world. His every public utterance would wind up being repeated by the media, even when his ideas were half-baked or alarmist. Later in his career, Hawking would warn humans of possible demise via asteroid or through a robot revolt, or even by a catastrophic change in a physical constant.

While Hawking was an extraordinary showman, his craving for recognition came with a price. At the very moment that he achieved worldwide fame, his first marriage fell apart under the pressure and his relationship with his children buckled under the strain. Though he was always surrounded by a bustle of nurses and fans, Hawking was often alone.

The singularity that is Hawking managed to garner an almost otherworldly status. He was considered to be an intellect in the company of Einstein and Newton and Galileo; a prophet of science who had climbed

Mount Sinai and then descended to share divine wisdom with the people; a philosopher-guru whose insights could inspire future generations; a man who refused to let a profound disability limit his achievements. It's a powerful image.

But even to those who've seen behind the public persona and know all of Hawking's flaws and foibles, there was something unique, something profoundly inspiring, about the man. Ray Laflamme, one of Hawking's former students—and caretakers—has a photo from Hawking's famous zero-gravity flight on the wall of his office. In the middle of talking about his former mentor, he pauses. "I have lung cancer, and the prognosis for lung cancer is not so great. But I'm alive right now, thank you to modern medicine," he says. Then he gestures to the photo on the wall, Hawking grinning broadly as he floats in midair. "That's why I have that picture there. And when I get discouraged, I go back and I say that bugger can make it for fifty years. I'm gonna ask only for twenty-five."[9]

CHAPTER 2

RIPPLES (2014–2017)

—————————

W hen the alarm clock went off at 2:40 a.m., Barry Barish swallowed his disappointment. "I assumed they had passed us over," he later recalled. But then his cellphone rang.[1]

The early-morning phone call is one of the clichés about getting a Nobel Prize. A scientist is supposed to awaken, blearily, to the realization that he or she is suddenly going to be an instant celebrity. It's supposed to be stunning, humbling—and above all, it's supposed to be a surprise. The first Monday evening in October is a restless night for many physicists of great renown, but never for a moment does even the greatest-egoed among them go to bed fully expecting to win the Nobel the next morning.

Except this time. October 3, 2017.

Two years earlier, a novel telescope had made a discovery so important that a Nobel was not only assured, but would come at the earliest possible opportunity. The only question was how the prize would be divvied up. Though there were hundreds upon hundreds of people who had worked on the telescope and its observatory, the rule was that a prize could be split at most three ways.

Barish, the director of the observatory, went to bed confident. Rainer Weiss at the Massachusetts Institute of Technology (MIT), who had spent decades designing the machine, was a bit more humble; he went to bed that evening thinking he had only a 20 percent chance of waking up to a Nobel. The third person who fell asleep that night in anticipation of a Nobel the next morning was California Institute of Technology

(Caltech) theorist Kip Thorne, Hawking's colleague and close friend for five decades.[2]

Like Hawking, Thorne had devoted his life to studying black holes, gravity, and time, and this new telescope was about to shed new light on precisely these subjects. For the telescope that was about to win Thorne, Barish, and Weiss the Nobel Prize was not the ordinary sort of telescope that gathers light from distant stars. Instead, this telescope, the Laser Interferometer Gravitational-wave Observatory (LIGO), was a tool designed to detect not light, but gravitational waves from colliding black holes. And with this tool, Thorne was able to begin testing the theories that Hawking, Thorne, and other physicists had developed in the late 1960s and early 1970s—a time so rich in theoretical discovery that Thorne dubbed it the Golden Age of Black Holes. "Among the nicest features of the Golden Age was the way we all built on each other's work," Thorne wrote. "Hawking laid the foundations, and one after another his compatriots built an edifice upon them."[3]

With his Nobel, Thorne belatedly fulfilled a promise that he had made on Hawking's sixtieth birthday fifteen years earlier: "I'm afraid it is more in the form of a promissory note than a concrete physics result," he had said. "Your birthday gift is that our gravitational-wave detectors [including LIGO] will test your Golden-Age black-hole predictions, and they will begin to do so well before your 70th birthday. Happy Birthday, Stephen!"[4]

• • •

By the time he reached his seventies, Stephen Hawking was the world's most famous living scientist, and had been for several decades. His first guest appearance on the animated show *The Simpsons*—a reliable indicator of apex pop-culture status—had happened nearly twenty years prior. And Hawking's most important research was two generations in the past.

In the last decades of his life, Stephen Hawking's fame was not really that of a scientist, but of a cultural icon. Though his celebrity was mantled in science, science had become almost incidental to Hawking's notoriety. It was mostly irrelevant whether his latest pronouncements about physics were valuable or not; just the fact that he would make them every so often was more than enough to maintain his status as an icon. The public simply didn't care all that much about the details of Hawking's scientific achievements or ideas. Yet Hawking wanted to be famous for his physics, not for his personality, or his condition, or anything else.

In the late 2010s, Hawking had hope that this would finally change. Hawking's work of thirty and forty years prior had suddenly become hot again. The most exciting physics of the day had to do with gravitational waves and black holes, areas of knowledge where Hawking's scientific work had had the most profound effect. After years upon years of labor, physicists around the world (including Hawking's best friend, Kip Thorne) were finally getting results from experiments that held the promise of testing a number of Hawking's decades-old predictions. If they confirmed some of his theoretical work, Hawking might finally achieve his wish to be known as a brilliant scientist first and a celebrity second. However, that dream would come to naught if he were left on the periphery when the Nobel committee came calling.

. . .

Stephen Hawking's research revolved around gravity; he spent much of his career trying to figure out how gravity behaves under some of the most extreme conditions that the universe can possibly produce, especially near black holes. And for the first time, the LIGO discovery began to give scientists a direct glimpse of matter being subjected to the enormous gravitational fields that Hawking had been studying for most of his life.

LIGO is so new at this point that scientists are still trying to understand precisely what the data can tell them, but it has opened an exciting new chapter in the story of gravity—a tale that starts a bit more than three hundred years ago.

The tale has a well-known introduction: in 1666, Isaac Newton, sitting under an apple tree, started to wonder why an apple plummeted to the ground rather than flying off into the air. From that moment, Newton began to solve a puzzle that philosophers had pondered for millennia, and the physicist's solution was as radical as it was powerful.

Newton built a theory that described a mutual attraction among all massive objects; his theory declared that any two bodies, no matter where they are in the universe, are somehow tugging on each other, invisibly, inexorably, and inexplicably. An apple on a tree feels the simultaneous pulls not just of every single other object on Earth, but also of every single star in the universe, no matter how distant. Newton's equations provided a beautiful and accurate description of the gravitational forces that act upon matter. Just how those forces work—what they're made of and how they can work at great distances—Newton couldn't explain. It would be another two and a half centuries before another physicist, Albert Einstein, could.

Einstein didn't set out to revolutionize our understanding of gravity. As a student and a young researcher, his work involved molecules and atoms and motes of dust rather than the motion of stars and planets; he was more concerned with the behavior of electric and magnetic forces than gravitational ones. But when he started picking at a loose thread in the equations that govern electromagnetic fields, he unwittingly unraveled everything that scientists thought they knew about gravity.

The equations that describe electromagnetic fields, known as Maxwell's equations, had a subtle flaw in them: motion mucks with the equations. Under certain circumstances, two observers moving at different velocities—say, one sitting still with the other zooming by in a train—could even get contradictory answers from the equations. One might predict that the particle in the experiment would be pushed to the right, while the other thought that it would be pulled to the left. Physics shouldn't work this way. The laws of physics—and the equations that encode those laws—should be the same for all observers, no matter how they're moving.

In 1905, Einstein realized that he could fix the flaw in Maxwell's equations by making a few changes in the underlying assumptions that physicists were working with. But the amended rules came with a lot of philosophical baggage. One of them, in particular—that a beam of light always travels at the same speed—seemed to violate common sense.

That is, Einstein was asserting that a light beam would always seem to pass an observer 186,282 miles per second, the quantity known by physicists as c. If you're sitting still, the light beam zooms by at c. If you're moving toward the source of the light at a million miles an hour, no matter; the beam zooms by at c. Even if you try as hard as you can to outrun the light beam, flying away at 99 percent of the speed of light, it won't change the outcome: the beam passes you at c, just as fast as it would had you been standing still. This makes no sense unless we tweak the way we think about speed.

Speed is just a measure of how much distance an object covers (such as miles) in a certain amount of time (such as per second). So any change to the concept of speed would automatically mean that something was wrong with our understanding of distance or time. Or, as it turns out, both. This was one of Einstein's greatest realizations.

Einstein understood that if all observers agree on the speed of light, it has to mean that those same observers will *disagree* about their measurements of distance and time. That is, a fast-moving observer's wristwatch will tick at a different rate (more slowly, in fact) compared to a stationary observer's. A stationary observer's yardstick will be a different size (larger) when he compares it to a fast-moving observer's.

This was a huge break with the way physicists thought about the universe before 1905. Science no longer held that the length of a yardstick was an objective fact. Observers moving in different ways will disagree about how long that yardstick is—and all can be right at the same time despite their mutually contradictory measurements. There was no "absolute" length, no right answer. Even stranger, time was no longer immutable, flowing at the same rate everywhere in the universe. There could be no "absolute" time, no way for everybody to agree about the moment at which an event happens. Time and space are malleable rather than rigid. This insight explains how all observers can see light zoom by at precisely c—the differences in their perceptions of time and space conspire to ensure that the speed of light is a universal constant.

Einstein's 1905 paper showed that time, length, and motion are linked in a way that Galileo and Newton had never dreamed of. The way one moves through space affects the way time flows as well as the distance between objects. Your speed changes not just *where* you are, but *when* you are. It was a heavy intellectual price to pay to repair Maxwell's equations. Yet these strange effects have been observed numerous times—fast-moving clocks slowing down and the like—in real life. It no longer made sense to talk about time and space completely independently; they are inextricably linked.

The concept of absolute space and absolute time were ingrained in the laws of motion. Newton's equations governing how objects moved when subjected to forces depended implicitly on the fact that time and distance were immutable. So when Einstein discarded the assumption of universal time, it meant that Newton's laws had to be subtly wrong. All of physics was broken on some level. Including the law of universal gravitation.

The first hint that something might be wrong with gravity came from another realization Einstein had in 1905: it's not just length and time that depend on the motion of the observer, but the mass of an object—its avoirdupois—also depends on motion. The faster it moves, the heavier it seems to be to a stationary observer. An object that is completely still will have less mass (its *rest mass*) than when it's in motion (its *relativistic mass*). By the end of 1905, Einstein had begun to understand the relationship between an object's energy, E, and its mass, m—which, of course, led to the famous formula $E = mc^2$.

Once you start fiddling with the nature of mass, however, that is naturally going to have consequences for gravitation. By 1907, Einstein realized that gravity is not a special, immutable property of massive objects. The same force can come about in other ways, through motion: when an elevator suddenly lurches upward and accelerates upward, you feel a force pulling you to the floor that can be thought of as a "gravitational" force akin to the Earth's downward pull. Like space and time, like mass and energy, gravity is tangled up in how objects and observers move about. Space, time, mass, energy, and gravity were caught up in a complex dance that made them impossible to separate from each other.

It took Einstein from 1907 through 1915 to turn that basic insight into a set of equations, a collection of mathematical rules describing that intricate dance. (In 2001, Stephen Hawking wryly noted that these break-throughs came as Einstein's first marriage was collapsing, leaving him estranged from his wife and children: "The fact that he spent the war years as a bachelor, without domestic commitments, may be one reason why this period was so productive for him scientifically."[5]) These *field equations* tie together mass and energy, space and time, and gravity into one package. Surprisingly, the field equations, which look like this—

$$G_{\mu\nu} = (8\pi G/c^4)\, T_{\mu\nu}$$

—are fundamentally a description of a smooth, curving sheet: in math-ematical language, a *manifold*. The right side of the equation encodes all the matter and energy in a region of space and time. The left side describes the curvature of that same region. And gravity is nothing more than a manifestation of that curvature. The mystery of the gravitational field, the questions about how distant objects invisibly pull upon each other over vast distances, is gone; gravity is just geometry.*

This is a mind-blowing concept that requires advanced mathematics to unpack fully, but there's a pretty good analogy that helps: Think of space-time as a mattress, or a rubber sheet, or some other springy surface. A dollop of matter and energy (say, a star) embedded in spacetime is akin to a heavy object sitting on that rubber sheet—it creates a dimple, curving the surface. If another body (say, a comet) passes by the star, its path will be deflected because of the curvature. Instead of moving in a straight line, the comet will swerve toward the star because of the dimple in the rubber sheet. That's gravity—the "attraction" between massive bodies is really the outward sign of how massive bodies cause spacetime to curve.

The rubber-sheet analogy goes only so far; the rubber sheet is two dimensional, when, in fact, the spacetime manifold is actually four dimen-sional, and three of those dimensions (the familiar up-down, left-right, front-back of space) have different properties from the fourth (which describes time). Mass and energy curve all four of those dimensions: a body passing near a star will not only swerve through space, it will "swerve"

---

* As physicist John Wheeler famously put it, "Spacetime tells matter how to move; matter tells spacetime how to curve."

through time as well—its clock will be affected in subtle ways by nearby mass and energy. Even so, the rubber-sheet analogy is really useful for describing general relativity and how it's different from Newton's description of motion and gravity.

That's the theory in a nutshell. However, a theory is merely an idea until it has experimental backing. Physicists can believe it; it can even get a large following of devotees. But it's really when a theory's predictions are tested by observations of the natural world that it becomes something more than a mere idea.

Einstein was lucky; it was easy to see a number of experiments that could test relativity theory. The differences between Newton's universal gravitation and Einsteinian gravity are usually subtle, but in certain circumstances, they can be stark. When objects move very, very fast—close to the speed of light—or come extremely close to monstrous gravitational bodies, such as stars, Einstein's field equations predict phenomena that Newton's laws did not. The moment experimenters could spot one of those relativistic effects, they could show that Einstein's equations were more than a mere mathematical mirage. Some of these experiments happened right away; others would take decades or more before they were within the realm of possibility.

The first big experiment came in 1919 with a solar eclipse. If spacetime were really curved, like Einstein's equations say, a beam of light moving near a massive object like the sun wouldn't travel in a straight line, but would bend, almost as if it were shining through a lens. This *gravitational lensing* effect should be visible: when the light from a distant star passes near the sun, the warping of spacetime would cause a tiny deflection in the star's apparent position in the heavens. That is, stars right near the edge of the sun should appear in the wrong places in the sky. During a solar eclipse, when the sun's light is temporarily blotted out by the moon, astronomers can measure the locations of stars around the sun and see if they appear in their usual positions (as Newton would have it), or if gravitational lensing makes them appear in slightly different places (as per Einstein). So in 1919, Sir Arthur Eddington, then the secretary of the Royal Astronomical Society, organized two expeditions to measure stellar positions during a solar eclipse. Lo and behold, Einstein was right! The stars had moved. Newton was overthrown.

Albert Einstein, almost overnight, had become an international celebrity.*

...

---

* In retrospect, it's not at all clear that Eddington's measurements were sufficient to prove Einstein correct, but that mattered little at the time.

Just shy of one hundred years after Einstein's rise to become the public face of science, Stephen Hawking was reigning in his place as the premier scientific celebrity. But the expectations of celebrity had changed a little bit in the interim.

"Analyzing data since the '66 World Cup, I have answered two of the biggest questions tormenting fans," Hawking told a gathered crowd of journalists in the basement of London's Savoy Hotel in 2014. "One, what are the optimal conditions for England's success, and two, how do you score in a penalty shootout?"[6]

The event was sponsored by Paddy Power, a Dublin-based bookmaker, best known for chasing publicity by offering bizarre attention-getting bets. ("As the oil disaster in the Gulf of Mexico enters its second month with little or no sign of abating," the company announced shortly after the Deepwater Horizon accident in 2010, "leading betting outfit Paddy Power are taking bets on the first species to become extinct as a direct result of the spill. Top of the bookies list at odds of 4/5 is the already critically endangered Kemp's Ridley Turtle."[7]) This time, Paddy Power was trying a different way of getting some attention.

"The technique I have used is called general logistic regression modeling," Hawking announced. But Hawking's "analysis" was anything but scientific. "Our chance of triumph can be worked out by looking at a number of variables. Statistically, England's red kit [uniform] is more successful."

For some reason, the British press seems inordinately fond of marketing ploys dressed up as nonsensical mathematical formulae—a formula for the perfect pizza to help drive sales for a pizza chain, an equation for the most miserable day of the year to encourage Britons to purchase a weekend getaway package from a certain travel agent, the formula for the perfect pancake to sell a supermarket chain's nonstick frying pans, and the like. To give the formula some credibility, the sponsor typically shops around for a scientist or a mathematician who's willing to accept a fistful of cash in return for lending his or her name to the whole silly endeavor. Typically, these scientists don't have enough of a reputation to damage it by producing nonsensical equations. "All are commissioned by companies as PR stunts and their value ends there," a science journalist wrote in *The Guardian*. "They are overwhelmingly drawn up by scientists whose names are unknown to any Nobel committee."[8]

This time was clearly different.

In his pitch to the press, Hawking seemed to be kidding around. ("As we say in science," he intoned, "England couldn't hit a cow's arse with a banjo.") Nevertheless, it was surprising that such an eminent scientist would lend out his name for such a ridiculous publicity stunt. It was a shock even to Paddy Power. A Paddy Power spokesperson later admitted that he never expected Hawking to agree to the bookmaker's request. "We thought there was a one percent chance he'd say yes," the spokesperson said. "But he did. I was totally surprised."[9]

When journalists asked Paddy Power representatives how much they had paid Hawking for his services, they got no answer. However, Hawking reportedly "said he split the fee between two charities, one devoted to saving children in Syria, and the other to motor neurone disease, the condition Hawking was diagnosed with as a student."[10]

• • •

Whatever you call it—motor neurone disease, Lou Gehrig's disease, amyotrophic lateral sclerosis (ALS)—it was an ever-present shadow over Hawking. Diagnosed with the disease when he was twenty-one, Hawking was expected to live another two or three years at most. Yet Hawking survived twenty times longer than the predictions allowed. And though he had quite a few close brushes with death, he made it into his seventies.

The disease had left him physically helpless, unable to speak, to eat, to hold his head upright—to move most of his muscles. He required round-the-clock nursing, and Britain's National Health Service provided nowhere near enough support to provide the care he needed to survive. And this care was damnably expensive. Even though Hawking made millions of dollars from the sales of his books—particularly his first popular book, *A Brief History of Time*, which sold more than ten million copies—he never seemed to have enough money to give him a lasting sense of security, says Al Zuckerman, Hawking's agent for more than thirty years.

A few years before Hawking died, Zuckerman says, "I was informed that Hawking badly needed money, and could I do anything to increase his income?" Zuckerman approached a number of people in the publishing industry with ideas—such as the possibility of turning some of his books into online courses. "And I approached some charities to support him and his research." But things just weren't jelling.[11]

• • •

Einstein and Hawking both occupied the same niche as celebrities—both uneasily bearing the title of world's smartest man—and it's not a coincidence that Hawking's area of study built upon Einstein's legacy. Hawking had become a master of Einstein's theory of relativity, and, like Einstein, he was able to predict phenomena that nobody had ever imagined before.

However, it wasn't easy to follow an act like Einstein. Not only did Einstein build a brand-new mathematical framework for gravity, space, and time, but he did it at a time when astronomers and experimental physi-cists could figure out ways to verify his predictions, showing conclusively that the young upstart had unseated Newton. Hawking's theories, unfortu-nately, didn't lend themselves so easily to experimental verification.

Gravitational lensing was just one of many new phenomena hidden within Einstein's field equations waiting to be verified by experimenters. It was just a matter of figuring out how. The 1919 eclipse provided one way. Physicists soon figured out another: general relativity predicted not just that light's path would be bent by strong gravitational fields, but that its color would be, too. Light climbing out of a deep gravitational dimple would be redder than equivalent light coming from an uncurved region of spacetime. In 1924, astronomers were convinced that they had seen this "gravitational redshift" effect. Einstein was right again.*

It took even longer—nearly three-quarters of a century—to measure an even subtler effect, relativistic *frame-dragging*. According to Einstein's field equations, a spinning massive body curves spacetime in a slightly different manner than a stationary one does. In the early 2000s, scientists detected this effect, first by observing matter spiraling around massive stars, and then by observing tiny changes in the motion of satellites in orbit around the spinning Earth.

However, the most radical, most profound prediction of Einstein's field equations—more important than gravitational lensing, gravitational red-shift, or frame-dragging—was *gravitational radiation*.

Relativity says that the matter and energy in a region of spacetime determine the curvature of that region. But what happens if the matter and energy in the region undergo sudden change—the matter and energy suddenly rearrange? Perhaps there's a supernova; perhaps two massive stars smash into each other. Perhaps it's not even a violent event, such as two massive objects orbiting each other, which constantly rearranges the distribution of matter in a small region. In cases where the distribution of mass and energy in a region is changing, the field equations say that these changes might cause ripples in the fabric of spacetime, ripples that propagate outward at the speed of light, carrying energy with them. (The rubber-sheet analogy is useful here, too. Imagine two massive iron balls

---

* As with Eddington's eclipse observations, these measurements, by astronomer Walter Adams, were in fact too error ridden to be good tests of general relativity. Later, however, more accurate observations verified gravitational redshift (and gravitational lensing) with high precision.

circling each other in the middle of the sheet; it's easy to visualize how those orbiting balls would cause undulations in the fabric.*)

Of Einstein's predictions, this might have been the hardest for experimentalists to verify. Gravitational-wave distortions are typically very small and subtle. Detecting the waves is therefore extremely difficult—a Nobel Prize–level endeavor. In the early 1970s, two astronomers, Russell Hulse and Joe Taylor, made an indirect observation: they watched two massive stars circling each other for half a decade and discovered that the orbits were decaying. Hulse and Taylor showed that the stars' orbital decay matched, beautifully, the energy loss predicted by Einstein's field equations. That is, the stars' dance caused the very fabric of space and time to squish and stretch, carrying energy away from the spinning stars, allowing them to sink ever deeper into each other's embrace. It was, as Taylor later put it, "a new and profound confirmation of the general theory of relativity," and won the pair a Nobel in 1993.[12]

That was for an indirect sighting of gravitational waves. What scientists *really* wanted, though, was to see gravitational radiation directly—to watch gravity waves ripple through spacetime, distorting its fabric. Those waves cause yardsticks to stretch and squish and clocks to speed up and slow down as they pass. But how to spot them? The effects are tiny—mile-long yardsticks change length considerably less than the width of a proton. But with lasers, and a lot of clever engineering, it's just barely possible to do. Lasers can be turned into exquisitely precise distance-measurers, and by arranging two of these laser yardsticks at right angles, scientists can spot gravity waves. When a gravitational wave ripples by, the stretch-and-squish effect of the changing curvature of spacetime is likely to stretch one yardstick while squishing another: the two yardsticks change length relative to one another.

This method isn't sensitive enough to detect the slow orbital decay of Taylor-Hulse-type stars; the rippling is too subtle. But for more violent events, with massive rearrangements of matter and energy in a small region of spacetime, an instrument built well enough could, theoretically, directly observe a passing gravitational wave.

What sort of events would be sufficiently violent? Well, when it comes to gravitational violence, it's hard to beat black holes, the most extreme objects in the universe. Denser and darker than other stars, they are more prone to extraordinary gravitational violence when matter approaches too

---

* But again, the "fabric" is really a four-dimensional manifold rather than a two-dimensional sheet, and includes time as well as space. In addition, the ripples are only created under certain circumstances, so the analogy is misleading in some respects.

close. And for more than thirty years, scientists had realized that this gravitational violence made them brilliant sources of gravitational radiation,
gravity waves ripe for detection.

This is what LIGO is all about.

LIGO is a pair of enormous L-shaped laser yardsticks designed to
detect the subtle distortion of a gravitational wave. They're so sensitive
that they are constantly being jiggled about by the gentle tremors of the
Earth, the low rumble of passing traffic miles away, and even the vibrations
caused by chainsaws felling trees in a nearby forest. And for thirteen years
after LIGO was turned on, that's precisely what the instrument detected.
Passing traffic, Earth rumbles, and chainsaws. No gravitational waves.

And then, early one September morning, the twin detectors saw a
wobble, a tremor that got faster and faster and faster and then, after a tenth
of a second, suddenly stopped, leaving only faint echoes behind. They were
gravitational waves. Two enormous objects, each roughly thirty times the
size of our sun—black holes—circled each other closer and closer and
closer together and slammed into each other, yielding an even larger black
hole. The gravitational waves emitted by that dance of death had traveled
over a billion light-years to the Earth. When those waves passed through
us, they stretched and squashed the very fabric of space and time—a distortion that, after decades of trying, we humans could finally detect. It was
the first time gravitational waves had been directly observed, and it was
the first experimental result to test the predictions about black holes that
Stephen Hawking and his colleagues had made during the Golden Age of
Black Holes more than forty years earlier.[13]

• • •

Even before LIGO detected its first gravitational wave—during that long
period when experimentalists still worried about whether the half-
billion-dollar experiment would ever break its long silence—it was an
exciting time for black-hole theorists. There was a fight brewing, a disagreement regarding one of the big lingering questions about black holes.
Forty years later, the debate was flaring up again, and Hawking was about
to enter the fray.

There are really two parts to the anatomy of a black hole. For the first
part of Hawking's career, he studied the very center of a black hole, a region
where the laws of physics break down: a singularity. Nobody quite knows
precisely what happens at the singularity, because the mathematical tools
we use to understand gravity—the general theory of relativity—stops
working. Relativity assumes that spacetime is a smooth sheet, a manifold.
A singularity is a point where that assumption breaks down, where there's

a puncture in the very fabric of spacetime. After spending the first part of his career becoming one of the world's experts on the properties of singularities, Hawking turned his attention to the second part of the black hole's anatomy. Every black hole has a region of no return, almost like a shroud surrounding the singularity. This region is known as the event horizon.

At a healthy distance, a black hole is like any other star or massive gravitational object: a spaceship can approach it, orbit around it, and fire its engines to zoom away. The closer the spaceship gets to the black hole, the harder the engines have to strain to break the black hole's grip upon the vessel and get safely away. However, the gravitational pull of a black hole is so enormous that if the spaceship gets too close, the pilot will discover to her horror that the energy needed to escape the gravitational pull is infinite. That is, no matter how powerful the spaceship's engines, no matter how much fuel is in its tanks, there's no way the spacecraft could expend enough energy to escape from the black hole. The spacecraft has crossed a point of no return, a region where nothing in the universe can resist the pull of the black hole's gravity. This is the event horizon—the ultimate boundary that makes black holes so special. No matter, no particles, no light, nothing at all, can return once it crosses the event horizon. Everything that ventures beyond is irrevocably cut off from the rest of creation; it is almost as if something falling past the event horizon has crossed into another universe entirely, for it is forever lost to ours.

The event horizon is more complicated than an abstract boundary might seem. For decades, physicists assumed there was nothing special to see near an event horizon. After all, the event horizon is a one-way portal leading straight down a black hole's maw, so the black hole couldn't shine in any way; any light emitted from inside the event horizon would be promptly swallowed again. The region near the event horizon would be completely black. Or so scientists thought.

Hawking's great scientific triumph was proving that wrong in the mid-1970s. Using mathematics to see where telescopes could not, he showed that right at the boundary of the event horizon, there's a faint glow. The boundary of no return wasn't as featureless as once thought.

Since then, cosmologists have been trying to figure out the properties of the event horizon. What, precisely, happens as a chunk of matter crosses the event horizon and falls into a black hole's singularity? What would an astronaut falling into a black hole see as he crosses the boundary? Unfortunately, scientists' answers even to the most basic questions about black holes tend to contradict each other. One of these contradictions led to Stephen Hawking's final scientific work, his ultimate burst of scientific creativity. It would be his very last attempt to plumb the secrets of the black hole.

The theory of relativity says that a freely falling observer shouldn't feel the effects of gravity at all.* This principle implies that an astronaut falling into a black hole would observe nothing special when crossing the event horizon—there wouldn't be any physical boundary or other sign marking the point of no return. Cosmologists dub this idea the "No Drama" postulate. Yet in 2012, four scientists—Joe Polchinski, Ahmed Almhieri, Donald Marolf, and James Sully—published an influential paper that argued—convincingly—that if Hawking was right about how black holes glow, then the No Drama postulate had to be wrong. In fact, the astronaut would encounter a tremendous "firewall" of radiation that would instantly burn him or her to a crisp. That's pretty dramatic.

Hawking didn't believe it. As an expert in relativity, he was firmly convinced in the bedrock assumption that an observer in freefall shouldn't be able to tell whether or not he or she is being pulled by a gravitational field. Yet a firewall would reveal the existence of the black hole's gravity field to the observer (a fraction of a second before the observer evaporates). To Hawking, this simply could not be the way black holes behaved.

However, Polchinksi and his colleagues had made a strong argument. Something was definitely wrong with the picture of black holes—the very picture that Hawking had helped build up over the past half century. There had to be a fix, and Hawking was determined to come up with it.

In late 2013, Hawking told a respectful but skeptical group of physicists that the way they had been thinking about black holes was fundamentally wrong. There was, he declared in his familiar computer-generated voice, "no event horizons and no firewalls. . . . [T]here can be no event horizon, as many of you assume." It was a baffling performance. Hawking seemed to be denying the existence of the very thing that makes a black hole a black hole: its event horizon. Yet few outside of the specialist field of black-hole cosmology had any inkling of what Hawking was suggesting. Even to fellow physicists, it wasn't clear quite what he was driving at.[14]

Hawking seemed to be arguing that there wasn't an impenetrable event horizon after all, but that infalling matter would be scrambled up and re-emitted by the black hole. The information wouldn't technically be lost. Unfortunately, Hawking didn't provide enough details for even specialists to understand his argument. "It sounds almost like he is replacing the firewall with a chaos-wall," Polchinski told a reporter. Another physicist at

* Except, technically, for tidal forces: a stretching pull that eventually turns the astronaut into spaghetti as he or she falls toward the singularity. For sufficiently big black holes, that tidal force should be negligible at the event horizon.

Caltech, Sean Carroll, added, hopefully, "It's very plausible Hawking has a better argument that he hasn't yet gotten down on paper."[15]

Hawking finally did get it down on paper in early 2014, but it didn't help much. He published an article on the Internet that was, mostly, a word-for-word transcript of his 2013 talk. It didn't have any more details that other physicists could begin to tackle. He had only added a few sentences, and one of them seemed calculated to catch the public's imagination. For Stephen Hawking, the master of the black holes, was suddenly declaring that "the absence of event horizons means that there are no black holes—in the sense of regimes from which light can't escape to infinity."[16]

News quickly spread around the world. "Stephen Hawking Says There Is No Such Thing as Black Holes, Einstein Spinning in His Grave," blared one grammatically challenged newspaper. "Stephen Hawking Stuns Physicists by Declaring 'There Are No Black Holes,'" insisted another. In terms of media firestorms, it was a mere ripple compared to what Hawking had triggered in his heyday, but media outlets from Bangladesh to Canada were once again reporting on the physicist's science. It was the first time in a decade.[17]

• • •

It's not like the media ignored Hawking. Far from it. It's just that the headlines were almost never about his science. They were about his pronouncements—or his personal life. No other scientist, not even Einstein, had a life story that so captivated the public, or had so many films made about it.

"Focus Features' much-awaited 'The Theory of Everything' world premiered at Toronto on Sunday night to the most rapturous standing ovation of the festival so far," gushed the Hollywood trade magazine *Variety*. "The Stephen Hawking biopic starring Eddie Redmayne and Felicity Jones (as his wife Jane Hawking) left not a dry eye in the house."[18]

It was not the first movie about Hawking, but it was arguably the best received. (It soon won the gangly, freckle-faced Redmayne an Oscar for his portrayal of the physicist as he grappled with the disease that robbed him of his ability to move.) At its heart, though, *The Theory of Everything* was a tear-jerker of a love story, one that Hawking wryly described as "broadly true."[19]

The story begins with the young Stephen as a geeky prodigy—a healthy one—beginning his studies at Cambridge University. Soon after meeting the love of his life, Jane Wilde, he is diagnosed with motor neurone disease and given two years to live. Naturally, he sinks into an angry depression.

Jane's love pulls him back from the slough of despond, and Stephen decides to make the best of the time he has left by studying time itself.

The love story is star-crossed from the start. Jane and Stephen spar throughout the movie about religion; a devout Anglican, she interprets his work—his attempts to "prove with a single equation that time had a beginning . . . one simple, elegant equation that will explain everything"— as alternately affirming and denying the existence of the creator who is so dear to her. And as Stephen's disease gets progressively worse, Jane turns to a widowed choirmaster, Jonathan Hellyer Jones, to help maintain the household and to act as a surrogate father for Stephen's children. Jane begins to fall for Jones, and when Stephen tells his wife that he "won't object" to Jones' presence in the household, Stephen's tortured expression makes clear the true nature of what that permission truly meant.[20]

But Jane refuses to act upon her feelings; the movie implies that her relationship with Jones is entirely chaste despite their mutual attraction. Only after Stephen falls for his nurse—leading to a weepy and poetic mutual parting of the ways—can Jane finally act on her suppressed feelings. Yet the two remain friends, and the movie ends with the pair looking lovingly at each other, hand in hand, as their three happy children play nearby.

Except for the few moments of despair and pain, Redmayne's Hawking always sports a twisted, wicked grin. He is affable and sympathetic even as he leaves his wife of twenty-four years for the nurse who seduced him. Felicity Jones' Jane seems peevish; the actress' attempts to look stoic and determined often come across as being annoyed at her husband's disability. This was almost certainly not what Jane had in mind when she agreed to sell the movie rights for her book.

The Theory of Everything was based on Jane Hawking's five-hundred-page memoir, *Travelling to Infinity: My Life with Stephen Hawking*, which in turn was a reworked version of her earlier six-hundred-page tell-all, *Music to Move the Stars*. In her books, Jane was able to control the narrative. She seemed somewhat shocked when the movie version didn't tell the story quite how she expected.

"The film really only shows part of our lives in Cambridge," Jane Hawking told *The Guardian*, explaining that the movie didn't give a sense of how much strain it was to take care of Stephen, especially with all the traveling he had to do. (She was rebuffed when she asked for the insertion of a video montage of frenzied packing and stowing and preparing for a trip.) Nor did she appreciate that she "didn't seem to have any friends or relations at all" in the movie. "I knew that if there were mistakes in the film that they were going to be immortalised, which they have been," she said, adding, "I found that very irritating and I didn't want it to happen. Don't ever believe what you see in films."[21]

Though the movie was based on his ex-wife's book, rather than his own autobiography, Hawking helped a great deal with the production. He spent time with Redmayne, and even lent the production crew the use of his distinctive robotic voice. "We'd been using this synthetic version of his voice which this company had drawn up for us," Redmayne told *Empire*, a movie magazine. "At the end [of the early screening] he gave us the copyright to use his actual voice."[22] *

Jane, as the author of the book that inspired the film, naturally got pride of place in the credits at the end of the movie; she is mentioned right after the director, producers, and screenplay writer, and in an equally large font.[23] † Then come credits for the actors and crew and thank-yous and acknowledgments for dozens more, including various artists, the locations where the filming took place, and the organizations that gave permission to reprint images in the movie. Even Jane's parents, who had been dead for years before the filming began, got a special thank you.

Nowhere acknowledged in the credits at all: Stephen Hawking.

• • •

Stephen had his own film projects, though nothing even close to the size of what his ex-wife had been working on. (*The Theory of Everything* grossed more than $120 million worldwide.[24] ‡) The professor and his robotic voice were a regular staple on cable-TV science shows.

Stem Cell Universe with Stephen Hawking, which aired on the Discovery Science Channel in April 2014, opened with a shot of the physicist in his wheelchair, his image superimposed upon a luminous spiral galaxy that wheels slowly around him as he narrates. "I have spent my life exploring

---

* The degree to which Hawking and his estate were and are in control of how third parties use his synthetic voice is an interesting legal question—it's a computer program, created by others, and can thus be used by others, yet it is associated with him just as surely as his face. While I have found trademarks associated with Hawking's name, I have not found any evidence of copyright on the voice or even an indication whether such a copyright could be possible.

† Font size is apparently a big deal in Hollywood circles.

‡ Even with a box-office success like this, the author won't necessarily make a huge amount of money. An SEC filing from 2006 revealed the terms of Jane's film-rights contract for *Music to Move the Stars*, in which she was paid an option price of $2,000 and promised 2.5 percent of the production budget and the same share of the net profits of the film should it ever be made. If her *Travelling to Infinity* contract was similar, she probably would have made at most a few hundred thousand dollars from the deal.

the mysteries of the cosmos, but there's another universe that fascinates me. The one hidden inside our bodies." With that, the galaxy suddenly contracts, collapsing into Hawking's midriff. And then, with a bang, a disk of little glowing indistinct blobs begins orbiting the physicist anew. "Our own personal galaxies of cells. Today we are on the brink of a new age in medicine. An age where we will be able to heal our bodies of any illness, all because of cells inside us which have special powers."[25]

Hawking can be forgiven for his hyperbole; after all, he is a cosmologist, not a biologist or a physician. The clumsy galaxies-to-cells sequence attempts to paper over Hawking's lack of expertise—and perhaps even real familiarity with—stem cell research. It doesn't matter; Hawking is science incarnate. His mere presence signals to the audience that what follows is serious, cutting-edge research. Hawking's six-part series, *Science of the Future*, which aired on the National Geographic Channel in 2014, dealt with such subjects as virtual reality, robots, urban design, and military technology—nothing anywhere close to the physicist's areas of study. Hawking had granted TV producers his voice and his name to give the show credibility; there wasn't much else he could (or needed to) contribute.

Unique among celebrities of the day, Hawking could, quite literally, lend his voice to a production. In the mid-1980s, when doctors performed a tracheostomy to save his life, Hawking lost the use of his larynx and his ability to speak. A team of engineers and software designers rigged his wheelchair with a computer system that he could operate despite his ever-diminishing muscle control. Embedded within that system was a speech synthesizer. Hawking could slowly compose a sentence, and then, with one final twitch of his muscles, send the text to the speech box, which would then attempt to pronounce the words that Hawking had keyed in.

Hawking's voice was truly disembodied; it resided in a little computer that could—and did—act independently of its master.

Hawking sometimes let others compose sentences for him, which could then be uploaded into his wheelchair computer. Hawking could then edit them—or not—as he saw fit. It was far more efficient than having to laboriously construct his own sentences from scratch on his computer.* The act of a twitch, a muscular assent, sent those foreign words through his own extended body, to be pronounced through his own voicebox, and become his own.

---

* At his peak, Hawking could only speak at about 15 words per minute; more typical was 3 words per minute or below. For comparison, a good rule of thumb for English speakers is about 120 words per minute.

The human in the wheelchair didn't even have to be present. By the time Hawking lent out his voice to the producers of *The Theory of Everything*, he had, on occasion, been allowing filmmakers to use his speech synthesizer for nearly three decades. Errol Morris, who directed the film version of *A Brief History of Time* in 1990, says that Hawking gave him a copy of his voicebox software so he could record Hawking's voice without the scientist's presence. "Theoretically, I could have Hawking saying anything. There's something quite absurd about that," Morris says. "You just type in a sentence and you record it, and you put it in the movie." At some point, Morris tweaked one of Hawking's statements and the scientist noticed the alteration immediately. "He said, 'You changed that,'" Morris recalls. "And then he said, 'But I like it better.'"[26]

Unlike a regular actor, Stephen Hawking could never flub his lines; if they were entered correctly into the computer, they'd come out as written every single time. And because Hawking didn't move his lips when he talked, a director with the full use of the physicist's voicebox could superimpose Hawking's speech on any image of the physicist sitting in his wheelchair, and it would seem like Hawking himself was intoning the words. No matter if Hawking himself hadn't composed, or even heard, what he was telling the audience. When Hawking outsourced his voice to a film production, he granted a director almost unheard-of control, an incredible money-saving boon. Shoot a few scenes of the physicist sitting in his chair (typically slow-moving spiral shots in a wood-paneled hall) and a few close-ups of his eye moving about or his head settled uneasily on his shoulder, and that's all that would be necessary for a production. The film editors could mix and match any of those images with whatever words were passed through the scientist's speech synthesizer.*

In 2016, Hawking put his name on an oddball series that aired on public television stations in the United States. Part reality show, part science documentary, Stephen Hawking's *Genius* asked contestants to tackle scripted challenges that illustrated scientific principles. (In one show, for example, a team was asked to melt a bucket of ice without an obvious source of heat; they wound up bending a metal bar back and forth to convert mechanical

---

* Even so, Hawking could be a difficult actor. One director of a 2005 documentary told sociologist Hélène Mialet how Hawking showed up extremely late to a shoot for a BBC documentary, then left early in a fit of pique. Later, attempting to get the Hawking voiceover, the director looked "on the Internet to try to find a kind of Stephen Hawking sound-alike, but we couldn't actually find one." Eventually, they convinced the physicist to "read" out the script that they had prepared. Hélène Mialet, *Hawking Incorporated: Stephen Hawking and the Anthropology of the Knowing Subject* (Chicago: University of Chicago Press, 2012), 96–98.

energy into heat energy.) The series consisted of six hour-long episodes. Yet the director needed less than four minutes of Hawking footage to carry through the entire season. The film editors used the same tiny library of shots over and over—cut and spliced in different places, digitally colored or otherwise altered in subtle ways, or even time-reversed to make them seem distinct. Only the most careful of observers would be able to tell that the exact same video footage showed Hawking talking about chemistry in one episode, evolution in another, and the expansion of the universe in a third.[27]

"You know, I would joke to him that he was the first non-talking talking head," says Errol Morris.[28]

• • •

To a slow, ambling beat, the electric guitar laments, wailing a gentle, feline wail, and then fades. Out of nowhere, a voice, a robotic voice, takes the lead. "Speech has enabled the communication of ideas, enabling human beings to work together to build the impossible," Hawking says, keyboard and guitar rising higher and higher in pitch as he continues his monologue. "Mankind's greatest achievements have come about by talking."

It wasn't Hawking's first appearance on a Pink Floyd album; two decades earlier, the physicist had provided vocals for the band's "Keep Talking." In 2014, Floyd decided to release its first studio album in twenty years, and it, too, included a Hawking-narrated track, "Talkin' Hawking." It was just as trippy, and Hawking's voice just as incongruous, as the first song.

Hawking was on his way to becoming a regular fixture in the world of rock 'n' roll. In 2015, he went on the road with the band U2—albeit virtually—on their iNNOCENCE + eXPERIENCE tour. At each show, fans were treated with a video in which Hawking declared, "One planet. One human race. We are not the same, but we are one." This line elicited cheers, but not as much as when he told the audience, "We give our elected officials their power, and we can take it away."[29]

By the time Hawking received payment for the U2 video, his foundation was up and running. At a gala dinner at the Royal Institution in London in the fall of 2015, Hawking and a variety of celebrities—including Eddie Redmayne—gathered to launch the physicist's new enterprise. Dedicated to promoting cosmology and helping people with ALS, the foundation began giving small grants to send young Britons to space camp, creating science tunes to help teach young children, and funding research into the very early universe. The foundation—overseen by Hawking's sister Mary; his friend Kip Thorne; and his Cambridge colleague Malcolm Perry—quickly started raising money.

In its first financial statement, the Stephen Hawking Foundation declared that it had earned £26,000 from a number of sources: rights from the U2 video, from marketing a set of postage stamps issued on the Isle of Man (one of the stamps featured the left side of Hawking's visage, perfectly mirroring the right side of Einstein's face immortalized on a neighboring stamp), and from the sale of scented candles. It wasn't a huge amount; straight-up donations from outsiders netted roughly twice as much.[30]

Given how little was flowing into the foundation's coffers, it's likely that Hawking himself was also receiving a relative pittance from trading on his name. Despite his agent's attempts to find more sources of funding, Hawking had decided to make a change.

"He had a meeting with me," says Al Zuckerman, Hawking's longtime agent. Zuckerman had started working with Hawking in the early 1980s, just as the physicist decided he wanted to write a popular book—the one that became *A Brief History of Time*. "I helped to make a lot of money [for him]," Zuckerman says with a grin. "Or should I say he helped me make a lot of money." Despite working with Hawking for so long, he had never signed the physicist to an exclusive contract, and he didn't handle all of the physicist's moneymaking ventures. "He had an office with a succession of different people, and they were besieged with requests for his appearances—some universities, and some conferences—and they were really in a position to make money for him," Zuckerman says. "But the people he had running [his office] were not equipped very well to do that, and so most of those requests did not come to me." It wasn't the most efficient arrangement, but it seemed to work for both Hawking and Zuckerman. So the meeting came as a complete surprise.[31]

"So he met me and [his lawyer], who decides to fire me. Which she did," Zuckerman says. "I think that he wanted more money than he was earning. And I set myself up to do a lot of the search for ways to get more income. But when push came to shove, he opted to go with this Brit guy." (This Brit guy being Robert Kirby, an agent who represented Hawking in his final two years and, at the time this book is being written, acts as agent to the physicist's estate.)

Roughly three years after that meeting, Zuckerman still seems a bit stunned. "He opted to go with Kirby. I have no idea why. I did a lot of work on trying to figure out how to find money for him, and I don't know what I said," Zuckerman explains. "I think I made a mistake, though, when I was there, because I was really mostly talking to his lawyer, who, you know, could click with and understand his speech and respond, whereas he was turned away and looking at a computer all the time. I should have realized that I should not have been looking at her but looking at him. But, uh," Zuckerman sighs. "Fortunately, you know, my life, my livelihood no longer depends on . . ." His voice trails off.

...

Stephen Hawking seldom wagers much money. It's a good thing, since he usually loses. Most of the time, it isn't by design.

This time, it looked like he had backed a winner. On March 17, 2014, a team of American physicists announced that they'd detected the subtle signal of gravitational waves—not from black holes crashing together, but from the Big Bang itself. When Hawking spoke on BBC radio the next day, he told listeners, "Yesterday a team from Harvard announced they had detected gravitational waves from the very early universe." Then, in an announcement that would be stunningly bathetic were it coming from anyone other than Stephen Hawking, he continued, "It also means I win a bet with Neil Turok, director of the Perimeter Institute in Canada." With that, Hawking's wager became part of the ongoing story. And a big story it was, as it seemed to be a snapshot of the very first moments after creation.[32]

A cataclysmic cosmic event, like the spiral-and-crash of two black holes, causes ripples in the fabric of spacetime, so it should come as no surprise that the biggest cataclysm of them all—the Big Bang and the early rapid inflation of the universe—sent shudders throughout the spacetime fabric of the cosmos. Those gravitational waves are so stretched and attenuated by the passage of time and the expansion of the universe that we can't detect them directly. Even LIGO's ultra-sensitive detectors are too weak and small to spot these waves. But those gravitational waves left their mark in the heavens.

Many billions of light-years away, boxing us in in every direction, we're surrounded by walls of light. These walls are invisible to the human eye; their ancient light has been so stretched by the expansion of the universe that only special microwave detectors can spot them. But they're there nonetheless, everywhere in the sky at once. They are the afterglow of the Big Bang, the light from a moment about four hundred thousand years after the birth of the universe when the glowing-hot clouds of gas that filled the cosmos cooled enough to suddenly become transparent, liberating the light that had been trapped within. That primordial light, now known as the cosmic microwave background (CMB), is ubiquitous—no matter which way you point your telescope, it's there. And it's the most distant object we can see. Across those walls lie the very young universe, the cosmos when it was younger than four hundred thousand years old.* From beyond the CMB, from beyond those walls, light simply can't reach us.

That's where gravitational waves come in. Even though we can't see light from the very early universe, gravitational waves aren't blocked by those walls. What's more, the rippling fabric of spacetime affects how those walls look—the gravitational radiation stretched and squished the primordial clouds of gas, and affects the nature of the CMB. Specifically, scientists were looking for telltale signs of gravitational waves in the polarization of the cosmic microwave background radiation.† And on March 17, 2014, when the researchers associated with a sensitive Antarctic microwave telescope, BICEP2, announced that they had found those signs, the physics community came alive with excitement. Much of the community, anyway.

"This is huge, as big as it gets," Marc Kamionkowski, a theoretical physicist, told the *New York Times* for its front-page story. "This is a signal from the very earliest universe, sending a telegram encoded in gravitational waves." Newspapers and press outlets were full of enthusiastic quotations from theorists who had studied the early universe, including Alan Guth (who came up with the theory of inflation describing the rapid expansion of the universe) and Andrei Linde (who helped devise the version of inflationary theory that is most popular today). Linde had even celebrated the discovery with a bottle of champagne brought over by a BICEP2 scientist. Physicists were even beginning to mutter about a crowning glory for the discovery; the chair of the Harvard Astronomy Department, theorist Avi Loeb, told the *Times* that if the results held up, "it's worth a Nobel."[33]

When Hawking went on BBC the following day, he was thrilled about the news; BICEP2's results implied not just that they had detected a signal from the very earliest moments of the Big Bang, but also seemed to show that Alan Guth's and Andrei Linde's description of the early universe—the

---

* Because light travels at a finite speed, looking at a distant object is like looking back in time. It takes about eight minutes for light to travel from the sun to the Earth, so when you look up at the sky and see the sun, you're seeing light that emerged from the surface of the sun roughly eight minutes ago. Light from the nearest galaxy, Andromeda, is more than two million years old.

† Light waves have a certain degree of directionality. They can be "waving" vertically, or horizontally, or in circles—think of all the different ways two children can wiggle a jump rope. By examining patterns of this directionality, this polarization in the CMB light, scientists could, in theory, figure out how gravitational waves were affecting the clouds of gas in the very early universe. For more information, see my book *Alpha and Omega*, 209 ff.

theory of inflation—was fundamentally correct. That's what Hawking's bet had been about.*

Hawking's former colleague, Neil Turok, along with several other cosmologists, such as Princeton's Paul Steinhardt and the University of Pennsylvania's Burt Ovrut, had been working on an alternative theory that did away with the rapid expansion of the early universe.† In their formulation, there wouldn't be any primordial gravitational waves rattling around the early universe, hence no imprint of gravity upon those hot clouds of primordial gas. Hawking, however, was all-in on inflation; not only was he close friends with Linde—and antagonistic toward Steinhardt—but Hawking had played a major role as midwife for the birth of the theory of inflation (see Chapter 11). So, a true believer in inflation, Hawking had bet Turok $200 that standard inflationary theory was correct—and that there had to be primordial gravitational waves. Or, just as importantly, Turok, Ovrut, and most especially, Steinhardt, were wrong.

Based upon the buzz in the theoretical community, it seemed that Turok would have to pay up. For once, Hawking had won a bet. However, Turok wasn't so sure. "I have some reasons for doubt about the new experiment and its results," he said later that day. "It's not entirely convincing to me that they have clearly seen what they have claimed to have seen." This wasn't just sour grapes. Nobody had yet had time to flyspeck the data, to assess for themselves whether the BICEP2 team had made a great discovery or just an embarrassing mistake.[34]

From a theorist's point of view, the BICEP2 observations were thrilling, because they suddenly gave direct support for an important segment of cosmology's theoretical scaffolding—inflationary theory. From an experimentalist's point of view, it was an exciting but technically challenging observation in which a lot of things could go wrong. Seeing a pattern of polarized light in the sky might well be a signal from the early universe; it might also be light bouncing off dust clouds. Telling the difference between the two is not so easy. And there was a concrete reason to worry: an expensive microwave-detecting spacecraft, *Planck*, hadn't seen anything of note. If, in fact, BICEP2 was right, *Planck* should have also spotted the signal— and there was no good explanation about why the Antarctic telescope was seeing something that the spacecraft didn't.

---

* Paul Steinhardt and his graduate student, Andy Albrecht, also deserve credit, which Hawking was loath to give.

† For more on this idea, known as the *ekpyrotic* or *cyclic* theory, see my *Alpha and Omega*, 196 ff.

Even if Hawking had enough time to read the BICEP2 paper carefully, as a die-hard theorist, he didn't have the experimental chops to find potential flaws in the analysis. Neither did Turok, for that matter, but his skepticism turned out to be warranted. Within weeks, eminent experimentalists began to poke holes in the BICEP2 team's analysis. It was a rather technical argument: when the BICEP2 team tried to subtract out the effects of dust using a mathematical model, they had done it in the wrong way. And time proved the skeptics right. More evidence—observations using multiple frequencies of microwaves rather than just one—showed that the polarized light came from dust, not from primordial gas clouds. By early 2015, the BICEP2 researchers had withdrawn their claim.

Hawking hadn't won the bet after all. But neither had he lost. That's more than can be said about his other bets, including the ones closest to his wheelhouse: the physics of black holes.

. . .

Perhaps the most striking property of a black hole is how featureless it is. The event horizon that shrouds the center of the black hole is completely opaque—nothing, no light, no matter, no energy, no information at all—can cross that invisible boundary and escape a black hole. This means that an observer outside the event horizon can tell nothing at all about what's on the other side of the horizon. The information blackout is almost total.

That's one of the most important conclusions physicists reached during the Golden Age of Black Holes. Hawking and others argued that because no information can escape the event horizon, an observer coming across a black hole could learn almost nothing about it: its age, its composition, what kind of matter it had swallowed since it was born—these were unfathomable mysteries hidden behind the veil of the event horizon. In fact, the physicists concluded, there were only three things that one could tell about a black hole from the outside: its mass, its charge, and how fast it spins. Nothing beside remains.

A black hole is almost totally devoid of distinguishing characteristics, a dictum that became known as the *no-hair theorem*.* However, starting in

---

* The term came to be when black-hole physicist Jacob Bekenstein, then a graduate student at Princeton, struck by a black hole's cue-ball-like featurelessness, exclaimed that "a black hole has no hair." His adviser, John Archibald Wheeler, popularized the phrase—but not everybody was happy with it. Physicist Richard Feynman "thought that was an obscene phrase," Wheeler once told an interviewer. "He didn't want to use it." Interview with John Wheeler, Part 84, "Feynman and Jacob Bekenstein," available at Web of Stories, www.web ofstories.com/play/john.wheeler/84.

the mid-1970s, Stephen Hawking came to realize that this featurelessness, which is required by the theory of relativity, causes immense troubles for theorists. A different set of laws—the laws of quantum mechanics—says just as firmly that black holes simply cannot be that featureless. Quantum theory says that somehow, the history of a black hole, the information about what it was made of and what it had swallowed, had to be preserved. And Hawking's own research implied that even a black hole can't hide such information from view forever; it either had to be destroyed (violating quantum theory) or emerge somehow (violating relativity). This contradiction—the *information paradox*—quickly became one of the biggest scientific puzzles of the day. It went to the heart of the nearly century-old fissure at the heart of modern physics, the mutual incompatibility of quantum theory and relativity, the two great physical frameworks of the twentieth century. Solve it, and it was possible that the solution would reveal the ultimate answer. It might show the way to an overarching, unifying theory that incorporated both relativity and quantum theory, something that described all the matter and energy and forces in the universe on all scales, large and small. Hawking had been groping for a solution to the information paradox ever since he described it, and toward the end of his life, he believed that the solution to the paradox was likely sitting right on top of a black hole's event horizon.

In 2014, when Hawking announced that "there are no black holes," it wasn't really a declaration of the nonexistence of the objects he had studied his whole career, but a salvo in his attempt to resolve the paradox by reexamining the physics of the event horizon. Though Hawking rejected the recent argument floated by Joe Polchinski and other physicists—that there was a firewall at the event horizon that incinerated infalling matter (and prevented information from getting lost)—he apparently agreed that something different was happening at the horizon, something that could resolve the paradox. Nobody understood quite what Hawking was driving at, not even Polchinski, who dubbed Hawking's nascent idea *chaos walls*. But the scientific community awaited further information.

Despite the headlines, Hawking never provided any more information. He soon abandoned the concept of chaos walls in favor of an idea he found much more exciting. A few months after publishing the chaos-wall paper, Hawking visited a ranch owned by a billionaire Texas oilman—George P. Mitchell—for a physics retreat. Mitchell was a major admirer of Hawking's; he had not only endowed a chair at Texas A&M in Hawking's name, but named a brand-new auditorium on campus after Hawking. Periodically, Mitchell would invite Hawking and some of his colleagues to gather and discuss the mysteries of the universe. This time, Andy Strominger, a physicist at Harvard who had known Hawking since the early 1980s, was there, and Strominger had some new results.[35]

Strominger had been working on a branch of general relativity that had been discovered—and abandoned—in the 1960s, and he was increasingly convinced that the ideas he was coming up with could help physicists understand what was going on at a black hole's event horizon. "And Stephen was very excited by this. I gave a seminar in the afternoon, and we stayed up until one in the morning," Strominger says. "He wrote, I haven't been this excited since . . . I feel like I felt when I discovered the area law.* And also his nurses said that they hadn't seen him that excited in forever. . . . He became very energized." And Strominger started working with Hawking and fellow Cambridge physicist Malcolm Perry to refine the idea. By late 2015, Hawking was publicly hinting that he, Strominger, and Perry were thinking about something new and exciting, promising that he was working on a "full treatment" of the concept.[36]

Collaboration with Hawking was always a bit challenging. "Malcolm and I would be writing at the board and getting cues from Stephen," Strominger says. "Stephen would sit there typing, and there was always a few minutes' delay." But the two were used to the awkward flow of a conversation with Hawking. More problematic was the difficulty of getting the three physicists in the same place at the same time. After several visits to the United Kingdom, Strominger invited Hawking and Perry to visit Boston. "There was always a money problem," says Strominger. "I mean, at some point some of his rich friends were offering their jets to fly him around, but he couldn't go on their jets anymore. He could only go on an ambulance jet, and not only that, his doctor insisted on one Swiss company, and it got very expensive . . . so I had to somehow find the money for him to come." Eventually the money came through—the visit would cost hundreds of thousands of dollars—thanks to one of those rich friends.[37]

But even with the money, the trip wasn't possible unless Hawking's doctor signed off on the trip. This time, the doctor reportedly said no. Hawking wrote a letter asking him to relent. "He wrote about how excited he was about the research and that it was potentially going to be a Nobel Prize," Strominger recalls. "I think his doctor was not very sympathetic to having him go off and meet celebrities, but when it came to his legacy and his science . . . and I think everybody could see that Stephen was invigorated by our new enterprise in a way that he hadn't been for quite some

---

* One of Hawking's key insights during the Golden Age of Black Holes.

years." And so in April 2016, Hawking prevailed; his doctor let him fly to Boston, where the three continued their work.

Despite all the difficulties, Hawking, Strominger, and Perry had a very productive collaboration. They soon delivered a series of papers arguing that the "no-hair theorem" wasn't quite right: black holes were, in fact, covered with "soft hair."

The idea is that when a charged particle falls toward the event horizon of a black hole, the rules of electromagnetism dictate that a bizarre, energyless particle of light—a "soft" photon—is born right at the black hole's event horizon. Such a photon not only stores information about the particle that created it, it will eventually escape its precarious position right on top of the event horizon. That is, soft photons record information about matter that has crossed the event horizon; at some point in the future, this information will be visible to an outside observer. A black hole isn't truly a featureless object, Hawking's argument went. It is not a void whose history is totally lost to the bare and boundless sands of time. It is a void with a recorded history, a history stored on soft photons at the event horizon. As no two black holes share the exact same history, their collections of soft photons must be different. Each one has its own unique head of "soft hair."

Hawking seemed convinced that, at long last, he had solved the problem of what happens to particles when they fall into a black hole—the question at the core of his scientific legacy. It would be his crowning achievement.

His collaborators were not so sure. "Stephen wasn't afraid of simplifying our work to get to the essence of it and to convey the excitement, but sometimes he said some things that were, you know, 'We've solved the black hole [problem],'" Strominger says. "Well, even worse, he'll often replace the 'we' with 'I.'"[38]*

Physicists outside the team were even less enthusiastic about the soft-hair idea. Hawking's former PhD student Raphael Bousso and a colleague argued that a "bad choice" in the team's mathematics had led to the wrong conclusion. "In other words, the soft hair is a wig," he wrote. Its roots didn't reach down beyond the event horizon—it was not connected to the innards of the black hole. The idea that this hair was preserving information that

---

* Strominger emphasized that he bore no resentment. "As I said to my friend, you write a paper with Stephen, you get ten thousand times as much attention as if you write it without Stephen. If I did 99 percent of the work, I'm still coming out ahead by a factor of a thousand," he said, laughing. "If you look at Stephen's stuff, he almost never gives credit to any of his collaborators when he's interviewed. I think he's given more credit to Malcolm and me than I've ever seen him give anybody. Sometimes he sort of suggests that we're working for him, but you know, I really don't feel cheated. I've had plenty of appreciation so I'm not about to complain, but it's a fact that he does this."

fell past the event horizon was merely an illusion. That is, soft hair bore "no relevance to the black hole information paradox." Another former student, Marika Taylor, agrees. "Stephen was so smart. He had to know that, maybe this kind of gave a contribution, but it was not solving fundamental puzzles," she says. "There's many different ways to see that this was not going to be the answer. And Stephen had to know that. But . . . he likes the adulation when he does something big."[39]

Even though soft hair was not the breakthrough Hawking seemed to claim it was, for the first time in more than a decade scientists were actively grappling with one of Hawking's papers—he had sparked a serious debate among his fellow scientists.

Once again, Stephen Hawking was at the center of a controversy; once again, the battleground was the event horizon, the boundary between the universe we know and the darkest unknowns known to humanity. However, this time, for the very first time, the experimentalists had begun to venture there as well.

. . .

Every year, rumors seem to fly faster than ever before. On social media, a quiet blurt can spread around the world at the speed of light. Thousands can hear it and amplify it within a matter of minutes. This is true even in the world of science. And on September 25, 2015, a stray comment on Twitter by a cosmologist in Arizona set off a bout of fevered speculation, leading once more to whispers about Nobel Prizes. This time, though, the outcome would be happier than it had been for the BICEP2 team.

"Rumor of a gravitational wave detection at LIGO detector," cosmologist Lawrence Krauss tweeted late that Friday afternoon. "Amazing if true. Will post details if it survives."[40]

It had been just a few weeks since the LIGO detector had finished a long upgrade and turned on once again. Physicists working on the project had been waiting nervously—LIGO had been unexpectedly silent since it had first turned on in 2002. Had its builders finally made good on the promises they'd made decades (and hundreds of millions of dollars) ago? The LIGO team said nothing, and after a brief flurry of blog posts and short stories in the specialist press, the buzz quieted down. For almost three months.

Then, Krauss struck again. "My earlier rumor about LIGO has been confirmed by independent sources. Stay tuned! Gravitational waves may have been discovered!! Exciting." This time, the rumors wouldn't go away. For almost a month, there was a steady drumbeat of speculation, which reached a fever pitch on February 8, 2016, when the collaboration declared a press conference three days hence. The team was coy about the precise nature of the upcoming announcement—merely saying they were going to provide an update, a "status report on the effort to detect gravitational waves." Even though it was the worst-kept secret in physics, the LIGO team itself was being extremely careful not to let things slip before the formal announcement. They almost succeeded.[41]

Sixteen minutes before the press conference was set to begin, a NASA astronomer used Twitter to post a picture of a cake decorated with a picture of two black holes spiraling into each other. "Here's to the first direct detection of gravitational waves!" it announced, in green icing. The rumors were true. A pastry had let the Einsteinian cat out of the bag.[42]

A few minutes later, Caltech's Kip Thorne was explaining the discovery to a crowd of journalists. "There was one regime in which general relativity had never been tested," said Thorne. For the first time, LIGO was giving a direct view of a place where the fabric of spacetime is not just extremely warped, but changing very rapidly—the conditions right near the edge of a black hole as it swallows a large lump of matter. "We have never had any tests in that regime," Thorne continued. "This observation tests that regime beautifully, very strongly, and Einstein comes out with beaming success."[43]

It was a success that Hawking could hardly have dreamt of when he was starting his career. Back then, and on through the Golden Age of Black Holes in the late 1960s and early 1970s, physicists couldn't even prove that black holes existed. The collapsed stars were the theoretical byproduct of the rules of general relativity; by following those laws to their logical consequences, Hawking and Thorne and numerous other colleagues were able to describe black holes in great detail. However, that wasn't the same as actually being able to point to a black hole in the sky. Finding a black hole—an object that absorbs light that ventures too close—was an incredibly difficult task. As Hawking put it, trying to find one was akin to "looking for a black cat in a coal cellar."[44]

By the time Hawking was starting his black-hole research, astronomers had seen a strange object in the constellation Cygnus—a mysterious mote in the sky that shone brightly with X-rays. Scientists suspected that it was a dead sun, and if it was, it was far too heavy to be any of the other kinds of collapsed stars that theorists knew about. It could just possibly be a black hole. Most astrophysicists came to the conclusion that it was a black hole, but the evidence was indirect. Indeed, Hawking's most famous wager—his 1975 bet with Thorne—was about whether this weird object in Cygnus was, in fact, a black hole. Hawking only conceded the bet in 1990.

Over the years, astronomers built up better and better evidence for the existence of black holes. Using X-ray detectors, they spotted more objects like the one in Cygnus. Using infrared and visual telescopes, they peered at the centers of distant galaxies looking for signs of massive black holes swallowing matter. (Astrophysicists now think that pretty much every galaxy has a black hole at its center.) They watched as stars at the center of our galaxy wheel around a massive invisible object. Astronomers are very certain that black holes exist, but using telescopes that detect different kinds of light—X-rays, ultraviolet light, visual light, infrared light, microwaves, radio waves—it's exceedingly hard to spot an object that doesn't let light escape, much less to probe its properties. The theorists with their equations were seeing far deeper into the abyss than the experimentalists with their instruments could possibly hope to go.

That's what LIGO had suddenly changed in September 2015. A brief shudder of spacetime heralded the cataclysmic collision of two massive black holes. And every few weeks, LIGO was detecting yet another such "merger"—one in October, another in December. Now, by using gravitational waves rather than light waves, experimentalists were not just detecting black holes, but beginning to gather data about the region not far from a black hole's point of no return: its event horizon.[45]

As two black holes spiral in toward each other, they orbit faster and faster and draw closer and closer, emitting gravitational waves all the while, setting spacetime all aquiver. In the final milliseconds before the collisions, the black holes' event horizons come into ever closer proximity. The gravitational waves that the holes emit in the very last split seconds before that colossal wreck, the last, violent tremors before all is silent once more, are signals from near those event horizons. Experimentalists were beginning to venture right up to the edge of the abyss—just at the border where they might be able to verify some key predictions that Hawking had made about black holes in the 1970s. If they could, Hawking would—after half a century of waiting—like Einstein, have an experiment demonstrating that his ideas were correct.

"He was very excited," Thorne says. "He wanted to know how well we can measure the masses and spins of the black holes in order to test his area theorem." Unfortunately, for technical reasons, LIGO isn't precise enough to provide a stringent test of any of Hawking's ideas directly: not his work from the 1970s, much less his recent fight over soft hair and firewalls. Even so, it was a stunning result. An observatory was for the first time scoping out the battlefield where Hawking had spent so much of his career. He certainly saw it as a victory. "Along with confirming Einstein's beautiful theory, the detections agree with predictions that I and other scientists have made about black holes," Hawking declared in late 2016. "The ripples of their work will flow through the field of astrophysics for many years to come."[46]

The following October, Barry Barish, Rai Weiss, and Kip Thorne won the least surprising Nobel in recent memory, "for decisive contributions to the LIGO detector and the observation of gravitational waves." At the moment of the announcement, the Nobel committee published copious information about the laureates' achievements. There was a press release announcing the basis for the prize; a popular account intended to make the science accessible to the public; and a dense, eighteen-page background going deep into the winners' work and its importance. In all of that material, with all of its dozens of references and kudos and historical explanations, one name was entirely missing.

Nowhere acknowledged at all: Stephen Hawking.

CHAPTER 3

MODELS (2012–2014)

T he stadium was bathed in an otherworldly blue glow, a color so vivid and pure that it seemed unnatural. At center, dwarfed by an enormous model of the moon perched on a tower high above, sat a tiny figure in a wheelchair. He was almost invisible, an immobile speck of dull brown in the middle of a gaudy, twinkling whirl of motion—a spectacle that had suddenly paused in anticipation. A hush fell over the crowd.

"Ever since the dawn of civilization, people have craved for an understanding of the underlying order of the world: why it is as it is and why it exists at all," Hawking's mechanical voice rang out over the loudspeaker. The physicist sat perfectly still as his speech synthesizer continued. "But even if we do find a complete theory of everything, it is just a set of rules and equations. What is it that breathes fire into the equations and makes a universe for them to describe?"[1]

As Hawking fell silent, a giant, glowing, smoking ball—resembling an atom with electrons circling about, or perhaps an armillary sphere representing the positions of the stars in the sky—descended from the heavens into a blue-lit void. Fireworks erupted in all directions as ponderous music poured from the loudspeakers. The 2012 Paralympic Games had begun with a Big Bang.

London's brand-new Olympic Stadium was nearly full to capacity. At over sixty thousand people, it was the largest assembly Hawking had ever addressed. However, even though he was in front of the biggest audience of his life—and was part of a production that had expended more money and was reaching more viewers than any of his prior appearances—Hawking had not gone out of his way to do anything special for the occasion. Completely unfazed by the pyrotechnics around him, he spoke as he always did, his voicebox pronouncing the same words that it had pronounced numerous times before.[2]

Hawking was often touted as one of the finest communicators of science in the world. However, he had to go through extraordinary lengths to make himself understood by the people around him. After losing his voice in the mid-1980s, he relied almost totally on the computer embedded in his wheelchair to speak and to write his books and speeches. Using a clicker in his right hand (and then, as his disease progressed, a sensor near his cheek), he navigated through a menu of options to select the words he wanted to utter, one by one. Only after Hawking had painstakingly assembled a string of words could he then instruct the voicebox to pronounce them aloud. At his peak, Hawking could get about fifteen words per minute out of the machine. But as his muscle control declined, his communication speed dropped as well.

"When I first met him, he was speaking in real time—maybe he spoke a little more slowly than the average person," Andy Strominger says. "Then, twenty years ago, you would talk about words per minute, and then it went to minutes per word, and then, you know, it just got slower and slower until it just sort of stopped."[3]

It was often easier for Hawking simply to call up an old sentence—one he had sent to his computer previously that was still stored in its memory bank—than to compose the same or a nearly identical one from scratch. As a consequence, his public utterances were often pastiches, bits and pieces of previous writings and statements fitted together like a mosaic. Even before Hawking's mechanical voice made this habit a labor-saving device, he seemed to delight in feeding the same bon mots to journalists over and over again. As one journalist recalled in the mid-1980s, shortly before Hawking's tracheostomy:

> "My goal is simple," he said, suddenly solemn. "It is a complete understanding of the universe, why it is as it is, and why it exists at all."
>
> I scribbled these thoughts into a notebook. When I looked up, Hawking was convulsed with laughter and there was a gleam in his eye.
>
> "Do those words sound familiar to you?" he asked.
>
> After an instant's reflection, they did indeed. A year earlier, I had written a story about Hawking that had appeared in a popular

American science magazine. In a prominent position early in the story, Hawking was quoted uttering those exact words, which he had spoken to me when I had met him the previous year.[4]

Not that it mattered to his audience at the Paralympic Games, but the other parts of Hawking's opening were also a quarter century old. "Ever since the dawn of civilization, people . . . have craved an understanding of the underlying order of the world," comes from his most famous book, *A Brief History of Time*, as does "Even if there is only one possible unified theory, it is just a set of rules and equations. What is it that breathes fire into the equations and makes a universe for them to describe?" Even Hawking's now-famous exhortation at the Games to "look up at the stars, not down at your feet" was repurposed from a few years prior.[5]

But there was one section of Hawking's address that was entirely novel, something that Hawking had never said before, at least in public. As his speech drew to a close, Hawking told the gathered crowd that the recent discovery of the Higgs boson at CERN was a triumph of a lifetime. "It will change our perception of the world and has the potential to offer insights into a complete theory of everything," he declared through his electronic voicebox as two glowing-visored DJs gesticulated wildly behind him. Hawking's voice faded and the sound system began to thrum with electronic dance music. Acrobats spun and gyrated, twirling red umbrellas. Actor Ian McKellen waved a placard to the beat, and backup dancers in wheelchairs pumped their fists in the air. Only about a month earlier, particle physicists had announced that the Higgs had been found, and already it had been repurposed for popular entertainment. Hawking sat, immobile, center stage, a striking contrast to the commotion surrounding him.

. . .

Hawking always enjoyed being at center stage, and in the last years of his life, he was used to being there. At least in popular perception, he inhabited the very center of the scientific universe. Like Newton and Einstein, Hawking was the shining star of physics, obscuring all others by his sheer brilliance.

Other pretenders to the throne would sometimes arise, other scientists who would be talked about almost as much as Hawking, at least for a short time. In 2012, for example, a massive particle accelerator on the French-Swiss border put Scottish physicist Peter Higgs in the headlines. For that multibillion-dollar machine was finally bringing to a close a forty-year-old quest to find a particle that Higgs (and others) had predicted: the Higgs boson. Or, as one Nobel laureate had dubbed it, the "God particle."

Unlike most of his peers, Hawking was a Higgs atheist; he didn't believe in the God particle, at least not until he was forced to. However, even had Hawking not had some unorthodox scientific views about particle physics, it was almost inevitable that he would get into a fight with Peter Higgs at Higgs' moment of greatest triumph.

Hawking wasn't afraid of controversy. His ideas were often unconventional, even heretical; even as he entered his scientific twilight, he and his collaborators kept trying to generate new theories to wield in scientific battles. (Sometimes Hawking's battles could turn nasty; he had gravely offended colleagues on occasion.) As a celebrity, however, Hawking had to deal with a whole other level of scandal, including tabloid gossip about his sex life. Of all the controversies that Hawking got involved with over the years, the one over the Higgs boson is a minor one. Yet for Hawking, it was the one that is most clearly born from sadness—sorrow that the universe wasn't turning out to be as interesting as the model Hawking had painstakingly built up in his mind.

...

It is almost as if the Higgs boson—the discovery that won Peter Higgs his Nobel—comes from a different universe than the one described by Einstein's general theory of relativity.

Einstein's cosmos is filled with a smooth, gently rippling manifold that stretches over almost unfathomable expanses of space and time. The force of gravity comes from the way matter and energy cause that manifold to curve. And while the curvature can be enormous on a large scale, in tiny, submicroscopic regions it becomes almost negligible. With the sole exception of a black hole—where the curvature is so extreme that the smoothness of the manifold breaks down—spacetime on the smallest scale is a continuous, quiescent, almost boring place.

The universe of the Higgs is not smooth manifold, but a frothy, discontinuous substrate that is ever more violent the smaller the scale. It is not a realm where forces are the result of how spacetime curves; instead, forces are carried by undetectable subatomic particles that pop into existence, carom off each other, and disappear once again into the void from whence they sprung. The ever-churning sea of the particle physicist looks almost nothing like the rubber sheet of relativity. This disconnect is the biggest problem of modern physics, and one that Hawking spent much of his career trying to understand.

Quantum theory describes how matter behaves on very small scales: how molecules and atoms and subatomic particles move and interact with each other. Embedded into the laws of quantum theory are some rules that are responsible for a froth, a churning, random activity of particles on the very tiniest segments of space and time. Like relativity, quantum theory was born at the turn of the century, and as with relativity, Einstein played a central role in its creation. But Einstein himself was repulsed by the randomness and the discontinuous nature of the universe of quantum theory. "God does not play dice with the universe," he declared. Einstein found the smooth manifold of general relativity to be much more comfortable than the roiling chaos of the subatomic realm of quantum theory.

The birth of quantum theory, not so coincidentally, coincided with the discovery of a zoo of subatomic particles. In 1897, Cambridge physicist J. J. Thomson discovered the electron, a tiny charged particle weighing in at a bit more than 1/2,000 the mass of even the smallest atom. A decade and change later, Ernest Rutherford, running the same lab—which featured prominently in the Hawking biopic *The Theory of Everything*—gave the name of 'proton' the much heavier charged particle he had discovered. Soon after, again in the same laboratory, James Chadwick discovered an uncharged particle almost as massive as the proton, the neutron. These were the three particles that, stuck together in various configurations, made up the atoms and molecules and all the everyday matter on Earth.

But the laws of physics had a surprise in store. In the late 1920s, Paul Dirac, also at Cambridge, came up with an elegant equation that fixed some of the difficulties that physicists were having with the electromagnetic force and how it behaved when combined with the principles of quantum theory. The equation worked beautifully, but it showed that the electron had a doppelgänger: a particle that weighed exactly the same amount, but carried an equal and opposite charge. (The particle was spotted shortly thereafter, and Dirac was rewarded with the Lucasian Professorship at Cambridge—the very one that Hawking would hold half a century later.)

Now there were more particles than chemists and physicists wanted, more than they needed to explain the composition of matter. In 1936 came the muon, a particle that behaved like the electron, but was quite a bit heavier and tended to decay after a short time. "Who ever ordered that?" quipped theoretical physicist Isidor Rabi. And the particles kept coming. Medium-weight particles like the pion and the kaon. Heavyweight ones like the lambda and the xi. The zoo of particles was expanding at an enormous rate. It was more than scientists could keep up with, or make sense of. At his Nobel lecture in 1955, physicist Willis Lamb joked, "I have heard it said that the finder of a new elementary particle used to be rewarded with a Nobel Prize, but such a discovery now ought to be punished with a $10,000 fine."[6]

It took until the early 1960s—at the same time that Hawking was beginning his studies at Cambridge—for scientists to begin to figure out how the myriad particles were related. There was a hidden structure underneath the chaos, an underlying order that, by the mid-1970s, had developed into what is now known as the *Standard Model.**

The Standard Model is a mathematical framework that describes all the matter in the universe and how it interacts.† This framework explains the vast and ever-expanding zoo of subatomic beasties by positing that there are a small handful of fundamental particles with very predictable behaviors.

On one side of the Standard Model, there are particles, the *fundamental fermions*, that describe the raw materials that matter is made of: the relatively massive quarks, which make up the protons and neutrons at the center of every atom; the lightweight electron, which tends to live in an atom's periphery; as well as other, more exotic species of particle, such as the neutrino, so light that its mass is almost undetectable. The other side of the Standard Model describes how particles interact with each other. We know, as just one example, that a negatively charged electron will repel a second negatively charged electron thanks to the electromagnetic force. In the Standard Model, this repulsion force is carried by yet another particle, a force-carrying one known as a *gauge boson*. This is a particle that interacts with both electrons and, essentially, causes them to feel each other's presence. Just as the Standard Model has a gauge boson that carries (or *mediates*) the electromagnetic force, it also has a gauge boson that mediates the *strong force* that glues the protons and neutrons together at the center of the atom. Bosons are also responsible for a third force, known as the *weak force*, that, on certain occasions, causes particles to interact by changing each other's identities. The Standard Model has not one, but three gauge bosons that carry the weak force.‡

---

* For more details about the Standard Model and its development, see my book *Alpha and Omega*, chaps. 8 and 9.

† There is reason to believe that there are kinds of matter that aren't accounted for by the Standard Model, but we haven't discovered them yet. See my *Alpha and Omega*, chaps. 7 and 10.

‡ One of the main triumphs of the Standard Model is to show that the electromagnetic force and the weak force are really, fundamentally, the same thing, even though they look very different. It's somewhat akin to how ice and liquid water seem very different until you heat them up to a high enough temperature. Similarly, the electromagnetic force and weak force become indistinguishable when temperatures are very, very high—as they were soon after the Big Bang.

In essence, the Standard Model explains everything about matter—
what it's made of, how it behaves, what forces it feels, and how it feels
them—with a mathematical framework that describes two kinds of particles, the fundamental fermions and the gauge bosons. The fundamental
fermions, in various combinations and permutations, explained the nature
of every single subatomic particle that we humans have yet observed. But
even more astounding, the Standard Model describes every possible way
those particles can interact with each other, every conceivable way that
they can feel each other's presence via three fundamental forces: the strong
force, the weak force, and the electromagnetic force. Building this framework was a tremendous mathematical achievement, the crowning glory of
late twentieth-century physics.

But there is a problem. A big one.

There aren't just three fundamental forces. There are four. In addition
to the strong force, the weak force, and the electromagnetic force, there
is—of course—gravity. And the Standard Model doesn't explain gravity at
all. Particles in the Standard Model don't feel each other's masses; there's
no gauge boson that explains how one particle might be gravitationally
attracted to another. With no gravity in the mathematical framework,
there's no way the Standard Model can be considered complete. It is not a
theory of everything, just a theory of most things under many conditions.

But it's not easy to add gravity to the Standard Model. It's built upon a
mathematical framework that works just fine with massless particles, but
real-world particles that feel gravity and have mass tended to break the
mathematical underpinnings of the model.

This is where the Higgs boson comes in. In the early 1960s, several
scientists, including the Scottish physicist Peter Higgs, figured out how to
bridge the gap between the massless particles the mathematical framework
produced and the massive particles physicists observed in real life. This is
what is now known as the Higgs field.

Perhaps the best way to visualize the Higgs field is by analogy: imagine
that the universe is filled with some sort of viscous fluid, like honey. Some
particles, by their nature, might be blunt-shaped, clumsy things; when
pushed, they have a hard time moving through the honey. These particles
resist motion—they are *massive*—because they feel the effects of the honey
very strongly. Other particles might be more streamlined and would zip
through the honey with very little trouble. These particles have less mass,
or may even be *massless*; they don't feel the effects of the honey very much
at all. In a vacuum, all particles, no matter their shape, would move about
in the same way; there would be no resistance. But once you add the honey,
that equivalence is broken; some particles interact more strongly with the

honey than others, and thus move with more difficulty. In the same way, a particle in some sense starts off massless, but it acquires mass depending on how strongly it interacts with the Higgs field.*

As a theoretical construct, the Higgs field worked beautifully. It didn't truly explain where mass came from; nor did it fix the disconnect between Einstein's smooth picture of gravity and the lumpy, discontinuous universe of quantum theory and interacting particles. Gravity still wasn't a part of the theory. But the Higgs field was an intermediate step that allowed physicists to build a mathematical model including real-world particles with mass—and three of the four forces affecting them. It wasn't a complete solution, but the Higgs mechanism was an important part of what became the Standard Model.

However, once you assume there's a Higgs field, there has to be a particle associated with it. So if you believed in the plain vanilla Standard Model, you were forced to accept that there was a so-called Higgs boson out there waiting to be discovered.† Fail to find it, and the Standard Model would be in trouble. And so the hunt for the Higgs began.

Physicists find exotic particles by pouring enormous amounts of energy into a very small space: by colliding particles together at very high speeds and looking through the spray of debris to see if they can find something new. The higher the energy of the particle collider, the more exotic the objects that can be found. Since nobody knew precisely how heavy or how rare the Higgs boson was, particle physicists had only a rough sense of how powerful their colliders had to be to spot a Higgs. Starting in the mid-1980s, they started seeing hints—ghosts—in their machines that seemed to indicate the existence of the Higgs. And like ghosts, they tended to be products of overactive imaginations. After building new collider after new collider over the course of several decades, theorists were becoming increasingly worried as the Higgs failed to show up.

On July 4, 2012, nearly half a century after the original predictions, scientists at CERN near Geneva announced that the search for the Higgs was finally over.

"Physicists Find Elusive Particle Seen as Key to Universe," blared the *New York Times* on its front page. A more understated *Washington Post*

---

* This is obviously an oversimplification, and it's somewhat unsatisfying for a number of reasons, not least of which is that most of the mass of ordinary matter comes from a different mechanism. It's also somewhat misleading: there are four Higgs fields (and associated bosons), and we can only observe one of them, for reasons that have to do with the same theory that ties together the electromagnetic and weak forces.

† It's a different variety of boson from the ones that carry the electromagnetic, weak, and strong forces, but has some properties in common.

declared that "Scientists' Search for Higgs Boson Yields New Subatomic Particle." "Scientists Might Have Found 'God Particle,'" exclaimed the *Daily Telegraph* in the United Kingdom. And, next to that, "Higgs Boson: Prof Stephen Hawking Loses $100 Bet."

Hawking himself had nothing to do with the discovery of the Higgs boson, or the nitty-gritty of the Standard Model that had led to the prediction of the Higgs. Yet on the day of the announcement, the BBC turned to Hawking to comment on the significance of the discovery. "If the decay and other interactions of the particle are as we expect, it will be strong evidence for the so-called Standard Model of Particle Physics, the theory that explains all our experiments so far," Hawking told the reporter. "This is an important result that should earn Peter Higgs the Nobel Prize, but it is a pity in a way because the greatest advances in physics have come from experiments that gave results we didn't expect." He then added: "For this reason, I had a bet with Gordon Kane of [the University of Michigan] that the Higgs particle wouldn't be found. It seems I have just lost $100."[7]

The story of the Higgs discovery, as important as it was, was an abstract and difficult story to tell. The story of Hawking's long-running bet gave journalists a way to put a human face on the narrative, to tell a story that readers would find compelling rather than confusing. Peter Higgs himself is a shy and retiring character, modest almost to a fault—and not very good copy. Hawking, on the other hand, was already a fixture of public life, a staple of the science pages, and a known draw. And it probably tickled Hawking to know that he was stealing a bit of the limelight away from Higgs.

• • •

"I didn't really quite appreciate how much Stephen enjoyed the limelight," says Peter Guzzardi, Hawking's editor for *A Brief History of Time*, the book that made Hawking a household name. "I mean, he just thrived in it." By the time Hawking was explaining the significance of the Higgs boson to sixty thousand people at the 2012 Paralympic Games, he had been an A-list celebrity for a quarter century. Most of the time, he relished the attention. But even when he did not, he couldn't escape the public glare. "I can't disguise myself with a wig and dark glasses," he wrote. "The wheelchair gives me away." And so his every dalliance was always in public, and this made his personal life fodder for the tabloids. And, generally speaking, tabloids aren't terribly interested in the arcana of particle physics. They preferred to publish Hawking's quirks. And his peccadillos.[8]

"Renowned Physicist Stephen Hawking Frequents Sex Clubs," trumpeted one tabloid in February 2012:

Renowned physicist Stephen Hawking is somewhat of a regular at a Devore, California sex club, RadarOnline.com is exclusively reporting.

According to a source who has been a member of Freedom Acres swingers club for nearly half a decade, Hawking, 70, shows up to the club with a bevy of nurses and assistants and has a naked woman grind on him.

"I have seen Stephen Hawking at the club more than a handful of times," the source revealed.

"He arrives with an entourage of nurses and assistants. Last time I saw him he was in the back 'play area' laying on a bed fully clothed with two naked women gyrating all over him."

The source notes that Hawking, who is wheelchair-bound due to a long-time battle with Lou Gehrig's disease, has his staff stand nearby watching while he's with the women.[9]

Britain's *Daily Mail* went so far as to get a comment from Hawking's employer, Cambridge University. "It is not true that Professor Hawking is a 'regular' visitor to the club in question," a spokesman reportedly said. "This report is greatly exaggerated. He visited once a few years ago with friends while on a visit to California."[10]

Freedom Acres is a swingers' club in San Bernardino, California, about an hour's drive from Caltech. Single men are welcome, but the club primarily caters to nudists and couples who participate in the swinger "lifestyle"—up to and including swapping partners for sex—and tries to attract a crowd that attends regular "play" parties. According to Russ Thomas, one of the club's owners, Hawking was at Freedom Acres "once, maybe twice, but certainly was not a regular." Indeed, given that Hawking used a wheelchair, merely navigating the club in the dark—with its numerous out-of-the-way nooks and exhibitionistic lofts and stages—would have been a challenge. "We are not equipped to cater to people with disabilities," Thomas admitted. "But Mr. Hawking had his own staff with him and they took care of all his needs."[11]

Hawking's sexual outings in California were known to many in his circles. After all, such adventures, by definition, are highly social events. Hawking's friends and colleagues are understandably reluctant to discuss them directly, but some of those in the know give the distinct impression, albeit obliquely, that these outings were very much a part of who Hawking was.

By this time in his life, Hawking's disease had progressed to the point where he had almost no control over any part of his body except his face.

His legs had long since been rendered useless, their muscles atrophied after not having been used for decades. He couldn't use his arms, or his larynx, or (increasingly) his fingers. Everything he did—attending to almost every single task he needed to attend to—required the intervention of a third party, or at least having someone in close proximity. This not only included the basic necessities of life, such as eating and dressing and cleaning himself and voiding his bowels, but also the other needs of a sensual being. Hawking yearned to savor good food, to sip fine wine, to listen to music, to gaze upon a naked body, to feel the touch of a paramour.* As Hawking's disease progressed, these sorts of private pleasures increasingly had to be mediated by other people—and there always had to be a nurse or carer available who could respond to an urgent summons. Even in his own home, he was never alone but in small company.

And Hawking was a sensual man. Director Errol Morris tried to portray this facet of Hawking's personality in the film version of *A Brief History of Time*. "I wanted to capture something about his ideas. His obsessions. His struggle. And the pleasure that he got," Morris recalled. "I mean, one unforgettable . . . just around the corner from here, Kendall Square, at Legal Seafood, Hawking ordered a lobster. Of course, he needed enormous help eating this lobster. And it was just one of the biggest messes that I had ever seen." Morris paused, smiling. "The table covered with lobster, Hawking covered with lobster. But here is a man really, really enjoying himself."[12]

These sensual moments were made possible by the sacrifice of someone else's time and effort. And it had been some years since Hawking had enjoyed the willing self-sacrifice of a loving spouse—his second wife, Elaine, had left him in 2007, and whatever intimacy he had once shared with his first wife, Jane, had long since become an unbearable burden for her. At least that's how she described it to the world.

In her first memoir, *Music to Move the Stars*, Jane described her sexual relationship with Stephen in harrowing terms—for her, the sexual act itself was "empty and frightening," a source of anxiety and repulsion:

> I had reason to fear that the effort involved in sexual activity might kill Stephen in my arms. . . .
>
> The functions I fulfilled for him were all maternal rather than marital: I fed him, I washed him, I bathed him, I dressed him, I brushed his hair and cleaned his teeth. . . . It was becoming very difficult—unnatural, even—to feel desire for someone with the body of a Holocaust victim and the undeniable needs of an infant.[13]

---

* Despite his disability, Hawking's sex drive was essentially intact.

When Jane released the second version of her memoir several years later, this passage was softened significantly, but the essential message was the same: she found the relentless sacrifices needed to fulfill Stephen's needs to be fundamentally incompatible with spousal affection.

With his divorces, Stephen had lost the most reliable mechanism for him to get his basic needs fulfilled. Hawking had to rely upon his friends, family, and confidants to look out for him. Otherwise, he had to impose his will on others through the sheer force of his personality—or with his money.

Conversely, others tried to impose their will on him. The Hawking name was extremely valuable, and a number of people close to Hawking have described sometimes bitter conflicts behind the scenes over the control of how he spent his wealth, his attention, and his time in the last few years of his life. Most of these conflicts did not erupt publicly; there was one notable exception, but even then, the details are scant.

This exception was the case of Patricia Dowdy, one of Hawking's nurses. Dowdy had worked for Hawking in the early 2000s, left, and then came back several years later. In 2016, at the behest of a number of Hawking family members, the UK Nursing and Midwifery Council opened an investigation into Dowdy, immediately suspending her pending an inquiry. Of the twenty-five professional standards in the council's code of conduct, Dowdy stood accused of violating nineteen of them. The investigation lasted for two years and wrapped up with a hearing shortly after Hawking's death; Dowdy effectively lost her nursing license* in 2018 for "multiple misconduct charges" related to her care of Hawking, including "financial misconduct; dishonesty; not providing appropriate care."[14]

The circumstances were irregular and rather bizarre: Though such hearings are typically open to the public, the council made an exception in this case and held the hearing in private. To this day, the council refuses to release documentation regarding the proceedings, making it all but impossible to understand the nature of the charges against Dowdy, much less to try to get a sense of her guilt or innocence. And though Hawking's own papers almost certainly reveal the circumstances of the conflict that led to the end of Dowdy's nursing career, those documents are in the hands of the Hawking family who accused her—and they're not revealing anything either.

The rougher edges of Hawking's personal life can only dimly be seen poking out from underneath the surface, given how much effort has been expended to keep them under wraps.

---

* Technically, she was "struck off" the nursing registry.

...

Despite the flourishing market for details about Hawking's personal life, he himself didn't publish a memoir or autobiography until 2013, when he was seventy-one years old. And when *My Brief History* came out, it was almost the diametric opposite of his ex-wife's tell-all. Reviewers were struck by its brevity—and the author's reticence, his "air of such studied incuriosity, particularly about the people in his life, that it feels inveterate."[15]

"As for his private life, his two marriages—swirling around them rumors of infidelity by Hawking and his wives and of physical abuse of the disabled scientist—and his globe-trotting experiences as a celebrity, the understated Hawking reveals very little," one reviewer wrote. "He admits that his failing health and celebrity status didn't help his relationships, but offers few details or insights."[16]

In *Nature*, the peer-reviewed journal where Hawking had published his famous suggestion that black holes should eventually explode, the review was even more brutal: "*My Brief History* does not do what we expect of a memoir. It does not take the reader behind any scenes. Hawking narrates his life non-introspectively, celebrating its triumphs and burnishing its sensitive moments. It is a concise, gleaming portrait, not unlike those issued by the public relations department of an institution."[17]

Without any real introspection, *My Brief History* wasn't likely to become any sort of critical—or, more importantly, financial—success.* Here, too, Hawking's inability to communicate at more than three words a minute played a role in limiting the amount of new material he could produce. Much of the book was recycled. His account of his childhood and college years was a lightly edited version of a speech he'd given in 1987, bulked up with additional material from a few years later. His account of the writing of *A Brief History of Time* had appeared in *The Independent* and *Popular Science* in late 1988 and early 1989. Other chapters on his scientific work were collages of previous lectures and writings. The all-too-brief introspective parts that reviewers (and readers) craved were the only really novel elements in the book. Yet producing novelty—writing something new from scratch, rather than cutting-and-pasting from prior utterances—was what gave Hawking the most difficulty.[18]

---

* Hawking's financial situation improved temporarily in 2013 when he won a $3 million prize sponsored by Russian billionaire Yuri Milner. Milner and Hawking jointly made headlines again three years later when Milner started funding, and Hawking publicly supported, a rather odd proposal to send tiny laser-propelled spacecraft to Alpha Centauri.

Even if Hawking had found it easier to write new chapters for his memoir, he may well not have been so inclined. He consistently tried to maintain a dignified, humorous, and self-deprecating tone in his popular writings. It wouldn't have been compatible with his image to try to boost sales by writing about the more intimate parts of his life, especially anything that might come across as controversial or scandalous. So it shouldn't have come as a surprise that *My Brief History* gave no insights into Hawking's love life, or even into his politics or his religious beliefs.

Hawking made no secret of his atheism. Though he frequently referred to God in his writings, typically as a metaphor for the underlying order of the universe, he made it quite clear on a number of occasions that he did not believe in a supernatural being or spiritual force.* Indeed, the version of the cosmological creation that he came to believe in ruled out the existence of a creator. Yet that didn't seem to satisfy the public's thirst for spiritual wisdom from the physicist; he was constantly asked about his religious views. The total absence of God in his memoir was striking.

There were also relatively few clues about his politics. Since his youth, Hawking had identified as a socialist—and he clearly leaned left. However, he tended not to weigh in publicly on political matters, except on occasions when there was a scientific angle, as with climate change or nuclear disarmament. And in those cases, he would use his scientific credentials to bolster his authority.

It was largely irrelevant that Hawking's scientific expertise often had little to do with what he opined about. His years of studying general relativity and particle physics didn't give him any special insight into the nature of artificial intelligence, the need for space travel, or the desirability of contacting alien species, for example. Yet he was more than happy to speak at length on these subjects, as if he were representing scientific consensus. This sometimes irked even his longtime friends, such as astronomer Martin Rees. "[A] downside of [Hawking's] iconic status was that his comments attracted exaggerated attention even on topics where he had no special expertise—for instance philosophy, or the dangers from aliens or from intelligent machines," he wrote in 2018. "And he was sometimes

---

* The only significant suggestion to the contrary was made by his second wife, Elaine. If she is to be believed, during their marriage she regularly took Stephen to services at St. Barnabas in Cambridge, where "the prayers, hymns, and bible readings often evoked a lot of quiet weeping from him." In addition, she said, "At home, he read the Bible. He usually asked for the Old Testament stories and [we] often prayed together." Elaine Hawking, "Baptism Testimony of Elaine Hawking," April 2018, https://archive.org/details /chipping-campden-baptist-church-513231/2018-04-01-Baptism-Testimony-Elaine -Hawking-Elaine-Hawking-46453311.mp3.

involved in media events where his 'script' was written by the promoters of causes about which he may have been ambivalent."[19]

This isn't to say that Hawking only spoke out about science-adjacent topics. He would opine about social services for the disabled, and toward the end of his life, he grew somewhat more outspoken about politics. He made comments in reaction to the rise of populism in Europe and the United States, and even backed a few Labour candidates for Parliament. He opposed Brexit, just as a few years before that, he opposed Scottish independence from the United Kingdom. Even so, with a few exceptions, even in purely political matters, he tended to couch his statements in scientific and technological terms. In 2016, he said that Brexit "would damage scientific research in Britain"; populism was, in part, "another unintended consequence of the global spread of the internet and social media"; and, at the same moment that he labeled Donald Trump a demagogue, he urged the new president to choose a different head of the Environmental Protection Agency.[20]

Even when he got into a scrap with the Tories over cutting funds for the National Health Service, Hawking took then minister of health Jeremy Hunt to task in an op-ed as if he were a fellow physicist.* "Hunt had cherry-picked research to justify his argument," Hawking wrote. "For a scientist, cherry-picking evidence is unacceptable. When public figures abuse scientific argument, citing some studies but suppressing others to justify policies they want to implement for other reasons, it debases scientific culture. One consequence of this sort of behaviour is that it leads ordinary people to not trust science at a time when scientific research and progress are more important than ever." Hawking rarely deviated from this pattern of argumentation from scientific authority, and when he did, it was typically big news.[21]

In mid-2013, Hawking was scheduled to appear at a conference hosted by the then president of Israel, Shimon Peres. This got the attention of the British Committee for the Universities of Palestine (BRICUP), an organization of academics in the United Kingdom devoted to fighting the occupation of Palestinian lands by promoting a boycott of Israel and through other means. Jonathan Rosenhead, the chair of BRICUP, managed to get two dozen other academics to write Hawking a letter declaring that they were "surprised and deeply disappointed" that he was helping Israel to "camouflag[e] its oppressive acts behind a cultured veneer."[22]

---

* Toward the end of his life, Hawking's worries that the Conservative Party was destroying the NHS drove him to become much more publicly partisan, to the point of endorsing a Labour candidate for Parliament in 2017.

Hawking promptly withdrew from the conference, couching it as his "independent decision to respect the boycott" of Israel. Hawking's employer, Cambridge University, was apparently caught flatfooted; its spokespeople promptly denied that their universally beloved professor had deliberately waded into one of the most divisive political topics of the modern era. Accusing BRICUP of "misunderstanding" the reasons for Hawking's withdrawal, one university spokesperson insisted that "Professor Hawking will not be attending the conference in Israel in June for health reasons—his doctors have advised against him flying." But it was Cambridge that had misunderstood; Hawking canceled because of the boycott.[23]

Hawking's controversial stand meant that he would receive some extremely negative publicity—something he had mostly avoided throughout his life. While Palestinians and boycott supporters praised the physicist, critics attacked him. Several leveled accusations of hypocrisy against Hawking because his speech synthesizer relied on an Israeli-made computer chip. One furious professor at the University of Haifa even suggested that Hawking should get "a free trip on the *Achille Lauro*," the cruise ship from which hijackers had infamously thrown a passenger in a wheelchair overboard.[24]

This was an extreme reaction to a rare exception. Hawking didn't shy away from controversy, but when he got into a scrap, it was almost always about something scientific—so it was generally only fellow scientists who got really annoyed with him. Scientists like Peter Higgs.

· · ·

In 2012, when the Higgs boson was found at the huge Large Hadron Collider (LHC) at CERN, newspapers happily reported that Hawking had lost his $100 bet with physicist Gordy Kane. What they didn't report was that it was his third bet with Kane—and Hawking had won the other two.

Back in the mid-1990s, the tunnels at CERN were home to a different kind of particle accelerator known as the Large Electron-Positron Collider (LEP). The LEP was weaker than the LHC, but at the time, physicists were predicting that it would find the Higgs. Hawking was skeptical; his own calculations seemed to imply that on the very smallest scales—when you zoom in far enough to see the frothy, discontinuous substrate where the Higgs particle belongs—tiny black holes were constantly popping in and out of existence. This may seem like a bizarre idea, but the weirdness of the subatomic world means that such things are eminently possible. These tiny black holes, Hawking argued, would mask the existence of the Higgs boson—they would make it impossible to observe any trace of a Higgs. Most scientists didn't buy Hawking's argument, although fellow physicist Malcolm MacCallum described it at the time as "wild and provocative."[25]

It provoked Gordy Kane, at the very least. A theoretical particle physicist at the University of Michigan, Kane was extremely sanguine about the prospects at the LEP. He thought not only that it would likely spot the Higgs, but that it would discover hints of new exotic particles whose existence lie beyond the predictions of the Standard Model. So he bet Hawking $100 that the LEP would find the Higgs.

By the year 2000, things were looking grim for the Higgs search. After roughly a decade of operation, the accelerator had found scant evidence of anything resembling a Higgs, and the LEP was scheduled to be demolished to make way for the more powerful LHC. At the very last minute, just as the LEP was about to be shut down, scientists claimed to see a blip that might be a Higgs-like particle.* The LEP's execution was stayed a month to see if scientists could gather more evidence for a discovery, but no dice. The Higgs had not been found—and theoretical particle physicists were beside themselves with frustration. Hawking proudly claimed his $100 from Kane.

The ordinarily retiring Peter Higgs snapped, rubbishing Hawking's theory that the boson would never be found. "It is very difficult to engage [Hawking] in discussion, and so he has got away with pronouncements in a way that other people would not," Higgs told *The Scotsman*. "His celebrity status gives him instant credibility that others do not have."[26]

"I am surprised by the depth of feeling in Higgs' remarks. I would hope one could discuss scientific issues without personal remarks," Hawking reportedly responded, before turning the knife in the wound. "Higgs has got it wrong. I did not bet that the Higgs Boson doesn't exist, just that it would not be discovered at LEP and I have already won the bet."[27]

Just as the LEP was shutting down, the Tevatron, a newly upgraded accelerator at Fermilab in the United States was starting up. Once again, Hawking and Kane took opposing sides of the wager: Kane, firmly believing in the Higgs, bet that the Tevatron would find it; Hawking, still doubting whether the Higgs would ever be spotted, bet against.

Running a particle accelerator is an extremely expensive affair, even once the equipment's already built and in place. It sucks up an enormous amount of power, not just to get the particles moving close to the speed of light, but also to power and cool the magnets that steer the particle beams, and to maintain a vacuum in the miles upon miles of tubing the particles

---

* In reality, it was a budgeton, an evanescent particle whose existence is fleetingly glimpsed by every accelerator that's in imminent danger of being shut down.

travel in. By the mid-2000s, the US Department of Energy, which ran Fermilab, was sending out strong signals that the Tevatron was going to be the last US collider on the frontier of particle physics. If the Tevatron didn't find the Higgs, that was it as far as America was concerned—no bigger, better collider was going to be built. And as the end of the decade drew closer, it became clearer and clearer that the Higgs wasn't going to reveal itself; there were strong hints of where one might be hiding, but nothing that physicists could remotely call a discovery. It was looking like the Tevatron would soon be shut down, and Hawking would once again win his bet with Kane.

The situation was looking desperate. But the game wasn't over yet. There was one last hope to find the Higgs, and one last bet for Kane and Hawking to make. The LEP had been torn down to make way for a more powerful accelerator in the same tunnels—the LHC. If the LHC didn't find the Higgs, then it was game over, at least for the foreseeable future. It was highly unlikely that another machine would ever be built that could succeed where the LHC failed. So in late 2008, with the LHC finally ready to be turned on, Higgs hunters looked on with anticipation—and trepidation.

"I think it will be much more exciting if we don't find the Higgs. That will show something is wrong, and we need to think again," Hawking told the BBC the day before the accelerator started collecting data. Because Hawking's synthesized voice doesn't inflect to show emotion, it was hard to tell whether Hawking was being impish when he added, "I have a bet of $100 that we won't find the Higgs." Given the heightened emotions in the Higgs community at the time, he almost certainly knew he would provoke a reaction.[28]

The next day, at a press conference celebrating the startup of the LHC, Peter Higgs lashed out, deriding Hawking's theory that the Higgs boson would never be found. "I have to confess that I haven't read the paper in which Stephen Hawking makes this claim. But I have read one he wrote, which I think is the basis for the kind of calculation he does. And frankly, I don't think the way he does it is good enough," Higgs grumbled. "My understanding is that he puts together theories in particle physics with gravity . . . in a way which no theoretical particle physicist would believe is the correct theory. . . . I am very doubtful about his calculations." Apparently, other scientists on the panel then "moved swiftly to cut off the discussion" before things got out of hand.[29]

It would be four more years before the LHC finally produced enough evidence of the Higgs for Hawking to concede his final bet with Gordy Kane. But in the interim, Hawking seemed to enjoy tweaking Higgs and those at the LHC who were searching for the Higgs boson. On several

occasions, Hawking half-heartedly joked that the accelerator could produce miniature black holes that would earn not Higgs, but Hawking, a
Nobel Prize.* And when he finally conceded that he was wrong in 2012,
admitting that Peter Higgs deserved a Nobel, he immediately followed up
with his statement that "it was a pity in a way" that the Higgs boson had
been found.

Peter Higgs would never say so, but on this point at least, Hawking was
right. Finding the Higgs boson marked the completion of the Standard
Model of particle physics; it was the last missing piece in a very successful
theory. But the theory is inadequate. Even though the Higgs allows particles to have mass, it doesn't explain gravity. For that, scientists need to
build a bigger, better model, something that goes beyond the Standard
Model to describe not just the strong, weak, and electromagnetic forces,
but the gravitational force as well. Physicists need to start seeing evidence
of phenomena that can't be explained by the existing Standard Model to
figure out where to extend it and improve it. It is precisely where the old
models break down that the new physics begins. If the old models don't
fail, there's nowhere to start.

It's for this reason Hawking had spent his life seeking out boundaries where theories clash and break down—at the edge of a black hole, at
the birth of the universe, because it is those boundaries that give scientists
fleeting clues to truths even more profound than the ones we already know.
By pouring enormous energy into a very small space, the LHC had the
potential to provide just such a boundary. But it failed to do so, even as
it succeeded in producing the Higgs. The accelerator didn't produce any
hints of beyond-Standard-Model physics, no clash of theories that begets
new knowledge. So, to Hawking, who had spent much of his life hoping
for a "theory of everything" that would reconcile the frothy discontinuous
subatomic realm and the smooth one of general relativity, the LHC was a
disappointment of a lifetime.

---

* Even after he conceded the bet, Hawking seemed to enjoy popularizing a theory that
measurements of the Higgs boson might imply that the universe is in some sense unstable.

CHAPTER 4

GRAND DESIGN (2008–2012)

_____

"**N**ow I realize that thinking in four dimensions is not easy . . ." says Benedict Cumberbatch, his lugubrious English voice substituting for Hawking's tinny, vaguely American-sounding, synthesized one. "But hang in there. I've thought up a simple experiment that could reveal if human time travel . . . is possible now, or even in the future." Of course, it's a cliffhanger—time for a commercial break.[1]

Into the Universe with Stephen Hawking was yet another Discovery Channel miniseries featuring the celebrity physicist. Episode 2 is devoted to time travel. Hawking's experiment is not what one might expect. As the commercial break ends, we see no lab equipment, no chalkboards, none of the sights an audience might associate with physics. Instead, the screen fills with balloons. A tray full of champagne glasses bubbles with anticipation. Plates of hors d'oeuvres sit on the table as an attendant pops another bottle of champagne of uncertain provenance; it seems to bear the blazon of Hawking's beloved Gonville and Caius College at Cambridge. "I like simple experiments and champagne," explains Cumberbatch-Hawking, "so I've combined two of my favorite things to see if time travel from the future to the past is possible." Cumberbatch's voice slowly morphs into Hawking's as the camera spies the physicist sitting among the festive decorations. "I'm throwing a party: a welcome reception . . ." A banner draped across the hall reads, "Welcome Time Travellers."

It was a clever stunt. Hawking threw a party, but didn't tell anyone of its existence. Only after the event did he publish his invitation:

You are cordially invited to a reception for Time Travellers
Hosted by Stephen Hawking
To be held at
The University of Cambridge
Gonville and Caius College
Trinity Street
Cambridge
Location: 52 12' 21" N, 0 7' 4.7" E
Time: 12:00 UT 06/28/2009
No RSVP Required

The four-dimensional coordinates on the invitation—three of space and one of time—gave a precise location in spacetime. But since the invitation was only published after the event ended, only someone from the future could make use of that information. Which meant that the only attendees, besides Hawking himself (and his film crew), would be people from the future who were somehow able to travel backward in time to attend the party.

Unsurprisingly, nobody showed. "What a shame," Hawking said. "I was hoping a future Miss Universe was going to step through the door."*

• • •

The time-travel stunt had Stephen Hawking written all over it; his wit and humor combined with his unflinching confidence about the way the universe worked would mesmerize an audience every time. Little doubt it was a moment of authentic Hawking shining through in an otherwise humdrum Hawking-branded Discovery Channel special.

It was getting harder and harder to find the authentic Hawking underneath the persona, to distinguish the scientist's original thoughts and words from those that were thought and written for him by cadres of script writers and coauthors. Even in his science, it was getting increasingly difficult to figure out where he ended and his collaborators began. Stephen Hawking was as much a brand as he was a person.

In this, Hawking was little different from many celebrities, even from many scientists, who receive (sometimes disproportionate) credit for the work performed by the assistants in their lab or the graduate students under their tutelage. Hawking's situation was more extreme because of his

---

* One possibility that didn't occur to Hawking was that the lack of attendance might be because the champagne was plonk.

condition; almost every function of his life—eating and defecating, much less writing and preparing lectures—was a collaborative effort. Hawking had the ability, and frequently the necessity, to outsource some of the basic functions of the self. It could be hard to discern the line between Hawking the person and Hawking the collective creation. The brand was so famous that it was sometimes hard to remember there was a human, and an extremely ill one, underneath it all.

• • •

Though Hawking's time-travel stunt was more P. T. Barnum than Albert Einstein, the idea did, in fact, have its roots in the theory of relativity. In 1905, Einstein inadvertently opened the door to time travel. When he discovered that space, time, and motion were inextricably intertwined—that one's motion through space affected one's motion through time—the idea of going into the past no longer seemed like the ridiculous fantasies of Mark Twain or H. G. Wells. Science had declared that time was malleable; it was just a question of figuring out how far it could be stretched and bent.

Even before Einstein made the crucial connection between gravity and the curvature of the fabric of spacetime, he and his fellow physicists had realized that there was something very odd about the way space was shaped. And the quirks in the shape of spacetime were key to understanding some of the strange rules Einstein was discovering about how space and time behaved.

Understanding motion in Newton's universe is a matter of simple geometry: the same geometry taught in high schools around the world, the same geometry of Euclid and Pythagoras and Archimedes. The formula for the distance between two objects in three-dimensional Newtonian space—what physicists call the *metric* of Newtonian space—is just the Pythagorean formula we all learned in elementary school. And as far as Newton is concerned, that's as complex as it gets; our universe is three dimensional, we only need three variables (x, y, and z) to pinpoint an object in space, and good old Euclidean geometry is sufficient to describe everything perfectly.*

This particular metric says that space is, well, boring. It's featureless, flat space with no surprises, the sort of stuff that even the ancient Greeks would have been perfectly capable of describing. Indeed, it's called

---

* That is, the distance, d, between two objects in Newtonian space is

$$d = \sqrt{x^2 + y^2 + z^2}$$

where x, y, and z represent the left-right/up-down/front-back distances of the two objects from each other.

Euclidean space because it obeys the rules of geometry that Euclid wrote down roughly two and a half millennia ago.

Einstein's universe is trickier than Newton's. To describe a location in Einstein's universe requires not just the three dimensions of space, but one of time as well. And this means that the formula for "distance" in space-time (what physicists call a spacetime *interval*) must incorporate a fourth variable—one for time, *t*—in addition to the *x*, *y*, and *z*. And the formula for distance in Einstein's theory is more complicated than the Pythagorean theorem. The Pythagorean theorem treats all dimensions as interchangeable. Not so with the formula for distance in Einstein's universe: time, *t*, behaves differently from the three space dimensions *x*, *y*, and *z*. In fact, if you look at the formula, there's a strange little minus sign next to the variable for time that's missing from the variables for space.* It's a subtle difference, but it has a huge effect.

To a mathematician, a metric like the one above is much more than a formula; it's a description of a *geometry*, an encapsulation of what space (or in this case, spacetime) looks like in the neighborhood around someone. The minus sign in the metric for spacetime means that the geometry of the spacetime that Einstein's laws describe—what physicists call *Lorentzian spacetime*—is much more complicated than the standard, featureless geometry of Euclidean space of Newton's laws of motion.† It is this geometry that leads to all the weird consequences of relativity—the speed-of-light limit, the slowdown of clocks moving very quickly. Everything that's counterintuitive about relativity theory is merely a manifestation of the non-Euclidean shape of spacetime. It's possible to get a sense of this with a diagram.

If a person starts at Trafalgar Square, picks a direction, and moves through Euclidean space for a thousand feet, she ends up somewhere on a circle (or a sphere, if she can fly or dig) centered on her starting point. No

---

* That is, the distance between two objects in spacetime. Technically, the *spacetime interval*, is given by the formula

$$s = \sqrt{x^2 + y^2 + z^2 - t^2}$$

where *x*, *y*, and *z* represent the distances in three dimensions of time and *t* represents the difference in time between the two objects.

† Technically, the terminology is a little more complicated than this, as scientists have different terminologies for the geometry of an object and the object itself. The Earth is an object that (more or less) has a three-dimensional Euclidean geometry at every point on its surface. This makes it a so-called "Riemannian" manifold. Einsteinian spacetime, with the minus-sign metric, technically has what's called a "Minkowski" spacetime geometry; the cosmos, which has this geometry at every point, is known as a "Lorentzian" manifold. I've elided over the distinction in terms for simplicity's sake.

matter which direction she picks, if she sticks with it, her journey is constrained; it must end somewhere on that shape.

The same person moving through spacetime away from Trafalgar Square at midnight on January 1, 2001, for a given interval of spacetime has a different sort of constraint. One can get a rough sense of this by looking at a *spacetime diagram*: a graph where one axis represents space and the other represents time. On this diagram, our spacetime traveler cannot move in just any direction; she can't even sit still. She's always moving upward: forward in time. And the dictates of Lorentzian spacetime say that after a given interval, her final destination will lie not on a circle, as in Euclidean space, but on a very different sort of curve, an open-ended shape known as a hyperbola, which is bracketed by lines at 45 degrees.

That hyperbola and those 45-degree lines constrain where a person can visit in spacetime. A traveler can travel faster and faster, traveling farther and farther away from Trafalgar Square, but as her endpoint must be on the hyperbola, her trajectory will never cross the 45-degree lines. For in a spacetime diagram, 45-degree lines represent the speed of light. Those 45-degree lines—known as the *light cone* emanating from the starting point—are the ultimate boundary to her travels. Unless the traveler can figure out how to go faster than light, no matter how hard she tries, she can never cross the barrier of the light cone. She is eternally constrained to remain within that cone, unable to visit any regions of spacetime that lay beyond.

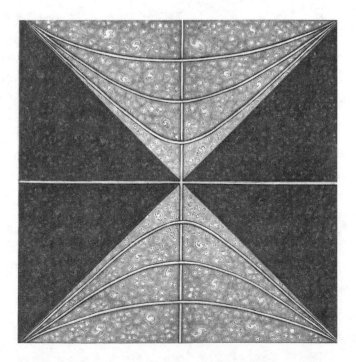

However, there's a loophole. If our spacetime traveler does somehow manage to travel faster than the speed of light, she can cross into the forbidden zone. Or there might be some flaw, some tunnel in the fabric of spacetime, that allows her to cross the boundary of the light cone. If she does, the mathematics of spacetime breaks down, and the rules get thrown out the window. She can communicate with people in the past—send them information that could change their destiny. She might even be able to move backward in time.

The ordinary rules of general relativity rule out both possibilities. Not only is traveling faster than the speed of light forbidden, but the fabric of spacetime is presumed to be smooth and unbroken, not full of holes or tunnels. But Hawking, like most other physicists, is confident that those rules don't always hold. Time travel is possible, at least in theory. And musing about the potential of time travel—and what sort of rules the universe must follow if time could be reversed—was one of Hawking's favorite activities.

It was also a favorite for his audiences. At the end of one of his public lectures, "Space and Time Warps," he would emphasize that "rapid space travel, or travel back in time, can't be ruled out, according to our present understanding." It was a thrilling statement, even though it was delivered in the flat affect of a robotic voice. Perhaps because it was delivered in the flat affect of a robotic voice.[2]

...

There were two things almost always true about a public Stephen Hawking lecture. First, it was very well attended—every seat taken, and if the fire marshals weren't a force to be reckoned with, large clots of people near the exits and in the aisles, craning their necks to get their first view of the celebrity physicist. The second was the hush that fell over the crowd as he wheeled out, accompanied by—or later in life, steered by—an attendant. There was typically a rustle of tension in the crowd as the attendant fiddled with the equipment and Hawking prepared to speak. And then came the first words, intoned exactly the same way each time, as if in a ritual: "Can you hear me?" The tension broke as the audience applauded, and the physicist triggered the voicebox to begin pronouncing the first words of his speech.

"Sometimes there was thirty, forty seconds of pure silence," says Christophe Galfard, a PhD student of Hawking's who, inspired by his adviser, become a science popularizer after he graduated. "For me, it was the silence within those forty seconds that made it so . . . that's what triggered my wish to pursue that road."[3]

Like many speakers on the public lecture circuit, Hawking had a set of stock talks that he would draw from, and he would occasionally tweak them or remix them, updating the science when appropriate, or adding new imagery. His talks were filled with fragments of text from his earlier books, and his later books contained lots of text from his talks. In fact, because all of Hawking's communications were mediated by his computer—the drafts of his books, the speeches he gave and the ones he never did, the answers to questioning by a colleague or an audience member— his utterances were never evanescent, but semipermanent, stored in digital memory. Every time he spoke or wrote, those phrases survived for decades, mutating, merging, and producing offspring every time Hawking searched his memory banks for something to say.

Hélène Mialet, a sociologist of science, wrote of the travails this situation caused for a librarian's assistant going through the physicist's papers. Typically, an archivist assesses the value of a document by its uniqueness:

> However, the material collected as part of the Hawking Archive is frequently characterized by its repetitiveness, by its iterability, as in electronic or published documents that can be reproduced at will *ad infinitum*. . . . And what does one do with all the speeches given in different places? Though slightly altered for each audience, they are essentially the same. It is thus difficult for the assistant librarian to trace their provenance and put them into context.[4]

Hawking was not the only public lecturer reusing his talks and papers—far from it. But the consequences of his illness were that any information in his brain that he wanted to transmit to the outside world had to flow through a digital channel—a channel that obscured the original thought that created that information in the first place. As Mialet put it:

> For years now, he has had to do everything through the computer. This is true not only for Hawking, but increasingly for all scholars. Hawking's particular condition . . . makes visible the practices of intellectuals who cut and paste from different documents to compose new talks, who recycle the same talk for different audiences, who transform their talks into articles or publish talks, interviews, and conferences as books. In other words, the drafting process that would

normally take us closer to the "origin" of the process of creation—that is, to "the author"—lead[s] instead to a hall of mirrors.[5]

With few exceptions, every speech Hawking gave was a reflection of a reflection of a speech or a book or an essay written years ago.

His audience cared not one whit. Hawking's speeches were always wildly successful, even more so when he was able to take questions, which were often submitted in advance or prescreened by assistants. (All the better to reduce the audience being subjected to long pauses as the physicist composes his responses.) Mialet describes a press conference in 1999:

> Interestingly, the questions are always more or less the same. . . . "Well, basically," the assistant explains, "what they are saying is: 'Do you believe that we're going to find a [theory of everything]?' And Stephen has given a public lecture, a number of times . . . called 'The Millennium Lecture,' which he gave to Bill Clinton, where he is saying in twenty-five years. What Stephen will probably do to answer that question is just take the last paragraph of his public lecture and say it."[6]

Hawking's assistants rolled their eyes at the shallow and predictable questions, but the esteemed members of the press were thrilled just to be in the great man's presence. "He'll probably take the last paragraph of his lecture, and go 'dadadadadada,'" one graduate student said. "And the journalists will be fans—they'll go 'Wow.'" It simply didn't matter if Hawking had said something a hundred times before; people relished the experience of being there in the room when he twitched his muscles to send the next block of text to his speech synthesizer. Novelty was beside the point.[7]

. . .

On the occasions when Hawking needed to produce something truly novel, he needed help. And he had no shortage of people who were more than willing to assist.

This was the natural way of things for an academic physicist—indeed, for a typical professor guiding students to their doctorates. Graduate study is an apprenticeship under a master with more ideas than he or she has time and energy to pursue. The professor uses doctoral students as tools to extend his or her ability to find new approaches, to think in new ways, to grind through tedious calculations, and as a means to greater glory—for the advising professor typically gets a significant share of the credit for the graduate students' projects.* Hawking usually had four graduate students working for him at any given time; as the Cambridge PhD usually took four years, each year one would graduate and be replaced from a pool of eager young candidates. This was standard in physics—even if Hawking, because of his disability, needed more assistance from his graduate students than normal: not just with physics, but with the physical.

A number of times over the years, Hawking had graduate students as live-in caretakers. Hawking described one such arrangement in the early 1970s: "He helped me with getting up and going to bed and some meals, in return for accommodation and a lot of academic attention." Like many of Hawking's students, Bernard Carr, one of his earliest graduates, was reverential, almost worshipful of his adviser. He told journalist Dennis Overbye that helping Hawking was "like participating in history." Overbye wrote, "At such comments, Hawking snorted in return that it was hard for a student to be in awe of his professor after he has helped take him to the bathroom."[8]

When it came to his popular books and speeches, Hawking also relied to some extent upon graduate students to help him, but when it really mattered, Hawking got professional help. And after not having produced a major book in a decade—and apparently needing the money—Hawking's next literary venture definitely mattered.

Leonard Mlodinow, a physicist at Caltech and popularizer of science, had worked with Hawking a number of times before. A decade prior, in the early 2000s, Hawking was discussing with his agent, Al Zuckerman, the possibility of reworking *A Brief History of Time*. "Hawking was looking for someone to collaborate with," Zuckerman says. As it happened, Mlodinow's literary agent, Susan Ginsburg, worked for the same agency as Zuckerman. And so *A Briefer History of Time* was born. "Mlodinow was not on there as a coauthor; I think it was just acknowledged. [But] Mlodinow certainly wrote *A Briefer History of Time*, and Hawking didn't like that," Zuckerman recalls, adding that it wasn't the lack of control that irked Hawking: "He gave Mlodinow 30 percent of the money. [Hawking was annoyed] because what sold the book was his name, even though Mlodinow did the writing." ("He continued to ask me to write other things for him, so he must have thought it was worth it," Mlodinow retorts.)[9]

---

* Sometimes that credit is arguably exaggerated or even misplaced. "I do think, for example, our most famous paper together, he should have let me publish it as a single author, because the main results were mine," says Marika Taylor, a former student of Hawking's. "He took credit for something which wasn't his." This is not an unusual complaint against thesis advisers, unfortunately.

The idea was to do something different from the books Hawking had published before. "I realized that [*A Brief History of Time*], *A Briefer History of Time*, *Universe in a Nutshell*, *The Illustrated Brief History of Time*, they were all pretty much the same topic, and that was his work in the '70s and '80s," says Mlodinow. "And he'd done a lot of stuff since then."[10]

This was the origin of *The Grand Design*. Unlike Hawking's last few literary ventures, it was a wholly new book—in the words of the publisher, it would be "the first major work in nearly a decade by one of the world's great thinkers."

Mlodinow, who would be a full-fledged coauthor, desperately wanted to get it right—though after a year's worth of labor, the two hadn't gotten the outline into perfect shape. "But then one day, [Hawking] just said, basically, 'We're done,'" Mlodinow says. "So I figured he means let's go to dinner or something. But no, he meant we're done with the outline. . . . He didn't base finishing the outline on the fact that we were done with it, but on the fact that he got tired of doing it."[11]

Hawking had a point: even without the perfect structure, the book was almost guaranteed to be a hit. "Repetition of the multi-mega-copy sales of *A Brief History of Time* (1988) can be safely predicted," warned *Booklist* in 2010. "Expect queues in stores and libraries for Hawking's latest parting of the veil to far-out physics."[12]

Sales prospects got even better thanks to the press attention before the book was released, mostly concentrating on a line in the manuscript exclaiming that "it is not necessary to invoke God to light the blue touch paper and set the universe going." This was sure to upset people who looked to Hawking for homey theological wisdom—the wisdom implicitly promised in the famous last sentence of *A Brief History of Time*, when Hawking suggested that constructing the ultimate laws of physics would mean that "we would know the mind of God." Indeed, the last line of *The Grand Design* echoed that line, but had displaced God. Now, wrote Hawking, constructing those laws would mean that "we will have found the grand design."

"Hawking: God Did Not Create Universe," blared *The Times* in London, while hundreds of similar headlines around the world expressed amazement and some discomfort at Hawking's atheism. Baroness Susan Greenfield, a neurobiologist and celebrity scientist in her own right, took to the BBC airwaves to explain how Hawking's "smugness" made her cringe: "Of course they can make whatever comments they like but when they assume, rather in a Taliban-like way, that they have all the answers then I do feel uncomfortable. I think that doesn't necessarily do science a service."* But those who criticized Hawking's statements about the lack of a need for God, or accused him of "perform[ing] a U-turn on the intellectual highway" to arrive at his atheism, clearly weren't sufficiently familiar with Hawking or his work.[13]

As more perceptive reviewers realized, *The Grand Design* was not an attempt to pit God against science. It was, at core, a pop cosmology book of the type that he and other physicists had written a number of times before, and sometimes with more skill.

· · ·

The title of *The Grand Design* evoked a hope that physicists had been harboring for more than half a century: sometime soon, scientists would figure out a simple set of equations; a beautiful, self-contained list of rules that governed how all matter and energy in the universe behave under all circumstances and on every scale. All the different kinds of particles, all the forces in the cosmos, all these would be described completely by these equations. This is the mythical "theory of everything"—the white whale of theoretical physicists since Einstein. This set of rules would essentially be the blueprint for the universe. It would be the Grand Design—the constraints that even the creator had to obey.

Physicist Richard Feynman once likened the search for the rules of the universe to trying to figure out the rules of a game of chess by watching it being played—but you're only able to see a few squares of the board rather than the whole thing at once. Given a short time, you might figure out that a pawn only moves forward, that each side has a bishop that stays only on the white squares and another that stays on the black ones, and a few basic rules of the game. But then, "you suddenly discover one day in some chess game that the bishop doesn't maintain its color, it changes its color. Only later do you discover the new possibility that the bishop is captured and that a pawn went all the way down to the queen's end to produce a new bishop. That could happen, but you didn't know it." Feynman explained. "And so it's very analogous to the way our laws are. They sometimes look positive, they keep on working, and all of a sudden, some little gimmick shows that they're wrong—and then we have to investigate the conditions under which this bishop changed color."[14]

---

* According to *The Telegraph*, Greenfield walked back the comments: "She later claimed her Taliban remarks were 'not intended to be personal,' saying she 'admired Stephen Hawking greatly' and 'had no wish to compare him in particular to the Taliban.'"

Physicists posit a set of rules about how the world works, and then they make observations, over and over again, to see whether those rules are holding as expected. Most of the time, they do—but when they don't, there's the possibility of getting a deeper, better understanding than before.

For example, by the end of the eighteenth century, scientists had divined a set of rules that seemed to be working pretty well. Atomic theory described the composition of matter: it was made of invisible, uncuttable atoms that had various properties. Newton's law of gravitation described how matter was affected by (and affected) the gravitational force, and Maxwell's equations described electricity and magnetism. In the early twentieth century, though, scientists were seeing that the atom itself wasn't uncuttable, but composed of pieces of different sorts (electrons, neutrons, and protons), and that a different sort of force altogether—the weak force—was involved in radioactive decays. By the 1970s, the rulebook had gotten yet more intricate; protons and neutrons were themselves composed of smaller particles, quarks, and were bound together with another kind of force, now known as the strong force.

It might seem that the rules of the game have been getting more and more byzantine at each iteration. But from a physicist's point of view, it's just the opposite: the rules are getting simpler, thanks to *unification*. Scientists realized that the electric force and magnetic force are really different aspects of the same thing; the weak force, too, at high enough energies, is also governed by the same fundamental laws. Similarly, the whole particle zoo, which seemingly produces an endless variety of different subatomic creatures, can be explained with a small handful of fundamental particles. As Feynman put it, "in the case of the chess game, the rules become more complicated as you go along, but in the physics when you discover new things, it becomes more simple." Perhaps it's a simplicity that's only truly appreciated by theoretical physicists, but the mathematical framework of the Standard Model is remarkably simple—with just a few assumptions, it's possible to describe all the kinds of matter that scientists have ever encountered, and how they are affected by three of the four fundamental forces: electromagnetism, the weak force, and the strong force. However, since gravity isn't included, at best, you can have a theory of *almost* everything.[15]

So physicists watch and wait, observing the universe through their particle colliders and their telescopes, hoping to see Nature breaking the rules so they can come up with newer, deeper, and hopefully simpler sets of rules that explain more. Perhaps, even, a theory of everything, a set of rules that would explain the behavior of all matter and energy in the universe. But such a theory would require a new mathematical framework—and new assumptions about the physical world that go along with that framework. What, precisely could that framework be?

By the time Hawking wrote *The Grand Design*, he had come to believe that a mathematical construct known as *M-theory* was the only real contender for that framework. M-theory wasn't really a theory—yet. It was a hypothetical construct, an inkling of a mathematical framework that theorists knew existed but couldn't yet describe very well. And to Hawking—as well as many other physicists—M-theory was the best shot at going beyond the Standard Model to explain all the matter and forces in the universe. However, M-theory—as potential theories of everything are wont to do— comes with some philosophical baggage that needs to be explained. For example, M-theory seems to imply the potential (or worse, the actual) existence of multiple different universes with different physical laws. Hawking spent a good portion of the book tackling this implication and explaining why we inhabit the universe that we do among all the myriad possibilities.*

Unfortunately, Hawking was fairly late to jump on the M-theory bandwagon, and there had already been several popular books written on the subject. So those reviewers who could look beyond the "Hawking denies God" storyline tended to be disappointed. "Even allowing for the need for the pleasures of digression, there is too much padding and too much recycling of long-stale material," wrote Graham Farmelo, a physicist and writer of popular math and physics books. "It gives me no pleasure at all to say that I doubt whether *The Grand Design* would have been published if Hawking's name were not on the cover."[16]

Nevertheless, Hawking's name was on the cover. Upon its release, *The Grand Design* immediately shot up to number one on the *New York Times* Best Seller list. Its sales were almost certainly helped by controversy. It wasn't just Hawking's stance on God that got critics clicking their tongues, but also his swipe at one of his perennial targets. "Philosophy is dead," said *The Grand Design* on its opening page. "Philosophy has not kept up with modern developments in science, particularly physics." It was a statement that seemed calculated to irk a certain segment of academia.

Though successful by any reasonable measure, *The Grand Design* was no *Brief History of Time*. For most authors, eight weeks on the *Times* Best Seller list would be the achievement of a lifetime. But *Brief History* had been there for more than two years solid. *Grand Design* just didn't come together

---

* By invoking the anthropic principle—roughly speaking, the idea that the universe is the way it is because if it were much different, we wouldn't be alive to observe it. More on M-theory and its problems in Chapter 7.

in the same way. Even Hawking's distinctive sense of humor—wry, some-times wicked, and yet self-deprecating—came across as flat. "Borscht belt," complained the reviewer for the *New York Times*, who seemed to pin the blame on Mlodinow rather than Hawking. Yet it was far from clear where Mlodinow's contribution ended and Hawking's began.[17]

. . .

At nearly seventy years old, Hawking was no longer taking new grad-uate students. Over his career, he had been the adviser on more than forty PhD theses, and his last advisee graduated in 2010. But Hawking kept working with his peers and publishing scientific papers, most of which were coauthored with two longtime collaborators, James Hartle at the Uni-versity of California at Santa Barbara and Thomas Hertog at KU Leuven in Belgium, and which mostly dealt with a subject that Hawking had been grappling with ever since his own PhD thesis nearly a half century earlier. It, too, had to do with time. Specifically, how it began.

The universe, as we now know, had a beginning. About 13.7 billion years ago, the universe came to be in a Big Bang. There was a moment of creation when the very fabric of space and time came into being. Since then, that fabric has been stretching—expanding, making galaxies rush away from one another. If you were able to travel backward in time, you'd be able to see the expansion running in reverse: the fabric of the universe would shrink and shrink, getting tighter and tighter until . . .

Until what? This is one of the questions that Hawking spent his career trying to answer. His PhD thesis came to the conclusion that there had to be a breakdown in the laws of general relativity at the moment of the Big Bang, that the fabric no longer could behave like the smooth manifold the laws of relativity require in order to work properly. However, for the last quarter century of his life, Hawking, along with Hartle, had come up with a clever dodge, a way to preserve the smoothness of the spacetime fabric even at the very moment of the Big Bang. Hawking was extremely proud of this solution, which came to be known as the *Hawking-Hartle no-boundary proposal.*

The no-boundary proposal was a powerful but controversial idea, and since the early 1980s, Hawking firmly believed that it explained what had happened to spacetime at the beginning of the universe. It also became the scientific foundation from which he felt he could dispense with the idea of a creator for the universe.

. . .

"I have not allowed my disability to stop me from doing most things," Hawking said during a prize acceptance speech. "My motto is, 'There are no boundaries.'"[18]

Hawking continued to work despite the increasing toll of his condition. With limited control over his musculature as well as a tracheostomy tube and a gastric tube that needed constant tending, he was regularly in and out of the hospitals, typically with respiratory infections. Many of them were quite serious. When Hawking decided to boycott the Israeli meeting in 2013, he did so from his hospital bed. Just weeks before his "Welcome Time Travellers" party in 2009, he pulled out of a scheduled conference in Arizona and rushed to Addenbrooke's Hospital in Cambridge, "very ill" with a chest infection, making headlines around the world.[19]

On January 8, 2012, on his seventieth birthday, Hawking was once again in the hospital, unable to attend the festivities. "All week at the scholarly symposium preceding this feast, there had been hopes that he'd appear. Maybe today? Maybe tomorrow?" wrote Kitty Ferguson, a longtime Hawking friend and erstwhile biographer who was in attendance. She recalled her shock when an a cappella group at the event quietly burst into song: "'Swing low, sweet chariot, comin' for to carry me home.' The crowd froze. Had someone chosen this way to announce Stephen's death? There were tears in the eyes of guests across from me." It was a false alarm. But it was a reminder of how Hawking had spent his entire life treading the boundary of the abyss—and refusing to let it constrain him.[20]

CHAPTER 5

CONCESSIONS (2004–2007)

———————————

S tephen Hawking's mechanical voice betrays no emotion. Every sentence is pronounced with the same flat affect, devoid of joy, of sorrow, of pain, of pleasure, of fear. He can't whisper, or murmur, or growl, or scream. He can only type.

When he typed "i cannot be left alone with her" to an assistant, reportedly as his second wife, Elaine, raged, there weren't any clues in his tone to give a sense of his inner thoughts. "please don't go. get someone to cover the shift."[1]

For a number of years, there were dark intimations—aired largely through the tabloid press, and strenuously denied by Hawking himself—that the physicist was being abused and manipulated. And it wasn't always clear who, if anyone, was doing the manipulation.

Elaine Mason had been one of Hawking's nurses starting in the mid-1980s, not long before *A Brief History of Time* made the physicist an international star. In 1990, Hawking left his then wife, Jane, and moved in with Elaine. It was not a movie-script-ready, dignified, mutual parting of the ways. Five years later, when his divorce was finally complete and Stephen married Elaine, there was still tension simmering underneath all of Hawking's familial relationships—with his ex-wife, with his three children, and with Elaine.

In many ways, Elaine was the diametric opposite of Jane. Where Jane was reserved, intellectual, steady, and very conscious of social norms, Elaine was fiery, volatile, demonstrative, and passionate. By all accounts,

Stephen's relationship with Elaine was tempestuous. "They would wind each other up," says one former acquaintance of Hawking's. "One would know that they were having arguments, because he would be in the office with his nurse and you would hear the arguments."[2] Some of his nurses and associates privately expressed concern about Stephen's safety.[3]

Those concerns began to erupt publicly in late 2000 and early 2001, when Stephen suffered mysterious injuries—a broken wrist, a split lip— and reportedly refused to explain how they had happened. Cambridgeshire police opened an investigation. Jane and her children discussed the matter with the authorities. However, the physicist refused to cooperate with the investigation, not responding to police calls and letters, and even threatening to sue for "harassment." The investigation was quietly dropped.[4]

But that wasn't the end of the matter. There were reports of other injuries: the physicist broke his hip in 2002, the result of a wheelchair crash. (As one newspaper put it: "Professor Hawking laughed off the incident, saying he had had 'an argument with a wall' and 'the wall won.'" One of his assistants at the time says that Hawking was going down a hill a bit too fast and was injured when he was thrown from his wheelchair.[5]) The following year, his daughter, Lucy, received a telephone call "report[ing] that the scientist had been burned after he was left out in the blistering sun in his garden. A nurse alerted police, who launched an investigation." Soon, the tabloids exploded with lurid details about how the professor was allegedly being abused, with the clear implication—sometimes the explicit claim—that his new wife, Elaine, was responsible. In *The Telegraph*, a nurse said that Elaine had bumped Stephen's wheelchair down the stairs, verbally abused him, and handled him in a rough manner. "She calls him a cripple and an invalid, which depresses him. The verbal abuse is unbelievable," the nurse recounted. "She would tell him no one cared about him. She is a very strong woman and can pick him up by herself. She does that a lot, and she can be very rough. It was all about humiliation, verbal abuse and rough handling. She would lift him up then drop him heavily into his chair." The same nurse told *The Times* of London that Elaine "let [Stephen] wet himself in front of his own mother. She refused to let anyone get the [urine] bottle." Several of Hawking's nurses reportedly made statements to the police detailing these claims and more.[6]

Not even Stephen's closest friends and family knew for certain what was happening. Some apparently believed nothing untoward was happening.[7] Others were just as firmly convinced that Stephen was, like many an abused spouse, covering up for a tormentor. "I believe quite strongly that these allegations are true, based upon what I have been privy to in the

past," Tim, Stephen's youngest child—who was twenty-four at the time—
was quoted as saying. "It makes me feel sick . . . and completely helpless.
I'm caught in the middle. He denies it every time I speak to him about
it . . . but I would hope he would respect me enough to tell me the truth."
Jane, too, expressed her disquiet. "The revelations have made me feel ill,"
she said in a statement. "He is a special man and vulnerable man but, when
his children see the aftermath of these events, they can only tell him he
must do something about it."[8]

Vulnerable Stephen might have been, as he could not directly control
his environment. But he was still able to exercise his will—and his will was
to support Elaine. "I firmly and wholeheartedly reject the allegations that I
have been assaulted," the physicist told the public. "The stories in the media
are completely false and I am profoundly disappointed by the circulation of
such personal and inaccurate information. My wife and I love each other
very much and it is only because of her that I am alive today."[9]

Once again, the police dropped their investigation. "The depth of my
inquiry has provided me with a great deal of information," the officer with
the Cambridge Constabulary in charge of the investigation said in a statement. "Based on what I have learned, I can find no evidence to substantiate
any assertion that anyone has perpetrated any criminal acts against Professor Hawking."[10] Then Stephen fired the nurses who had complained to the
authorities: "We were got rid of by being made redundant, saying that he
didn't need nurses anymore," one such nurse recalls. "He could make do
with 'carers,'" caretakers without the depth of training that nurses had. No
charges were ever filed.[11]

"It was a huge blow for us all when they dropped the inquiry," Stephen's
middle child, Lucy, told an Australian newspaper. "I really felt for Mum. It
was hard for her." And with Stephen firmly in Elaine's camp, Lucy said, "It's
been made quite difficult for us to see him and so to maintain an ongoing
relationship."[12]

• • •

It's one of the greatest ironies of Stephen Hawking's life: he was touted
as one of the world's foremost communicators of science, yet in his
communications, it was difficult to perceive the authentic human being
underneath the layers of mediation between him and the outside world.
Staff, nurses, family, and students were not always reliable conduits for his
thoughts—for his will. Even his computer and voicebox could be—and
were—occasionally manipulated by others who wanted to express their
own ideas in the professor's voice. Sometimes, this was benign: like other

celebrities, he would read advertising copy in exchange for cash or endorse a product for other consideration. Sometimes the only damage would be the risk to the Hawking brand, such as when he allowed publishers to cash in by producing sub-par books bearing his name.

The allegations of abuse in the early 2000s were the starkest example of the difficulty of trying to divine what was going on in the head of the world's most famous physicist. But as Hawking would prove time and again, he was anything but passive—and he was able to surprise even those who knew him best.

• • •

E ven as his personal life was in turmoil—his ex-wife and children all but estranged, police looking into allegations of abuse at home, frequent visits to the hospital as his health waxed and waned—Hawking found solace at work, at the Department of Applied Mathematics and Theoretical Physics at Cambridge University, which he visited almost every day. His office, decorated with his awards and with a large poster of Marilyn Monroe, offered escape. As he liked to say, "Although I cannot move and I have to speak from a computer, in my mind, I am free."[13]

Hawking bristled at being considered vulnerable—unable to move his muscles, he shaped the world around him through sheer force of will. And that will was formidable, if sometimes hard to live with. "If you are a disabled person and you are going to do something with your life, then there is a measure of egocentricity. If you are reliant on other people all the time, you've got to have that force of ego that will persuade people to do things for you," his daughter, Lucy, recalled. "But then, if I were in that situation, I doubt whether I'd have the determination to do as much as my dad does and to keep going. He is incredibly tenacious; a very stubborn person. Once he sets his mind on something, he's like an ocean liner: he doesn't change course."[14]

But change course he could—and did, even on fundamental matters. For much of his career, Hawking had plowed on, full steam ahead, with the conviction that there was a fundamental problem with one of the two great physical theories of the twentieth century, quantum theory. Until he convinced himself that he was wrong.

Quantum theory, like relativity, was born at the turn of the century, and its origins had to do with equations that weren't working as they should. And once again, the equations had to do with light.

In the 1800s, physicists were firmly convinced that light was a wave. That is, light behaved like smooth, undulating ripples on the surface of a pond rather than like a collection of tiny, discrete objects—corpuscles or particles. And the equations that dictate how light moves have exactly the same form as equations that describe the motion of waves in the ocean.

However, by the turn of the century, physicists began to realize that treating light as a wave didn't always work. For example, in 1900, an eminent British physicist, Lord Rayleigh, used wave equations to calculate how light would behave when trapped in a reflective box at a given temperature. He found, to his chagrin, that no matter the temperature, no matter the size of the box, the wave equations dictate that the cavity inside contains an infinite amount of light energy. That simply couldn't be. Something was going wrong.

Nobody knew it yet, but this was the end of the era of classical physics; a new and bizarre theoretical framework was slouching toward Germany to be born.

That framework was quantum physics. In December 1900, a forty-two-year-old German physicist, Max Planck, was attacking roughly the same problem that Rayleigh was, but from a slightly different direction. Instead of imagining an empty box, Planck started with a lump of matter and tried to calculate what sort of light it could absorb and emit. The equations that resulted gave an answer that baffled Planck at first. They seemed to imply that the lump of material could only absorb or emit light that came in discrete packets—packets that had a minimum size—rather than as a smooth, continuous wave. Light wasn't behaving like a smooth, continuous wave. Planck soon realized that with this assumption about light, this "quantum hypothesis," his calculations suddenly matched what scientists were seeing in the laboratory when they measured how much light was emitted by an object at a given temperature. The equations of quantized light had succeeded where the equations of continuous wave-like light had failed.[15]

Planck later described his quantum hypothesis as an "act of desperation," a mathematical kludge that would make the equations of physics start working correctly again. He didn't realize at the time that he had stumbled upon a fundamental property of the universe on very small scales: nature wasn't continuous, but discrete. And in 1905, patent clerk Albert Einstein explained a bizarre property of metals—how certain colors of light would cause them to spit out electrons when others would not—by applying the quantum hypothesis. But unlike Planck, Einstein suggested that the quantum hypothesis held for all kinds of light, everywhere in the universe, not just the sort of light that shines from a body at a given temperature. Then, in 1913, Niels Bohr realized that the quantum hypothesis applied not just to light, but to matter, too: by assuming that the electron in the hydrogen atom could only take on certain energies and not others, Bohr was seemingly able to explain, with great precision, the colors of light that hydrogen would absorb and emit. The quantum hypothesis had quickly become the key to understanding the behavior of objects on the very small scale; there was something fundamentally discontinuous about electrons and atoms and packets of light. Planck, Bohr, and Einstein all won the Nobel Prize for their quantum-theoretical insights.*

However, it wasn't until the mid-1920s that physicists figured out just how weird a quantized universe really was. Planck's desperate act had loosed upon the world a totally different conception of the fabric of reality.

Quantum theory is founded on our inability to predict with certainty precisely how an object is going to behave. This inability, this uncertainty, is woven into the very fabric of the mathematics of the theory. On short time scales and in small regions of space, that mathematical framework paints a picture of nature that is unpredictable and chaotic. Questions that have simple answers in classical physics—Where is an object? How fast is it moving? How much energy does it have?—can have fuzzy answers or even no accessible answers at all in the quantum realm. Or, another way of looking at it: in quantum theory, nature limits the information it does and doesn't reveal to an observer.

This is a very strange property, making quantum theory unlike any scientific theory that came before it. And it has a flip side: just as the mathematics of quantum theory treats information about the natural world as something that can't always be extracted, it treats that information as inviolable. That is, this information can be transferred from place to place, it can be split up and scattered, it can be transformed, but it can never truly be lost. A neutron never can lose touch with its "neutronness" even if it gets split apart or transformed (say, into a proton, an electron, and an antineutrino, as often happens). If a scientist painstakingly collected all the parts—the proton, electron, antineutrino, and energy—in theory, she would be able to reconstruct the neutron from the debris it left behind when it decayed.

It's not just a matter of different philosophies, or different mathematical approaches. It's a fundamental incompatibility between quantum theory's picture of how nature works and relativity theory's. In relativity, the canvas of spacetime is smooth and gently curved. In quantum theory, the canvas is rough and foamy on small scales. Quantum theory dictates that information about an object is never truly lost. Relativity makes no such promise. Indeed, the so-called no-hair theorems seem to imply otherwise. This is the issue at the heart of the black-hole information paradox.

---

* Einstein never won for his relativity theory, but for his quantum work regarding what happens when light hits metal objects.

The phrase "black holes have no hair" is physicist shorthand for the assertion that black holes are almost entirely featureless; other than the collapsed star's mass, charge, and rotation rate, there is nothing that an outside observer can determine. It is impossible to measure the black hole and figure out what it was made of—how much of its mass came from swallowed neutrons or swallowed protons and electrons, how those neutrons or protons and electrons were moving before they fell into the black hole, or anything like that. All that information is lost; once the matter falls past the event horizon, it is gone. Not preserved, like quantum mechanics said it was, but totally destroyed. To the people who studied general relativity—Stephen Hawking chief among them—a black hole was a cosmic eraser that permanently wiped out information.

This was a belief that struck at the heart of quantum theory, a belief that Hawking had held for three decades. And yet, even as the controversy about his home life swirled about him, the ocean liner Hawking had made a sudden and drastic U-turn.

• • •

"**C**an you hear me?" The physicist's wheelchair sat in the shadows stage right, inconsequential compared to an enormous screen glowing with a white PowerPoint slide. It read, simply, "The Information Paradox for Black Holes," with his name underneath in a smaller font: "Stephen Hawking." Technical talks at general relativity conferences are seldom packed full, but this one was. Every seat was taken, and latecomers—mostly journalists and grad students—packed themselves along the walls of the lecture hall. Most of those assembled had little idea what Hawking was going to say. Even the organizers of the conference didn't know precisely what he had in mind. "He sent a note saying 'I have solved the black hole information paradox and I want to talk about it,'" the chair of the conference's scientific committee told *New Scientist*. "To be quite honest, I went on Hawking's reputation." But the rumors had spread quickly, and people from around the world had gathered to this relativity conference in Dublin to hear what Hawking had to say.[16]

"I want to report that I think I have solved a major problem in theoretical physics that has been around since 30 years ago." For three decades, Hawking had been the chief advocate of the picture painted by the equations of relativity—that black holes destroyed information. In his view, the identities of particles falling beyond the event horizon were irretrievably lost.[17]

This view was by no means universally held. The physicists who came at the problem from the particle-physics/quantum-mechanical point of view, like Caltech's Lenny Susskind, pointed to the fact that the very mathematical framework of quantum theory forbade the loss of that information; it had to be preserved. "If the standard laws of quantum mechanics are correct, for an observer outside the black hole, every little bit of information has to come back out," Susskind said.[18]

By the late 1990s, the weight of scientific opinion was against Hawking, but in 1997, he and his friend Kip Thorne defiantly took up a bet against Caltech physicist John Preskill:

> Whereas Stephen Hawking and Kip Thorne firmly believe that information swallowed by a black hole is forever hidden from the outside universe, and can never be revealed . . .
>
> And whereas John Preskill firmly believes that a mechanism for the information to be released . . . must and will be found in the correct theory of quantum gravity,
>
> Therefore Preskill offers, and Hawking/Thorne accept, a wager . . .
>
> The loser(s) will reward the winner(s) with an encyclopedia of the winner's choice, from which information can be recovered at will.[19]

Flash-forward to Dublin in July 2004. Hawking's presentation at the conference was hard to follow, even for the specialists in attendance; he made the argument only in thumbnail, and they were based on a mathematical framework that was somewhat mistrusted by the audience—it made the results hard to interpret. It was clear that Hawking had reversed his belief; he had come to the conclusion that black holes don't, in fact, act as information erasers. Even though a black hole might have no hair, it somehow stores information about the stuff that falls past the event horizon and eventually disgorges it again into the universe. It might scramble that information, but it doesn't destroy it. "If you jump into a black hole, your mass-energy will be returned to our universe," Hawking said, adding that what gets returned will be "in a mangled form which contains the information about what you were like, but in a state where it cannot be easily recognized."[20]

Hawking had convinced himself. Working for more than a year with the assistance of his graduate student Christophe Galfard, he was now firmly in the information-preserving camp.* But he hadn't convinced anyone else. Not even Galfard, who thought that Hawking's arguments weren't airtight.

---

* Galfard, in fact, thinks that Hawking didn't actually reverse himself: "My true belief is that Stephen never believed in the information paradigm that information was lost. He just wanted to find where it had gone."

"I didn't believe that Stephen was right," Galfard says. "We had a funny aside the day before, because I really did not want to be on stage. It was the first time I spoke in public, and you can imagine how nervous I could have been with all the people and cameras and all of these things, and I was disagreeing with Stephen's results on top of that." So Galfard went to Hawking with his worries. "And [Hawking] said, 'That's absolutely fine. You can go on stage and disagree. You just have to prove it afterwards,'" Galfard recalls. "Yeah, I went and backed him up; I didn't manage to prove him wrong overnight."

Judging from the reaction of the scientists at the meeting, Hawking's arguments didn't have much better effect on them. The particle physicists in attendance already thought that black holes preserved information; those who didn't weren't swayed by an abstruse set of calculations condensed to fit in a brief presentation. Notably, Kip Thorne wasn't sold by the new argument—unlike Hawking, he refused to concede the wager. But the flashbulbs popped as Preskill, somewhat sheepishly, took to the stage to claim a baseball encyclopedia from Hawking.

Andy Strominger, Hawking's sometime collaborator, describes talking to Preskill soon after: "I said, 'John, why do you think the problem has been solved? Why are you accepting the encyclopedia? You know the bet's not settled,'" Strominger recalls. "And he said, 'I know, but this is my fifteen minutes of fame! No matter what I do with the rest of my career.'" Strominger laughs. "'Stephen wants to give him [a victory], what is he going to say? No? You gotta be kidding."[21]

Worse still, even if one accepted Hawking's presentation as a proof that information was preserved, the quirkiness of Hawking's mathematical framework obscured *how* that information managed to survive getting eaten by a black hole or how it wound up getting spit out. It was a performance that left scientists in the field somewhat confused.

If Hawking were truly to lay claim to solving the long-standing paradox, he would have to flesh out the details of the argument that he was making. He would have to write a deeply detailed mathematical exposition tracing how information evolves when swallowed and vomited up by a black hole. It took more than a year of work after the conference, but in late 2005, the paper was finally ready. It was . . . underwhelming.*

Instead of a detailed argument, it was three pages of dense text making a semi-qualitative argument rather than the quantitative treatment the subject seemed to call for. The entire article had but three equations.

---

* Kip Thorne put the blame on Hawking's "not . . . terribly strong student," presumably Galfard. Leonard Mlodinow, *Stephen Hawking: A Memoir of Friendship and Physics* (New York: Pantheon), 193.

Despite all the hoopla about Hawking's conceding the bet, his "solution" of the black-hole information paradox wouldn't even be in the top twenty-five of Hawking's most important works. For a man who had produced so many seminal physics papers, it was a disappointment.[22]

• • •

Perhaps even more disappointing were the sales of his brand-new book. *A Briefer History of Time*, which came out in late 2005, was supposed to be a more accessible, updated version of his 1988 best seller. And unlike the original, Hawking had formally contracted with someone who could help with the prose. Leonard Mlodinow, the Caltech physicist/author, wasn't given full coauthorship, but right underneath Hawking's name on the cover (in a smaller font, naturally), it stated that Hawking had written the book "with Leonard Mlodinow." Unfortunately for both men, the reviews were far from stellar.

"Long story short: the only way Hawking could have dumbed it down any further would have been to enlist Paris Hilton as a co-author, which would have cost him a certain amount of academic cred," snarked one wit in *Maclean's*. "But then again, it would have made the chapter on the Big Bang far more enrapturing." A more thoughtful review, in *Nature*, had a different take: "With the briefer version, I feel the baby has been thrown out with the bathwater," as the book lacked the "charming incomprehensibility" of a glimpse inside the mind of one of the world's leading physicists, the reviewer wrote. "It is just another run-of-the-mill popular science book on modern physics," with many of the topics in the book having been covered more skillfully by other authors.[23]

But the review that probably stung the most was the one in *The Times*. "It is interesting that the impetus for *A Briefer History of Time* was partly to make it more accessible by removing the maths of the original. This would be fine, except that Hawking is not very good at words. His language lacks the clarity and explanatory force of masters such as Brian Greene and Paul Davies."[24]

A Briefer History of Time didn't crack the best-seller lists. If it didn't sell, then his other books in the works didn't have much of a chance.

Shortly after *Briefer History* came out with his usual publisher, Bantam Books, Hawking started working on *God Created the Integers* with Running Press, an imprint out of Philadelphia. The idea of the book was to give readers a sample of key mathematical texts through the ages—such as Euclid's *Elements*, Archimedes' *Sand Reckoner*, and Descartes' *Geometry*. Hawking would organize and edit the works and write short introductions to each to put the mathematics and its author in context. It probably wasn't the sort of book that would sell great quantities under the best of circumstances, but there was a chance that Hawking's name-draw would generate some sales and make him and his publisher a little bit of money.

However, if *A Briefer History of Time* wasn't Hawking at his best, *God Created the Integers* was Hawking at his worst.

The book was carelessly edited and the core texts had lots of minor issues; the introductions had more significant errors that in some cases seemed to be the result of patchwriting.

Patchwriting is a form of plagiarism in which an author cleaves very closely to a text, changing some words around so that it's not "copied" in the ordinary sense of the word—but is still an improperly cited and badly digested regurgitation of someone else's work. It's not always easy to spot—and there's some latitude in interpretation—but patchwriting leaves some telltale signs: bizarre mistakes that stem from a slight misunderstanding of the original text; odd leaps of logic or in the focus of the narrative that are caused by omissions or distortions of the source text; phrasings that parallel the source material for long passages without explanation; attention to details that seem quirky or irrelevant outside of their original context. And the introduction-and-context parts of *God Created the Integers* suggested that large sections had been patchwritten from a number of different sources.

In the Descartes section, Hawking described the philosopher's death:

> Preferring his own treatments to anything Christina's doctors might prescribe, Descartes tried to cure himself with a concoction of wine flavored tobacco designed to make him vomit up the phlegm. He soon got worse, falling into a delirium and dying two days later on February 11, 1650.[25]

A biography on Descartes published a number of years prior by an Australian historian reads:

> [Descartes] preferred to rely on his own cure, wine flavoured with tobacco, to make him vomit up the phlegm. Coming out of a delirious state on the seventh day of his fever, his condition seemed to improve slightly, but he suffered a fatal turn on the ninth day, and died at 4.00 a.m. the next morning, 11 February.[26]

There appear to be distortions of fact that seem to come from a misreading of the text (it's wine flavored with tobacco, not wine-flavored tobacco; he came out of the delirium two days before dying, rather than falling into it.)

In his seven-page introduction to the works of George Boole, *God Created the Integers* pays a disproportionate amount of attention to the logician's living arrangements:

> For the first two years after their marriage, they lived in a house called "College View," which was about a ten-minute walk to the college. It was convenient, but soon became too crowded for the Booles. . . . [T]hey moved to a rented house in the village of Blackrock, four miles from the college but only half a mile from the railway station. The Booles loved this house. It had a wonderful view of Cork's magnificent harbor.[27]

This level of detail is much more appropriate to a full-sized biography of Boole, such as one written by an Irish math professor:

> For the first two years of their marriage, they lived, probably in rented accommodation, in a house called "College View." . . . Here Boole was within ten minutes walking distance from the college in an area much favored by the rich merchant classes of the city. . . . Early in 1857, the Booles decided that their expanding family needed more room so they moved to a modest rented house in Castle Road, a short distance from the village of Blackrock, some four miles from Queen's college. The house stood overlooking the sea with a good view of Cork's magnificent harbor. . . . [T]he little Blackrock station was about half a mile from the house.[28]

There are lots of other examples from biographies and other similar (uncited) sources—*The Dictionary of Scientific Biography* seemed to be the inspiration for more than one section. None of these are damning on their own (or even, arguably, damning in sum), but the writing process was inarguably below the quality that one would expect from a major author. The errors that crept in, the carelessness of phrasing, the seeming slavishness to a small handful of sources: all of these led to a book that not only failed to make Hawking very much money, but risked tarnishing his reputation as a communicator of science.

However, it is almost certain that Hawking himself did not write the passages in question. Merely holding a book to patchwrite from was beyond the physicist's capability by the time he "wrote" *God Created the Integers*; it's hard to imagine Hawking composing text at three words a minute while having an assistant help him pore over the source material. Far, far more efficient would be to have one of his many assistants do the work. Hawking would then put his name on it in hopes of selling more copies. "Well, the Running Press books, he didn't write those books," says Al Zuckerman, Hawking's agent at the time. Running Press had gotten Hawking's name on not just *God Created the Integers*, but also a similar (and rather better-executed) biographies-plus-primary-sources book about physics, *On the Shoulders of Giants*, a few years prior. "Those books were compendiums, for which he wrote the introductions, and he in fact didn't even write those introductions. . . . Hawking's name was put on the books to make money."

Gil King, who ghosted five of the biographical essays in *On the Shoulders of Giants* for Hawking, got the impression that Hawking wrote the introduction and little else.* "I think Stephen Hawking wrote an original introductory essay, but they said, 'You know, he's not really going to be able to write the biographies of Galileo and Newton,'" King recalls. "They needed just like a basic introduction of who these physicists were. They had me working with a real physicist . . . [and] I basically just did all the biographical stuff. There was really no need for me to interact at all with Hawking; I don't really even know how much he wanted to be involved with this project. So it was sort of like a way to get a product out there, I guess. And so that's what I did." King only heard from Hawking's people once or twice during the entire process. "It was sort of a work-for-hire thing. I finished it fairly quickly. And I didn't really interact. I think the most I ever got was an email from Stephen Hawking's assistant. It was a basic question about, you know, deadline or word count or something like that." After completing his essays, King forwarded his work on to the publisher to get Hawking's name stamped on it.[29]

In this, Hawking is little different from many modern artists—such as Andy Warhol or Jeff Koons—who have had other people manufacture their work. ("I'm the idea person," Koons told the *Journal of Contemporary Art*. "I'm not physically involved in the production.") When it came to patchwriting in *God Created the Integers*, it seems likely that one of Hawking's ghostwriters let the physicist down.[30]

Hawking took sole credit for *God Created the Integers*: his name was the only one on the cover and in the author biography. ("Stephen Hawking is considered the most brilliant theoretical physicist since Einstein . . .") In the acknowledgments, he does, however, extend special thanks to a number

---

* King later won a Pulitzer Prize for his work on Thurgood Marshall. I have not detected any patchwriting in the sections I know King is responsible for.

of people, some of whom were ghostwriting for him.* But whoever was responsible for the major problems with the book, Hawking himself cut dangerous corners when he agreed to put his name on it.[31]

The Hawking name was a valuable commodity. The physicist was more than just a person: he had become a brand, a symbol. Something to monetize. And Stephen Hawking himself only got a small fraction of what his brand was worth.

Not long before *A Briefer History of Time* came out with Bantam, a different publisher, New Millennium, published an embarrassingly slim volume: *The Theory of Everything: The Origin and Fate of the Universe.* As with all his books, Stephen Hawking's name was plastered on the front in huge type—he had long been one of those rare authors whose name was always a stronger selling point than the title of the book. And as with most of his books, its publisher tried to extract the maximum value out of it by remixing and rebranding it; within a year, a bulkier "Illustrated" version— still under two hundred pages—was hitting the shelves. "Although the book is short, it is full of information," wrote one reviewer. "And despite Hawking's efforts to keep it simple, the going gets pretty dense sometimes."[32]

But Hawking didn't write the book, at least not in the ordinary sense. Despite the huge "Stephen W. Hawking" emblazoned on the front cover, Hawking had not written anything for New Millennium; he didn't want them to publish a book in his name, and they did it over his objections.†

The Theory of Everything was old Hawking material—transcripts of a handful of fifteen-year-old-and-older lectures he had given at Cambridge University. New Millennium was mostly an audio publisher and had apparently acquired the rights to publish the tapes of the lectures. Unfortunately, the original contract—signed in 1988, at the height of *A Brief History of Time*'s success—contained a clause allowing the publisher to produce "in written form the text of said recording." Al Zuckerman told a reporter at the time that "maybe I was a little negligent in letting [the audio publisher] put that language in there," but added that he thought he was just yielding permission to package a transcript of the audio along with the recording— not to publish a separate book with Hawking as an unwilling author.[33]

---

* Among the people he thanks is Leonard Mlodinow, who assisted Hawking not with the little biographies that had the problems with patchwriting, but with the overarching introduction to the book, which didn't seem to have the same issue. (Leonard Mlodinow, personal communication with author.) In our discussion, I did not indicate to Mlodinow that there were patchwriting issues, much less where the problems were, so his unprompted description of the sections he contributed would seem to rule him out as a culprit.

† In the interest of full transparency, I should reveal that I, too, had an unpleasant fight over intellectual property that involved New Millennium.

But that's precisely what happened, and Hawking couldn't stop it. He tried complaining to the US Federal Trade Commission (FTC). "It was professor Hawking's concern that, if the Cambridge lectures were published in book form, his large fan base would purchase the new book based solely upon his name," his lawyer wrote, "only to find that the work was not authorized by Professor Hawking, that the quality of writing was not what they would expect of him, and that its content was not new, but instead merely a repackaged presentation of the material contained in 'A Brief History of Time.'"[34]

The FTC didn't pursue the matter, and Hawking didn't file a lawsuit. "I just don't have the financial resources and energy to take on this guy in court," Hawking's lawyer recalled the physicist telling him. New Millennium went ahead and published its book in Hawking's name and made its money.[35]

The physicist, in the meantime, could only publish an outraged note on his website:

> It has come to our attention that the book **"The Theory of Everything: The Origin and Fate of the Universe"** has been published. . . . **We would urge you not to purchase this book in the belief that Professor Hawking was involved in its creation.**[36]

Hawking's name, like any other commodity, could be traded and used in ways that the original owner never intended. Most of the time, though, the physicist was cagey enough to get some benefit out of its use.

• • •

With a few years' worth of hindsight, Hawking described his relationship with Elaine as "passionate and tempestuous," with "ups and downs." In 2006, it was taking its final plunge, with both Elaine and Stephen filing for divorce in November. One employee of the court remembers that Stephen's lawyers filed first, and then, within about an hour, Elaine's solicitors appeared with their own papers. The couple were closed-lipped about the reason for the breakup, but there were rumors that the discovery of an affair had sunk the marriage. Only after Hawking died did Elaine comment publicly. ("Stephen was the love of my life, and I was his. Neither of us had been as happy or as loved before," she told a congregation a few weeks after his death in 2018. From Elaine's point of view, she made the difficult decision to separate from him because she "wasn't tough enough to cope with the constant undermining from one particular quarter of our marriage.")[37]

According to Hawking's agent, the terms were apparently harsh, at least financially. "Of course, when his second wife left him, she demanded a big settlement," Zuckerman says. "And he was doing a book for a half a million dollar advance and the contract was signed, and they began working on it, and it was, I don't know, maybe halfway into [the time] before it was to be delivered. I got an email that he wanted to change the deal to a million dollars."* Incredulous, Zuckerman passed the demand on. "I thought the publisher would tell him to go to hell. But they agreed to it. That was because he had to pay her off."[38]

The divorce with Elaine was not just a breakup with his wife, but with a key person who was in charge of his physical well-being. Since the mid-1980s, Elaine had been caring for Stephen, and soon she had become the most important person in the team of nurses who gave Stephen the round-the-clock care he needed. The breakup robbed Stephen of the care of a person who, by his own account, "saved my life on many occasions."[39]

Hawking would never remarry, though soon after the divorce, he had acquired a new live-in companion whom he referred to as his "house-keeper." And, despite the turmoil, the breakup with Elaine did have one major benefit for Hawking. It removed a major source of discord that had damaged his relationship with his children and prevented any sort of reconciliation with his ex-wife.

During the marriage, Lucy, in particular, seemed to be very hurt by her father's choosing to stand by Elaine, and while she didn't consider herself estranged, she admitted that the relationship was strained. And she made plain what she thought the fundamental problem was: "There was one person who was always more important than everyone else collectively and there is still this assumption. In the wider world, to my dad's colleagues and some of his family, he's like a Monopoly hotel to one of our houses," she told a newspaper in 2004. "Yes, of course, he's important. But we matter, too."[40]

As Stephen's marriage with Elaine deteriorated, he entered a new phase of his relationship with his daughter. Three months before Elaine filed the divorce papers, Lucy, a novelist, accompanied Stephen on a trip to China. At a press conference during their visit, Stephen and Lucy made a surprise announcement: they would be writing a children's book together. "It's like Harry Potter meets *A Brief History of Time*," Lucy said. Stephen echoed the comparison to Harry Potter, but it would be "about science, not magic," he added.[41]

---

* Zuckerman was unable to recall precisely which work it was, but the timing and the size of the advance imply that it was *The Grand Design*.

In September 2007, Lucy published *George's Secret Key to the Universe*.
As always, the word "Hawking" was emblazoned prominently across the
cover. But for the first time in his life, Stephen Hawking conceded top bill-
ing as the author of a book. In smaller type, above the large "Hawking," it
read "Lucy & Stephen." (On the title page, but not on the cover, underneath
the coauthors Lucy and Stephen Hawking, there was the admission that the
book was written "with Christophe Galfard," Hawking's recently graduated
PhD student, who himself went on to publish several pop-sci books.)

Like the Harry Potter series, *George's Secret Key* begins with a young
boy suffering a miserable home life. But the misery wasn't because of mean
foster parents; it was because of his actual mother and father:

> Wanting to live a purer, simpler life, they washed all their clothes by
> hand and didn't own a car and lit the house with candles in order to
> avoid using any electricity.
>
> It was all designed to give George a natural and improving upbring-
> ing, free from toxins, additives, radiation, and other evil phenomena.
> The only problem was that in getting rid of everything that could pos-
> sibly harm George, his parents had managed to do away with lots of
> things that would also be fun for him. George's parents might enjoy
> going on environmental protest marches or grinding flour to make
> their own bread, but George didn't.[42]

George escapes the clutches of his Luddite parents with the help of a (surprisingly short-tempered) scientist neighbor named Eric, his daughter
Annie, and their miracle-working computer.

The same year, Jane Hawking, Stephen's ex-wife, published her memoir
of her time with Stephen, titled *Travelling to Infinity*. It was essentially a
lightly edited version of her earlier memoir, *Music to Move the Stars*. Some
of the harsher (and franker) passages had been softened or eliminated, but
it still painted a fairly dismal picture of life with the physicist. This was the
book that, seven years later, would become the Oscar-winning *Theory of
Everything*.

...

By this time in his life, Hawking had been the living embodiment of
scientific intellect for almost two decades. When he spoke on matters
touching upon science, his words were uniquely persuasive. Nobody else
could match the implicit authority about space, time, physics, or science in
general that Stephen Hawking carried. Such a strong voice attracts power.
And money.

It doesn't take a lot of money to make a theoretical physicist very happy; unlike experimentalists who need laboratories with large staffs and expensive equipment, theorists can make do with a blackboard and a graduate student or two. This makes them a cheap date for multimillionaires looking to draw scientific superstars into their orbit. And over the years, a number of wealthy people—mostly men—did just that, cultivating relationships with leading theorists for reasons of their own.

Some were smart but lacked formal education, and relished in surrounding themselves with the smartest people on the planet. In the 1970s and 1980s, Werner Erhard, who made his money with a controversial self-help program called est, not only forged ties with leading scientists around the country but also hosted intimate conclaves where they gathered to discuss topics of interest. (Hawking attended one of them; more on this in Chapter 12.) But at this point in Hawking's life—the early 2000s—the new undereducated billionaire on the scene was Jeffrey Epstein.

Epstein would later become infamous for his pedophilia and sex trafficking. By then, a number of celebrities, including Prince Andrew of England, lawyer Alan Dershowitz, and computer theorist Marvin Minsky, stood accused of improprieties with Epstein-procured underage girls. (Both Prince Andrew and Dershowitz have vigorously denied the accusations; Minsky died in 2016.) In March 2006, a few months before Epstein's first indictment hit the headlines, he hosted a small conference on gravity in the Virgin Islands. Hawking was there—and apparently on Epstein's nearby private island along with some of the other attendees of the conference: Kip Thorne, Nobel laureate David Gross, and Harvard theorist Lisa Randall. There's also a picture of Hawking peering out the window of a submarine, which Epstein reportedly had modified to allow Hawking to enter. Given the date, Hawking was likely unaware of the allegations hanging over Epstein. There is no evidence that Hawking did anything improper or, unlike a number of other scientists, had any sort of contact with Epstein after the accusations became public. For Epstein didn't adopt Hawking in the way some other billionaires had.[43]

In 2002, Hawking met oil baron George Mitchell and began a decade-long relationship that ended only when the magnate died in 2013. "He's basically the guy who invented fracking," says Hawking collaborator Andy Strominger. "So he set up these two, three week retreats—he was a huge fan of Stephen—which were basically designed to create an environment where Stephen can work." (Mitchell would go on to name an auditorium at Texas A&M—his alma mater—after Hawking.) "George P. Mitchell was a remarkable individual who combined vision with wisdom and persistence," Hawking said after the billionaire died. "Through sheer hard work and dedication, he leaves behind an extraordinary legacy. It can be said of very few people that they changed the world—but George Mitchell is among those few."[44]

Another billionaire that Hawking singled out for praise was Yuri Milner, a Russian oligarch and Internet tycoon who helped invest hundreds of millions of the Kremlin's dollars into US corporations. "Yuri Milner is something of a visionary," Hawking enthused in 2016. "He sees that while there are many good causes and pressing problems, ultimately our chances of thriving as a species depend on tending and feeding the precious flame of knowledge."[45]

Milner met Hawking at a conference in Moscow in 1987—Milner has a background in theoretical physics—and helped to tend and feed Hawking a quarter century later. In 2013, Milner awarded Hawking a "special" version of the billionaire's new "Breakthrough Prize," which came with a hefty $3 million purse. And Milner was, in fact, the "rich friend" who funded Hawking's 2016 trip to Harvard to work with Andy Strominger on soft hair. "You know, it cost half a million to fly him there," recalls Marika Taylor, one of Hawking's former students. "Yuri Milner gave him the half a million, so here he had to believe that this was going to be a scientific breakthrough . . . but in reality, it wasn't a scientific breakthrough, right?"[46]

The relationship between the billionaire and the physicist got closer over time. Hawking put his name to two of Milner's hugely expensive and somewhat daft scientific projects—a search for alien radio transmissions, and an attempt to send ultra-tiny spacecraft to the nearest star—either of which would likely have gotten little attention without a big name like Hawking giving them his imprimatur.

"[Hawking] was very much conscious that if he wanted to keep in the public eye, he needed sponsorship; you know, he needed support," says Taylor. And that support—and Hawking's public profile—was to a large degree dependent upon making headlines. "He was often asked to comment about topics where he should have said, 'You know, actually I don't want to comment about this; this is not something I have the expertise for,'" Taylor says. "He would admit this privately, but he said it in public because . . . there may have been sponsorship involved or he may have been doing it because keeping him in the public eye helped the sponsorship."

Though Milner's projects were arguably a little kooky, they probably appealed to Hawking on some level. It didn't take much for Hawking to swallow whatever discomfort he felt about straying from his expertise to put his voice behind them. "Many of these things were not things he wanted to talk about," says Taylor. "But he felt he should."[47]

For example, Hawking's support of the Milner spacecraft project was helped by the fact that Hawking was personally an advocate of spaceflight. He had argued on numerous occasions that the long-term survival of the species depends on our eventually being able to leave the planet. Of all of Hawking's out-of-left-field stances, it was this one that led to the most benefit for Hawking. Including one of his most iconic moments, again, sponsored by the largesse of a titan of industry.

In October 2006, Hawking met multimillionaire Peter Diamandis, a California businessman who himself had a great interest in space and space technologies. Six months earlier, Diamandis had signed a contract with NASA to use the shuttle landing pad at Kennedy Space Center.[48]

Diamandis' new company, the Zero Gravity Corporation, needed the landing pad to perform zero-gravity flights with its Boeing 727. That is, the jetliner would take off and climb to twenty-four thousand feet. And then, in a set of stomach-churning maneuvers, the plane would suddenly lurch up and zoom to thirty-four thousand feet, then tip over and plummet back down again. The idea was to behave essentially like an object in freefall—if the pilot did the maneuvers just right, the passengers would feel no gravity at all for about thirty seconds at a time. For years, NASA used a similar plane to train NASA astronauts; it was dubbed the "Vomit Comet" for all-too-obvious reasons.

Diamandis offered Hawking a zero-gravity trip. "And he said, on the spot, 'Absolutely, yes,'" Diamandis told an audience some time later. In April 2007, Hawking—with medical professionals on standby—took the plunge. Flashbulbs popped as two handlers helped him float into the air. For many of his fans, this was the only image that they had ever seen of the physicist out of his wheelchair. Emaciated, his hands curled at unnatural angles, he was a spectral figure—with a look of delight on his face.[49]

The headlines across the world about Hawking's zero-gravity flight certainly didn't hurt the Zero Gravity Corporation's prospects. And as it happened, at the time of Hawking's flight, the corporation had bid on a contract to provide zero-gravity flight services for NASA—a contract that the corporation won in January 2008—and that NASA later regretted.* The corporation's performance was often terrible—at times, its pilots failed to do the maneuvers properly more often than they did them right, meaning that the flights were all but worthless—yet all the public ever knew

---

* Even without the Hawking stunt, the Zero Gravity Corporation would almost certainly have won the bid; of the four companies in the running, it was the only one that owned its own plane.

of the corporation, if they heard of it at all, was the Hawking flight. From the physicist's point of view, a free zero-gravity flight for a little PR was a no-brainer. Upon landing, he gushed about the experience. "I could have gone on and on," he said. And then he exclaimed: "Space, here I come!"[50]

For Hawking had his eye on an even more ambitious target than a free zero-gravity jaunt: a free trip to outer space. He had already acquired the ticket.

In 2006, Richard Branson, billionaire owner of the Virgin corporation, was famously trying to set up a service to launch tourists out of the atmosphere and (hopefully) bring them back to Earth in one piece. At the time, the service, Virgin Galactic, was scheduled to begin in 2009. When Hawking appeared on a BBC radio show in November 2006, perhaps emboldened by his encounter with Diamandis, he made a not-so-subtle appeal. "My next goal is to go into space," he told the host. "Maybe Richard Branson will help me out."

Branson, more than happy to oblige, promptly promised to give Hawking a ride to outer space. "With delight I found myself with what I understand is the only free ticket for a Virgin Galactic space flight that Richard has ever handed out," Hawking later said.[51]

But a promise is not a trip, and Hawking eagerly awaited the opportunity to cash in that free ticket. Branson asked nothing in return. In the meantime, the billionaire's engineers spent years working out kinks and overcoming delays while trying to get their spaceship into shape. Just when it looked like Virgin Galactic was on a clear path to commercial flights, in October 2014, a Virgin Galactic ship undergoing spaceflight testing broke up high over the Mojave Desert. One of the two test pilots was killed. Branson's endeavor, already long delayed, was on the ropes.

After a year of investigations, redesigns, and retrenchments, Branson was nearly ready to unveil the redesigned spaceship. So, in December 2015, the Virgin team stopped by Cambridge University for an inspirational chat with the professor. Hawking fondly recalled his appearance on the BBC show when he expressed his desire to go into outer space. "[The host] asked me whether I was worried by the prospect of death. I replied that as my death, according to the medical profession, has been predicted many decades earlier, it did not overly concern me, but that there were still a few things left on my bucket list. Near the top of that list was the desire to experience space for myself," he said. Accidents will happen and people will die, but space travel is not for the weak-hearted. *Dulce et decorum est pro astra mori.*[52]

When Branson unveiled the new prototype spaceship in February 2016, Hawking was unable to travel to the event, but he regaled the crowd with a recorded message: "We are entering a new space age, and I hope this will help to create a new unity," he declared, naming the new vessel. VSS *Unity*, gleaming, sported a painted banner on its side bearing a gigantic image of Hawking's eye.

Though the physicist himself never got to ride, that rocket will always bear Hawking's name.

• • •

Hawking was known as a scientist of the first rank and one of the most celebrated popularizers of science of our age. Yet in the last decade of his life he did little science of note—certainly none considered to be of high import—and very little science communication. He was neither scientist nor communicator as much as he was a brand.

The authentic Hawking, the man who had devoted his life to physics, and who had a passion to be understood not just by his peers but also by the public, is barely visible behind the image, the commercial product that he had become. It's a vexing, almost paradoxical situation: Hawking's celebrity had almost completely obscured the very elements of Hawking that had made him a celebrity in the first place. To find them, one must travel further backward in time.

PART II

IMPACT

_____

"I hardly move, yet far I seem to have come."
"You see, my son, time here becomes space."

—RICHARD WAGNER,
Parsival

"You'll have in your possession absolute perfection
upon this earth. No-one is so rich he may vie with your
wealth, if you have given the question its due."

He said: "I did not ask the question."

"Alas that my eyes see you," said the grief-laden maiden,
"since you were too daunted to ask the question! But you
saw such great marvels there—to think that you should have
refrained from asking then! There you were in the presence
of the Grail."

—WOLFRAM VON ESCHENBACH,
Parzival

CHAPTER 6

BOUNDARIES (1998–2003)

J ournalists shifted uncomfortably, waiting for the big arrival.

Under ordinary circumstances, the room would have been large enough. Only a handful of beat reporters—perhaps bolstered by a TV crew or two from the local broadcast stations—would typically attend a press conference like this at an American Physical Society meeting. But this year's meeting was special; 1999 was the society's centennial, and thousands upon thousands of physicists from around the world had descended upon Atlanta for the occasion. And the headliner was Stephen Hawking, and he had promised to field a few questions from the press. Now, dozens of eager reporters shoehorned themselves into a way-too-small conference room in the hope of hearing the great man speak.[1]

Earlier in the day, journalists had been asked to submit their questions on little slips of paper. His graduate assistants had sorted through the stack of slips and chosen a few questions for the physicist to answer. The press conference organizers looked around nervously as latecomer journalists tried to find room to stand.

Then there was an excited whisper as Hawking whizzed in, zooming in his wheelchair to the front of the room, seemingly heedless of the journalists trying to scramble out of his way. It wasn't easy to interpret the expression on his face, but the physicist certainly didn't look happy to be there.

The press conference went as a Hawking press conference usually does: the same shallow, predictable questions that the physicist has fielded dozens of times already, the same respectful hush from the crowd in anticipation, the same snippets of speeches repurposed as answers, the same muttered appreciation of the physicist's wisdom. The crowd seemed entirely oblivious to Hawking's clear desire to be almost anywhere else besides playing the sage to a bunch of awestruck journalists.

There was only one moment when Hawking's responses broke new ground: new astronomical observations had just forced him to abandon a belief that he had harbored for decades. "I now think it is very reasonable that there should be a cosmological constant," he told the reporters. It was couched in scientific terms, but the consequences of the new position were quite dramatic. "The universe may keep flying apart forever."[2]

• • •

It was a time of revolution. In the late 1990s and early 2000s, a series of astronomical observations—particularly of distant supernovae and of the cosmic microwave background—were forcing cosmologists and gravitation experts to revisit one of the basic assumptions about how the universe behaved. For those observations were showing that spacetime was acting in an unexpected way: it was as if there were some mysterious substance, "dark energy," that was slowly, inexorably stretching the fabric of the universe. Once cosmologists understood what these observations meant, they would not just be able to pinpoint when the universe began; they would also divine how it will end.

Stephen Hawking was not a part of this revolution. Back in the 1970s and 1980s, his work on black holes and singularities and the very early universe put him at the very center of the swirl of scientific activity in cosmology and general relativity. By the late 1990s, he had found himself on the sidelines, trying to stay relevant. Not just scientifically. A Brief History of Time was more than a decade old, and he hadn't published any major popular work since. Hawking's story of the universe was at risk of getting stale—or, worse, he might lose his audience entirely.

As he turned fifty years old, Hawking's struggle had taken on a new dimension: he was no longer at the very center of physics or even of the popularization of science. He would have to fight to avoid drifting to the periphery.

• • •

The science of cosmology is largely about how the universe began and how (whether!) it will end. Both of these questions, fundamentally, are about geometry: about the shape of the cosmos. For, as Einstein's theory shows, the shape of the universe describes how it behaves not just through

space, but also through time. The fabric of spacetime encompasses everything in our universe, no matter how far distant in space or in time it might
be from you—including the very start of the universe and the finish, if it
ever happens. Just as surely as the Fates of Greek mythology could predict
any event by examining the warp and weft of the fabric they wove, cosmologists can figure out the origin and ultimate fate of the universe by
understanding its shape on the largest scale.

Just like any other geometric shape, the curve of spacetime looks rather
different on different scales, even though they're all part of the same object.
The surface of the world that we live in is a good analogy. How should we
describe its shape? On the smallest scale—looking at, say, the surface right
beneath our feet—it's generally pretty much flat. Zoom in far enough, and
the surface looks like a smooth, flat plane, even if you're standing on a little
hillock or in a shallow pit. You can describe a location on such a plane with
two coordinates: call them x and y or N and E or something else, but two
numbers are generally sufficient to describe the location of any terrestrial
object. Also, with very good accuracy, on this small scale, you can figure
out the distance between any two points on the plane with the Euclidean
metric—the Pythagorean formula. In some sense, the world is flat; it's a
Euclidean plane.

Things get a little more complex as we zoom out. The world still looks
more or less like a flat plane—a two-dimensional paper map does a pretty
good job at describing how far apart two buildings are. But the countryside
isn't perfectly flat. Subterranean forces have warped the surface, creating
hills and valleys. A paper map of hilly San Francisco might not give you
a very good sense of how long it will take to walk from one building to
another across town; unlike a hike in nearly flat Kansas City, you have to
add time to the journey to account for all the scrambling up and down hills
that aren't typically pictured in a flat map. On the middle scale, the flatness assumption begins to break down; local features like hills and valleys
can distort the distance you would expect to travel based upon a typical
two-dimensional map.

Zoom out farther still, and you realize that the world isn't really flat at
all: it's round. You don't notice this in everyday life; you typically feel that
you're living in a flat(ish) world, interrupted only by hills and valleys. You
could spend your life blissfully unaware that you live on a ball rather than
on a flat pizza pie—if you're stuck on the planet, it takes some relatively
sophisticated measurements, such as observations of visual distortions of
tall objects at a great distance, or differences in solar angles at widely separated cities—to prove that we live on a sphere rather than a plane (or a

donut or a saddle or a bowl, for that matter). Even though on the smallest scale our *geometry* is a flat Euclidean sheet, the *topology* of the world as a whole is a sphere.*

The same sort of analysis holds for our universe, according to the theory of relativity; it's just that we're dealing with four-dimensional spacetime instead of two- or three-dimensional space. On the smallest scale, our universe is described not by the Euclidean metric of flat space, but by the more complicated Lorentzian spacetime metric. That's the fundamental shape of the fabric of spacetime, the spacetime we move through on a daily basis. Just as our little corner of the world appears fundamentally like a copy of simple, smooth, flat Euclidean space, our little corner of spacetime appears like a copy of Lorentzian spacetime. This space is "flat" in some sense: even though it's four dimensional and includes time as well as space, it isn't distorted by creases or pits (at least on the scales we encounter locally). This is spacetime undisturbed by gravity or other distortions: the spacetime of special relativity.

Zoom out a bit, and on the middle-level scale—say, looking at a solar system or a galaxy as a whole—we can see that spacetime isn't totally "flat." As the equations of general relativity predict, bundles of mass and energy, like planets or stars or solar systems or galaxies, cause spacetime to curve. For about a century, scientists have been able to spot that curvature by watching a star's light get pulled as it passes near the sun during a solar eclipse, or by spotting the shapes of distant galaxies being distorted as their light passes near huge lumps of mass. And with LIGO, scientists have just begun to detect the ripples of gravitational waves disturbing the flatness of our little neighborhood of spacetime. This is the spacetime of general relativity: a dynamic fabric that's still smooth but not quite flat, as it is distorted by mass and energy and gravitational waves.

Then there's the third level. Zoom out to consider the cosmos as a whole. What is its topology? Is it flat like a (higher-dimensional) pizza? Round like a ball? Or a donut or a saddle or a bowl? These are questions of cosmology: among the biggest questions that cosmologists can possibly answer.

Because the shape of the universe encodes time as well as space, knowing the topology of our manifold tells us about the universe of the future.† A

---

* This is really the essence of a "manifold": it's an object that might (or might not) have a complicated topology, but if you look at it on the smallest scales, it has a regular, smooth geometry.

† And, of course, the universe of the past, though that's a bit more complicated. More on this shortly.

sphere is a compact object—it doesn't stretch in all directions for infinity—and must be bounded. This means that if our universe is "spherical" our cosmos will have a well-defined end. The universe will expand for a while, all the galaxies and mass and energy flying apart as it does. The expansion will slow down until the collective gravity forces everything to start contracting again, stars and galaxies get pulled together again, and everything finally ends in a "Big Crunch." If, instead, our universe is flat like a pizza, or warped like a saddle, it expands and expands without limits either in space or in time. The universe keeps flying apart ad infinitum.*

The equations of general relativity don't tell us about the topology of the universe; we have to gather data to figure it out. Just as it took smart measurements and sophisticated reasoning for the ancients to understand that the world was round, it took astronomers a tremendous amount of observation and calculation to get a handle on what the overall shape of the universe might be. And in the late 1990s, after many years of effort, they finally began to succeed.

Two groups of astronomers were studying distant supernovae—stellar explosions so bright they are visible half a universe away—when they each noticed something very bizarre. The more distant supernovae were fainter than expected, which seemed to imply that the expansion of the universe was not slowing down fast enough to lead to a collapse; there was no way that the universe was going to stop expanding and start contracting again into a Big Crunch. It was an amazing discovery: the universe was expanding so fast that it would have to expand ad infinitum. For the first time, cosmologists could say something about the topology of the cosmos: it wouldn't collapse like a spherical cosmos would. This alone would have been worthy of a Nobel Prize (which Saul Perlmutter, Brian Schmidt, and Adam Riess won in 2011). But there was more to their discovery.

Not only was the expansion of the universe not slowing down very fast, but it seemed to be speeding up. This was forbidden by the field equations of general relativity—unless you added a little mathematical kludge known as a *cosmological constant*. The cosmological constant, in some sense, is an energy that's a part of the fabric of spacetime itself, an energy that, like

---

* Technically, the relationship between topology, the fate of the universe, and whether the cosmos is infinite or finite in extent in space and/or in time is a bit more complicated than this, once you allow for a cosmological constant, or if you don't rule out some of the more exotic topologies. So this is a bit of an oversimplification, even though the general principle holds.

other matter and energy, makes spacetime curve.* But it has the opposite effect as gravity; it pushes things apart rather than drawing them together. This mysterious, antigravity-like *dark energy* contributes to the shape of the cosmos—a shape that scientists were increasingly certain was infinite in extent. The universe was not curved like a ball and primed for recollapse; it had to be flat or saddle-shaped, and would expand forever and ever as galaxies fly ever farther away from one another.

The two supernova-hunting teams published their results in 1998, sparking a storm of excitement in the scientific community. The idea of an accelerating universe, of a cosmological constant, of dark energy, was so unexpected that it totally upset the pet theories of most of the cosmologists in the world, Hawking included. For Hawking had spent many years pushing a beloved theory, the no-boundary proposal, that featured prominently in *A Brief History of Time*. Unfortunately for Hawking, the no-boundary proposal seemed to imply that the universe would end in a Big Crunch—precisely the result ruled out by the new supernova observations. The observations were so compelling that Hawking had to admit that a Big Crunch was ruled out. So in the 1999 Atlanta press conference, when Hawking said that it was "very reasonable" that there should be a cosmological constant that would make the cosmos "keep flying apart forever," it seemed like he was finally giving up on a theory that was one of the cornerstones of his career.

As with many things Hawking, the truth was quite a bit more complex.

• • •

One year earlier, in March 1998, Hawking gave a grand address at the White House. President Bill Clinton, First Lady Hillary Clinton, and assorted political and scientific luminaries were in attendance. After the lecture, the audience peppered Hawking with questions. "Within the past month, we have seen evidence suggesting a strong, repulsive force in the universe—an anti-gravitational force causing the universe to expand, surprisingly, at an accelerating rate," Vice President Al Gore said. "How surprised were you by this finding? What are its most important implications?"[3]

Hawking made his skepticism clear. "What the Vice President is referring to is some observational evidence that suggests that there may be an anti-gravitational force that would cause the universe to expand at

---

* Because this is a book primarily about Hawking and not cosmology, this description merely skims the surface of just how weird—and how revolutionary—this discovery was. For details about the supernova observations, the cosmological constant, the cosmic microwave background, and the implications for cosmology, see my *Alpha and Omega*, chaps. 4 and 5.

an increasing rate," Hawking said. "The existence of such an anti-gravita-tional force is very controversial. Einstein first suggested it might exist, but later regretted it and said it was his greatest mistake. If it is there at all, it must be very small." He was still convinced that a Big Crunch was coming. "But don't worry, the Big Crunch won't come for at least 20 billion years," he said.

Even though it was nominally Hawking the scientist who addressed the audience, it was Hawking the celebrity who got most of the attention. Hillary Clinton read out a question submitted over the Internet. "How does it feel to be compared to Einstein and Newton?"

"I think to compare me to Newton and Einstein is media hype," Hawk-ing declared.

"I must say, you did look good at the card table," Clinton responded.

The card table in question featured prominently in Hawking's lecture earlier in the evening. At the very start of the talk, the physicist played a short clip from *Star Trek: The Next Generation* in which the physicist had been conjured in the twenty-fourth century to engage in a battle of wits—poker—with none other than Newton and Einstein. "All the quantum fluctuations in the world will not change the cards in your hand," Einstein declares. "I call. You are bluffing, and you will lose."[4]

"Wrong again, Albert," Hawking gloats, a huge grin on his face, as a motorized arm reveals his hand. Four of a kind.

Hawking may have considered it "media hype" to treat him as the intellectual equal of Einstein and Newton, but it was an image that was central to his public identity. In May 1999, viewers of the prime-time car-toon *The Simpsons* were visited by "Stephen Hawking, the world's smartest man." Armed with a gadget-filled flying wheelchair and an IQ of 280, the physicist makes fools of the Springfield locals. But he does bond with one character: the buffoonish, pastry-gobbling Homer Simpson. ("Your idea of a donut-shaped universe is intriguing, Homer," the animated Hawking says over a round of beers at the local pub. "I may have to steal it."*) Even in Hawking's appearances on TV commercials—such as his 1999 ad for Specsavers opticians, which reportedly earned him £100,000—his words were a conscious echo of Newton's. "For me, physics is about seeing fur-ther, better, deeper," his voice declares. Onscreen, a CGI spaceship and a new pair of glasses make the shoulders of giants entirely unnecessary for his transcendent vision.[5]

---

* Hawking would appear on *The Simpsons* several more times. He also played a recurring role a number of years later in the geeky sitcom *The Big Bang Theory*.

The jacket copy on Hawking's books did nothing to dispel the media hype either. "Stephen Hawking is the Lucasian Professor of Mathematics at the University of Cambridge and is regarded as one of the most brilliant theoretical physicists since Einstein," read the author biography in *Universe in a Nutshell*, published in 2001. The Lucasian Professorship at Cambridge is arguably the most prestigious academic post in the world—made famous by one of its previous occupants, Sir Isaac Newton. Hawking was the heir apparent to the legacy of Newton in more ways than one.

It was sometimes a heavy mantle for the physicist to wear. Moments after dismissing the comparison to Newton and Einstein as "media hype," Hawking suggested another reason that he might be considered the world's smartest man. "I fit the popular stereotype of a mad scientist or a disabled genius or, should I say, a physically challenged genius, to be politically correct," he said. "I am clearly physically challenged, but I don't feel I am a genius like Newton and Einstein."[6]

Nor did his colleagues put him in that category. In 1999, the magazine *Physics World* asked some 250 physicists around the world to name the five physicists, living or dead, who had made the most important contributions to physics. Einstein topped the list with 119 votes, followed by Newton, and James Clerk Maxwell with 67 votes. Galileo was sixth, Richard Feynman seventh, Paul Dirac eighth. Gerard 't Hooft and Stephen Weinberg, who each helped understand the weak force, and Charlie Townes, inventor of the laser, got two votes each. Hawking—along with a couple of dozen others, such as Martin Rees and John Wheeler—wound up at the bottom of the list, each with a single vote. It was an honor merely to be included, but the poll left little doubt that, generally, physicists didn't think his work to be in the same league as a Newton or an Einstein or even a Dirac, no matter what impression the mass media gave.[7]

Hawking knew that his disability deeply affected the way the public perceived him, transforming him from a mere human being into a living metaphor. The image of a man trapped in a quadriplegic body soaring, intellectually, to the ends of space and time; the idea of a person who is unable to speak, who can only, with heartbreaking amounts of labor, struggle to utter a few words a minute, becoming one of the world's best-selling communicators of science—the physicist's story was powerful and deeply ironic. And Hawking was keenly aware of that irony.

"He loves that *Hamlet* line," says director Errol Morris, "the 'nutshell' line. And why wouldn't he?"[8]

In Act II of Shakespeare's play, a courtier suggests to Hamlet that Denmark is too small for him, and restricts his mind. The doomed prince responds, "O God, I could be bounded in a nutshell and count myself a king of infinite space, were it not that I have bad dreams." Says Morris: "And in the Shakespeare itself, there's a very strong irony, which he is play-ing on, clearly." The transcendent idea, the beautiful image of destroying all limitations with the power of the mind, is snatched away by attendant nightmares. Hawking was stoic—and closed-lipped—about whatever bad dreams pulled him back to uncomfortable reality.

...

The Universe in a Nutshell was Stephen Hawking's first major book since his groundbreaking best seller, *A Brief History of Time*, published more than a decade prior. It was the first serious literary attempt to capitalize on his stature as the reigning monarch of physics popularization. But in the intervening time, there had emerged some new pretenders to the throne. Most notable was Brian Greene, a theorist at Columbia who studied hypo-thetical subatomic objects known as *superstrings*. Greene's book, *The Elegant Universe*, sold at a steady clip since it was published to rave reviews in 1999, and wound up on the paperback best-seller lists for several months the following year. The market for science nonfiction books was still there; the question was just whether Hawking could capture it once again.

Hawking lamented that many of his readers "got stuck" in the begin-ning chapters of *Brief History* and never finished the book.[9] *Nutshell*, unlike *Brief History*, would be lavishly illustrated, all the better to make it easier to understand and reach a broader audience. However, it would be a struggle for Hawking to produce new material for the book. This time, he would have help from the get-go. It would come not just from a graduate stu-dent (Thomas Hertog) but also from science writer and erstwhile Hawking biographer Kitty Ferguson who recalls:

> The first I saw or heard of *The Universe in a Nutshell* was as a bundle of typewritten pages . . . prints-outs of public and scientific lectures and papers, most of them recent, some easy to understand, some full of equations and the language of physics, repetitious of one another in places and occasionally of previous Hawking books—not resembling a coherent book at all. . . . Could it possibly make a book, [editor] Ann Harris wanted to know?[10]

Though the raw material was a total mess, it would have been "unthink-able" for Harris to turn down a proposal from Hawking. Ferguson did her best to even things out, but in the end, even the heart of the book—the illustrations—wound up being somewhat haphazard. One of Hawking's assistants, Neel Shearer, was asked to generate an idea for an image in the book. Shearer created a Monty-Pythonesque mockup of God shooting a lightning bolt and creating an explosion in space, and turned it in. "I hadn't appreciated that it was an image that was going to be published in the book," he says. "I imagined that it was an image that was going to be passed to the publisher's illustrator to be made into something proper." It was a great surprise to Shearer when he saw the completed book and his original illustration was in there.[11]

When *Universe in a Nutshell* came out in November 2001, it immediately shot up the best-seller lists, rapidly reaching number 4 in the *New York Times*. Reviews were mixed. Readers praised the beautiful illustrations, but, as one reviewer complained, "your £20 buys you a lot of repetition of things you first read in *A Brief History*. . . . [T]he images make this book a more handsome object, but also cloak the brevity of the text, a scant 100 pages of unadorned print. That means that quite a lot of things which should be explained are skimped."[12]

There was a good deal of overlap with *Brief History*, but there were definitely new elements, such as an explanation of the holographic principle, an increasingly important area of exploration in cosmology and black-hole physics. And some reviewers praised his exposition—fellow physicist Joe Silk called the book "a delight to read" because of "Hawking's caustic asides and his infallible optimism." But many of the reviews were lukewarm or harsh, such as one written by physicist and science popularizer John Gribbin—who himself wrote a biography of Hawking.[13]

> Whereas previously the party line was that *A Brief History* was all his own work (which anyone who has seen the early draft knows cannot be true),* this time Hawking thanks Ann Harris and Kitty Ferguson, "who edited the manuscript." I don't know the work of Ann Harris, but some of the less technical parts of the present book certainly read like the work of Kitty Ferguson. . . .
>
> Somebody at Bantam should have had the guts to tell Hawking that his jokes aren't funny, and Hawking, assuming he hasn't begun to believe his own publicity, should have told them in no uncertain terms to tone down the blurb. "Great" is an adjective that should be used sparingly, and when used in science reserved for the likes of Albert Einstein and Richard Feynman.[14]

Perhaps more shocking: in a few respects, Hawking seemed to be lagging some years behind the latest developments in cosmology. One of the

---

* In fact, Gribbin had seen an earlier draft. More on this later.

great scientific triumphs in recent times had been to nail down the age of the universe with great precision. By the time *Universe in a Nutshell* came out, cosmologists knew that the Big Bang had happened about 14 billion years ago, give or take a few hundred million years; within a year or so, they would peg it to 13.7 billion years, give or take a few tens of millions. Yet *Universe in a Nutshell* described the universe as 10 to 15 billion years old, a number that cosmologists had used for decades before the rash of observations in the late 1990s and early 2000s allowed for greater precision.[15]

Despite the griping, *Universe in a Nutshell* was a success in more ways than one. It lingered on the *Times'* Nonfiction Best Seller list for twenty weeks and won the Aventis Prize, a prestigious award for science popularization. It just wasn't the same sort of success as *Brief History*.

Plus the *Nutshell* project—which took roughly a year from concept to finished work—seemed to reinvigorate Hawking's love of publishing. In October, less than a month before *Nutshell* hit the bookshelves, Hawking's publisher announced that the physicist—with the assistance of Leonard Mlodinow—would be rewriting *Brief History* to make it more accessible. The new book was originally going to be titled either *A Brief History of Time for Children* or *A Brief History of Time for Young Adults*. However, Hawking eventually settled on *A Briefer History of Time*. And though the new venture was also supposed to be finished in a year or a bit more, it would be almost exactly four years before the book was complete. It was more than he had bargained for. Even if Hawking could indeed be king of infinite space, time was the one thing that he didn't ever think he had in reserve.[16]

• • •

Stephen was not the only one who had been busy writing. Shortly after her marriage to Stephen had fallen apart in 1990, Jane had decided to embark upon a "new project, a project of my own, through which I could prove to myself, if to nobody else, that I had a brain which was both capable and inventive," she wrote. She wanted to write her memoirs, not about her life with Stephen, but about her experiences as an Englishwoman setting up a second home on the Continent. "It would consist of amusing anecdotes and practical information, aimed at the considerable market of British buyers of homes in France."[17]

However, for some reason, her literary agent at the time was unable to generate much interest. One publishing house made an offer to print the book if Jane would commit to writing her memoirs as well. Insulted, Jane refused, and after feuding with her literary agent, resolved "to publish my French book myself, whatever the cost, and to deprive him in perpetuity of any commission on any other book that I might write." Once her contract with her agent expired in 1994, Jane self-published *At Home in France*, and in 1995, shortly after her divorce from Stephen was finalized, she accepted an invitation by Macmillan to write her memoirs, her tale of a "quarter of a century of living on the edge of a black hole," married to Stephen.[18]

The result, *Music to Move the Stars*, hit the bookshelves in 1999 to a flurry of press attention. One reviewer called it "a small book about a small life, bounded by the all-consuming needs of another, occasionally tinted with martyrdom and alarmingly empty of dreams." But the physicist himself was no longer the sole author of the narrative of his life. The woman he once loved—and who had devoted herself to his care—was painting him as a ruthless and selfish tyrant, "an unruly, demanding, and assertive child who needed my protection both on account of his physical helplessness and that peculiar naivety, born of his hyper-intelligence, which can blind the famous to the Machiavellian subtleties of personalities and motivations."[19]

For Jane, a central narrative of her marriage—and much of the reason for its dissolution—rests upon a central irony: "Her Christianity gave her strength to support her husband, the most profound atheist," as one journalist put it. Jane had inherited a deep faith in God from her mother "which had sustained her through the war, through the terminal illness of her beloved father and through my own father's bouts of black depression." Stephen's atheism offered only a "bleakly negative influence which could offer no explanations, no consolation, no comfort and no hope for the human condition," and, despite her faith, pulled her away from God. It wasn't until after Stephen left her, Jane wrote, that she "became convinced that God does exist as the ultimate power of goodness." For Jane, the separation was almost preordained by the Almighty, who had recognized that she was in the wrong marriage and would not allow her to find true happiness until she found the right helpmeet.[20]*

Jane's version of events struck a nerve—it was a classic narrative trope—and it soon provided fuel for a production at the Theatre Royal in Bath in late 2000. The play, *God and Stephen Hawking*, was no *Doctor Faustus*, and the dramatist, Robin Hawdon, was no Marlowe. He was, however, a veteran playwright who had written a wildly successful comedy, *The Mating*

---

* Though it was clearly a source of tension, religion does not figure prominently in what little Stephen wrote about his marriage to Jane or its failure. He implied, laconically, that the breakup was due to another cause altogether: "I became more and more unhappy about the increasingly close relationship between Jane and Jonathan [Hellyer Jones]," a local choirmaster who had taken up residence with the Hawkings. "In the end, I could stand the situation no longer." Stephen Hawking, *My Brief History* (London: Bantam Books, 2018), 87.

Game. And while Stephen had refused to respond to Hawdon's attempt to engage him in the project, Jane had made some "quite helpful suggestions." It was precisely the sort of thing to get the literati talking Hawking.[21]

God and Stephen Hawking pitted the two titular characters in a battle of wits—all about whether the almighty truly exists or was a figment of human imagination. Jane was a divine instrument, a "great weapon" that God used to try to win the battle that Stephen wasn't even aware he was engaged in:

GOD: Let me pose you a question. Which would you rather? A conventional life to suit your upbringing—few traumas, steady husband, family life—a peaceful end. Or something far more perilous?

JANE: Perilous?

GOD: (indicating the sky) Traumatic, exhausting—hugely dangerous. But with the chance of glory.

JANE: That, naturally.

GOD: Not naturally at all.

Both Jane and Stephen are tempted to betray each other, but only Stephen succumbs, running off with his nurse, Elaine. After pronouncing her closing line, "All the same, whether you want him or not, God be with you," Jane leaves, never to be heard again, ceding the stage to Stephen and his creator to battle it out amongst themselves.[22]

"The repetitive squabbling over religion is both inadequately written and ultimately fruitless," complained one reviewer. Another argued that the "extraordinarily misconceived" play "subjects Hawking to the humiliation of taking part in a rigged, bogus, clumsily written debate." Stephen himself found the play "deeply offensive and an invasion of my privacy," he told *Physics World*. "I could probably have got a court order against it, but I have never approved of public figures using [threats] of legal action and it might have just attracted more attention to a stupid and worthless play." Martin Birkinshaw, a cosmologist who helped Hawdon with the science in the play, didn't see it that way. "I rather feel that it is up to the playwright to decide who to write about. Stephen has been in the public eye for so long, and through such a wide range of television programmes and books, that surely this play can make little difference?"[23]

It was unusual for Hawking to complain about how he was portrayed. For example, in 2000, a website with the stylings of gangsta science rapper "MC Hawking"—complete with synthesized voice fronting hip-hop beats—went viral. In response to lyrics such as:

> Then up ahead cold chilling in the street,
> six motherfuckers from MIT. . . .
> I wait till I'm sure they can see my face,
> then I bust out slugs to the beat of the bass.
> The streets sketched out in the full moon light,
> MIT punks dying left and right.
> There's nowhere to run don't even try,
> cause all my shootings be drivebys.[24]

Hawking's assistant sent the author an email stating that the physicist was "flattered." But as crude as MC Hawking's lyrics were, the tongue-in-cheek portrayal of Hawking as a badass gangsta for science didn't conflict with the image that Hawking had been enjoying in popular culture. It was another thing entirely to paint him as needy, arrogant, manipulative, or—worse— vulnerable. That was not a narrative of his choosing. Nor was it one that he could control.[25]

The counternarrative gained strength in 2000, when the first reports began to emerge that Stephen had sustained mysterious injuries—and that the police had begun an investigation into the circumstances. Stephen and Elaine maintained their silence. But there were dark rumors that got even worse just before his sixtieth birthday in 2002, with the incident in which he broke his hip. Though Hawking insisted that he had run into a wall, and his assistant affirmed that the accident was because of excessive haste ("He was late for a meeting and running on Hawking time, as ever"), the questions about abuse rumbled just underneath the surface. It accompanied him and Elaine to scientific conferences and public lectures. And it brought not just Hawking's physical dependence on others into the public arena, but also his sexuality.[26]

The sexual elements that contributed to the end of his first marriage, and made their appearance in Jane's tell-all memoir, had brought Stephen's libido into the sphere of public discourse. There was no longer any way to maintain the image of the physicist as the ultimate force of rationality, a being who, as sociologist Hélène Mialet put it, "can do nothing but 'sit there and think about the mysteries of the Universe,' this intellect liberated from his body and seemingly emancipated from everything that clutters the mundane mind." Stephen Hawking was a flawed human being with an ego and a libido and who could be snickered about behind his back. It was about this time that the press started getting interested in Hawking's habit of visiting nudie bars, a habit he might have picked up in the States.[27]

"At Caltech, there was this tradition of going to strip clubs, I believe quite a bit," says physicist Neil Turok, who collaborated with Hawking a number of times before going on to run the Perimeter Institute of Theoretical Physics in Waterloo, Canada. "And Stephen, when he was here [in Waterloo], his assistant says to me one day, 'Neil, do you know if there are any gentlemen's clubs?'"[28]

"And I said, 'What do you mean gentlemen's clubs? I have no idea what you mean.' So she said, 'Gentlemen's clubs.' And I said, 'You mean like in London, posh clubs, you know, for upper-class people like in Mayfair?' I literally had no idea what she was talking about. And she finally says, 'Strippers!'" Turok's tone lowers, and he speaks between clenched teeth. "And I said, 'If you dare take Stephen to a strip club in Waterloo, I will explode! This, it will ruin our name as a scientific . . . we cannot . . . So don't take him there. Even if there is one, don't take him.'"

Because Hawking was so dependent on others for his mobility, this yen for strip clubs (and even more risqué entertainment) naturally had the potential to put his hosts, his nurses, and his students in an awkward position.

The press first noticed Hawking's predilection for nude bars in 2003, when Hawking was spotted with actor Colin Farrell in Stringfellow's, a London strip club. According to Peter Stringfellow, the club's owner, he spent more than five hours with a stripper named Tiger—but he didn't reveal what the couple did or discussed. Ever after, Stephen's visits to gentlemen's clubs and eventually sex clubs—along with the nitty-gritty of his spousal relations—were grist for the tabloids, which seemed to delight merely in the image of a nearly immobile physicist in a wheelchair becoming sexually aroused.[29]

Despite the best efforts of the tabloids, the prurient stories never reached the vast majority of Hawking fans. He stayed almost entirely aloof; his image was still of the same *Star Trek*–worthy intellect, the same symbol of rationality, the same stoic hero who, uncomplaining, overcame the most profound adversity to understand the workings of the cosmos in a way no other mortal could.

• • •

It was a time when the workings of the cosmos were becoming more visible by the day. In 2000, the supernova observations were still new, and cosmologists were just beginning to grapple with the possibility of dark energy, a totally unexpected substance that stretches and curves the fabric of spacetime. And there was another set of observations just around the corner—observations that had the potential to be just as revolutionary for cosmologists' understanding of the shape, origin, and fate of the universe. In retrospect, Hawking would later say that the ongoing discoveries were the most exciting developments in physics during his career.[30]

Flash back thirty-five years, to 1965, when two engineers at Bell Labs in New Jersey—Arno Penzias and Robert Wilson—were having trouble with an antenna that was designed to pick up microwaves designed for satellite communications. Wherever they pointed it in the sky, the antenna hissed with noise. It seemed like the equipment was malfunctioning; no matter how they tinkered with it, no matter how carefully they controlled where they pointed the antenna, they couldn't seem to eliminate the noisy static.

At the same time, a few miles away, at the Princeton campus, physicists Bob Dicke and Jim Peebles were doing some calculations having to do with the origin of the universe. If the universe was really born in a Big Bang—a sudden explosion that created the manifold of spacetime all at once—the tiny, early universe must have been very, very hot and glowed brightly with highly energetic light. But as the cosmos expanded, it cooled down, and the light—which was everywhere in the universe—stretched out. As light stretches, it becomes less and less energetic, changing from gamma rays to X-rays to ultraviolet to visible light to infrared to microwaves. So, the Princeton scientists reasoned, billions of years after the Big Bang, that remnant light should be visible as microwaves coming from all directions in the sky.* A microwave antenna pointed upward should be able to detect that light—which would be a background hiss that wouldn't go away no matter where the antenna looked. They were planning to build a microwave antenna to see if they could spot that hiss when they got a call from the engineers at Bell Labs asking for help with an odd little static problem they were having.

This was the discovery of the cosmic microwave background (CMB) radiation. It's hard to explain this sort of omnidirectional, almost uniform hiss of microwaves unless you assume that the manifold of spacetime was once very small and very hot, and expanded to give us our present-day universe. Penzias and Wilson received the Nobel Prize for their discovery; Dicke and Peebles, the theorists, got a nice thank-you.

---

* This is a bit of an oversimplification. The remnant light is actually from a time roughly 380,000 years after the Big Bang, when clouds of electrons and protons—which were opaque to light—cooled down enough to form hydrogen gas.

But there was another layer of discovery just out of reach of 1970s technology. Even though the microwave static looked the same in every direction, the theory says that it can't be *quite* uniform; as scientists like to say, it was *anisotropic.* That is, if you look hard enough, you should be able to see differences in the microwaves at different scales, hot patches that had more energetic light than average and cold regions that had less energetic light. The fabric of spacetime in the early universe was rippling with energy, squashing and stretching out clumps of matter at various times. This, in turn, made the spacetime of the early universe curve on various different scales. If the Big Bang model were correct, those ripples and curves in the early universe should have caused fluctuations in the energy of the CMB, hot spots and cold regions of various sizes. Penzias and Wilson's antenna wasn't good enough to see those variations. But a better instrument should be.

That instrument was the Cosmic Background Explorer (COBE), a satellite launched in 1989, a quarter century after the discovery of the CMB. The COBE team quickly proved that the CMB did, indeed, have hot and cold regions, just as the theory had predicted. This discovery, too, won a Nobel Prize. However, COBE's instruments weren't good enough to resolve those spots with very much detail. It was like a nearsighted person looking at an eye chart without glasses; it's not hard to recognize that there's something written on the chart, and perhaps even to read the largest few letters at the top. But the chart is a blur, and it's impossible to gain enough information to figure out what the smaller letters are saying. COBE was looking at the CMB with blurry vision, unable to resolve, or extract information from, smaller spots in the radiation. And encoded in the size of those smaller spots was nothing less than the age of our universe—and its shape.

The theory that describes the hot and cold spots caused by the rumbles of energy in the early universe also implies that there should be a characteristic spot size—a maximum breadth for the hot or cold spots caused by the rumbles in the early universe.* These maximal spots were predicted to be about a degree across in the sky. And this gave cosmologists a way to measure the curvature of the universe.

If we lived in a spherical universe, distant objects would seem large in the sky, magnified just as if we were seeing them reflected in a bowl-shaped shaving mirror. If we lived in a saddle-shaped universe, the opposite would

---

* Because the CMB was released roughly 380,000 years after the Big Bang, the spots that are created by these so-called acoustic oscillations in the early universe are limited by how far a given dollop of energy and matter can stretch or shrink in 380,000 years, which itself is limited by the speed of light.

be true. Only if the universe were flat would those distant objects be undistorted, their apparent sizes in the sky neither magnified nor reduced by the curvature of spacetime.

All scientists needed to do to pin down the curvature of the universe was to measure the size of those maximal hot and cold spots in the CMB. If they were larger than expected, it was evidence of a spherical, closed universe. If they were smaller than expected, it meant that we were in a saddle-shaped one. And if they were precisely the size theorists said they would be, well, that would be a very strong indicator that our universe was flat. It was a measurement cosmologists had been hoping to perform for decades—but microwave antennas weren't sophisticated enough to get the smaller-than-one-degree resolution needed for the requisite comparison.

Flash-forward again to 2000. After years of effort, scientists had finally managed to improve their microwave detectors enough to start resolving hot and cold spots about a quarter of a degree across. Data began to trickle out, and as best as cosmologists could tell, those hot and cold spots were precisely the size that theory predicted.

The universe seemed to be flat after all.

It would expand forever. There would be no Big Crunch.

. . .

Hawking had gone most of his professional life as a theoretician without any experimental tests of his theories. That had been the norm for cosmology and general relativity for more than half a century. In the 1960s and early 1970s, when Hawking did his most important work on black holes, there wasn't a confirmed observation of even a single black hole. (One of Hawking's most famous bets, made in 1974, was about whether or not a newly discovered object would actually turn out to be a black hole.) Experiments and observations that could advance the debates over the origin and fate of the universe were incredibly hard to come by. Other than a few blurry pictures of the CMB, scientists had to make do with error-ridden and difficult-to-interpret measurements of galactic properties and distributions to get a sense of the shape of the universe and its age. In the 1990s, the best guess was that the universe was somewhere in the neighborhood of 12 to 15 billion years old, and there was no real clue about the overall shape of the universe. By the early 2000s, thanks to new observations, astronomers had figured out that it was 13.7 billion years old, give or take a few tens of millions of years, and was dead flat, or very nearly so. This incredible increase in precision suddenly meant that theorists could finally begin to use experimental evidence to challenge ideas that had sat around, untested, for decades. Among them was one of Hawking's own—one of the ones he held most dear: the no-boundary proposal.

The no-boundary proposal was an attempt to figure out where time and space came from at the very beginning of the universe, a question that Einstein's theory simply can't answer. General relativity describes how a smooth manifold of spacetime behaves. It says nothing at all about how such a manifold might have come to be. If you were to look at our expanding and cooling universe and try to run the equations backward—the universe shrinks and gets denser and hotter—the mathematical framework of relativity has to break down at some point. (In fact, Hawking's first major contribution to physics was proving exactly this, and it became the basis of his PhD thesis.) Even a fully functional quantum theory of gravity, which would explain the behavior of gravity even in the tiniest, most chaotic, roughest patches of space and time, would still have to contend with describing all the matter and energy in the universe compressed into a single infinitesimal point—a singularity—and explaining how that singularity came to be out of the (presumed) nothingness that came before it.

In the early 1980s, though, Hawking, along with fellow physicist Jim Hartle at the University of California at Santa Barbara, used a clever trick that, they argued, allowed them, to some extent, to dodge the messy singularity at the beginning of time. The idea was to look backward not in time, but in a mathematically related quantity known as *imaginary time*. This concept will be explained more fully in Chapter 10, but for the moment, suffice it to say that with this change in perspective—looking through the lens of imaginary time rather than regular time—a space-imaginary-time universe has a different sort of topology from the ordinary spacetime universe we're used to. It seems like a nice, smooth, compact, self-contained sphere-like object with neither beginning nor ending. From our perspective, as creatures moving through regular time, it seems like time and space suddenly exploded out of nowhere in a Big Bang. From an imaginary-time perspective, nothing was born, nothing died. There were no boundaries to the universe, no places on the spherical universe where one could say "the universe begins here" or "the universe ends here." It is a completely smooth universe without boundaries. Hawking himself described his no-boundary cosmos in *A Brief History of Time*:

> There would be no singularities at which the laws of science broke down, and no edge of space-time at which one would have to appeal to God or some new law to set the boundary conditions for space-time. One could say: "The boundary condition of the universe is that

it has no boundary." The universe would be completely self-contained and not affected by anything outside itself. It would neither be created nor destroyed. It would just BE.[31]

If you find this hard to understand, you're not alone. The Hawking-Hartle no-boundary proposal was controversial from the start; the shift from regular time to imaginary time made it hard to interpret, and it wasn't at all clear that the mathematical calculations Hawking and Hartle were doing to describe their picture were valid. As physicist and former Hawking student Don Page wrote, "although at first sight the proposal looks conceptually clear (at least to those who understand the concepts, so long as they do not worry about details), when one looks at the details one finds that the proposal is not yet mathematically precise." This made it hard for many physicists to engage seriously with the concept. "The idea has never been accepted," says physicist Neil Turok, who helped Hawking explore the proposal in the 1990s. "I would say 90% of cosmologists or theoretical physicists don't even form an opinion. Of those who do, 90% of them would say they probably don't agree with it, or they thought there was some problem with it." Nearly four decades after Hawking first lofted the no-boundary proposal, physicists are still fighting over its validity.[32]

While this idea might seem extremely complex to some, to others, there's a certain elegance, a certain beauty to the way Hawking dodged the problems inherent in asking about the origin of the universe. And Hawking, like many other theoretical physicists, believed that the true laws of the universe possess a certain spare beauty. As a consequence, he became very attached to the no-boundary proposal. "That just rang true to him. Whether that was for cultural or religious or whatever reason, I couldn't say, but I think he thought it was true just because it's simple," Turok says.

The no-boundary proposal featured prominently in *A Brief History of Time*, but when Hawking wrote the book in the late 1980s, the proposal required that the imaginary-time-universe be rather symmetric—the Big Bang at the beginning of the universe had to be matched by an equal and opposite Big Crunch. And because our universe would end in a Big Crunch, it couldn't be flat. Moreover, for technical reasons, the no-boundary proposal seemed to imply that there was no cosmological constant—that it is "not necessarily zero, but zero is by far the most probable value."[33]

And so things remained until the supernova data in 1998, the high-precision CMB data in 2000, and, finally, the results from the successor to COBE, the WMAP satellite, which were released in 2003. We finally had a definitive answer; the universe was flat or very nearly so, and it would keep on expanding forever. A Big Crunch was no longer a possibility. The cosmos was open, not closed. And there was a significant cosmological constant.

As theories are wont to do when subjected to experimental pressure, the no-boundary proposal began to wiggle and transform. In 1998, Neil Turok, who was then a professor at Cambridge, suggested to Hawking that, hidden inside his sphere-like, finite, imaginary-time universe, there might reside an open, infinite, real-time universe rather than only a closed one that would collapse in a Big Crunch. "It's such a surprising thing. If you start with a finite universe, inside it, you can create an infinite universe, so that's very, very paradoxical," Turok explains. "Inside a small piece, you can form this infinite spatial universe which lasts forever." No longer was a Big Crunch an absolute prediction of the no-boundary proposal. But there was another problem.[34]

"The truth is, his proposal wasn't really working. It predicted an empty universe; you can have this picture of the universe coming from nothing, but the prediction is that it's empty," says Turok. Even after tweaking the model to force it to produce at least one galaxy, the model produced exactly that—a universe with just one galaxy. Nothing more. And nothing that in any way resembled our own universe. "We predicted an empty universe or one galaxy; it's not very good. Stephen said, 'You know, don't worry about that too much. It's still progress. . . .' So he was pretty happy-go-lucky about it. I was actually amazed."

Hawking was able to keep holding onto his proposal even as some of its predictions failed and the mathematics frayed around the edges. The details simply weren't working out. But the details weren't really the point in the first place. In the 1980s, Hawking had presented a beautiful idea—a seeming way to avoid many of the messy questions about how the universe came to be. To Hawking, the no-boundary proposal had the ring of Truth with a capital T, and that was enough, even if he and his collaborators couldn't work the kinks out of the mathematics. It just felt correct to him, and to Hawking, instinct was a powerful driver. The big beautiful idea was the important thing, not the details.

And Hawking had become less and less able to work out the details himself. As he himself admitted, since he'd lost the use of his hands in the early 1970s, he couldn't do mathematics in the same way everyone else did. He couldn't write complicated formulae, doodle diagrams, or even store a transient thought in an efficient way. That made it hard for him to manipulate a lot of the necessary mathematical formalisms or to build the equations that breathe concrete details into a beautiful and innovative, but skeletal, idea. "It is difficult to handle complicated equations in my head," Hawking told the audience at the White House Millennium Lecture in 1999. "I therefore avoid problems with a lot of equations or translate them into problems in geometry. I can then picture them in my mind."

"He has gradually trained his mind to think in a manner different from the minds of other physicists: He thinks in new types of intuitive mental pictures and mental equations, that, for him, have replaced paper-and-pen drawings and written equations," Kip Thorne wrote of his longtime friend. "Hawking's mental pictures and mental equations have turned out to be more powerful, for some kinds of problems, than the old paper-and-pen ones, and less powerful for others." But, according to Andy Strominger, Hawking's disability definitely limited his reach as a physicist. "He did, of course, develop amazing abilities to do things in his head that anybody else would have needed a paper and pencil for. But even so, he couldn't do as much in his head as other people could do with a paper and pencil," Strominger says. "And, you know, and it got worse as the years went by and whenever it would get to a detailed calculation, you know, he would go as far as he could in his head, and then, you know, the rest would be up to his collaborators."[35]

"Working with Stephen doesn't mean working with equations. It means working with words and with concepts. That's how he leads my research," Hawking's graduate student Christophe Galfard told a BBC film crew making a documentary about Hawking in the early 2000s. Galfard was helping Hawking not with the no-boundary proposal, but with the black-hole information paradox—the question that led to Hawking conceding his bet with John Preskill in 2004.[36]

Quantum theorists and string theorists, unlike general relativity specialists—and unlike Stephen Hawking—had always been fairly confident that information wasn't permanently lost to a black hole, and by the mid-1990s, Hawking was losing the argument. And then, in 1996 and 1997, a rash of papers by physicists Juan Maldacena, Ed Witten, Andy Strominger, Cumrun Vafa, and others seemed to settle the question fairly decisively against him—or so the quantum theorists believed. "Beyond a shadow of a doubt . . . information would never be lost behind a black hole horizon," wrote Stanford physicist Leonard Susskind in his book about his "war" with Hawking over the information paradox. "The string theorists could understand this immediately: the relativists would take longer. . . . Although the Black Hole War should have come to an end in early 1998, Stephen Hawking was like one of those unfortunate soldiers who wander in the jungle for years, not knowing that the hostilities had ended." Yet the Black Hole War continued to claim casualties.[37]

"Andrew Farley turned out to be one of the two best graduate students that I have ever had," writes Peter D'Eath, one of Hawking's early PhD students who joined the physics faculty at Cambridge shortly thereafter. Under D'Eath's tutelage, Farley wanted to examine certain quantum-mechanical effects after the collapse of a black hole: effects that strongly implied that information would *not* be lost when falling past the event horizon.[38]

The department's rules were that a student couldn't formally enter the PhD program until the fourth semester, when two assessors reviewed his or her work and plans and gave the go-ahead for the student to register. Unless the student were performing poorly, the assessment was mostly a rubber stamp—with the added benefit of having experts give the student some feedback on potential weaknesses and point out avenues for further exploration. D'Eath, quite naturally, picked Hawking to be one of Farley's assessors. "Of course, in retrospect, this caused a serious amount of trouble," D'Eath notes. It was unexpected: D'Eath had a collegial relationship with his former mentor, and frequently invited him and Elaine over for meals. However, after Hawking began looking at Farley's work, he no longer seemed willing to visit the D'Eaths. "For the first time, Stephen had either declined or put off any acceptance indefinitely."

The assessment of Farley's work was delayed for several months, and there were some very odd changes of the rules—for example, the assessment wound up using three rather than two assessors, and D'Eath himself was "interrogated" about Farley's work in May 1999. ("Indeed, I have never heard of a supervisor being required to take part in such a process," D'Eath writes.) And when the assessment finally happened, it was clear that Hawking was very displeased with Farley's project. Hawking went on the attack and eventually recommended to the university that Farley be kicked out of the program.

Though the head of the department overruled Hawking, and Farley was allowed to continue with his studies, that wasn't the end of the matter. D'Eath says that he later discovered that "Stephen had continued to try to have Andrew Farley de-registered via the Degree Committee of Mathematics, then the Board of Graduate Studies, then the Vice-Chancellor himself. . . . Each time, Stephen was acting *ultra vires*, and each time his application was denied."

Farley himself didn't elaborate on his adviser's account, writing, "I have nothing material to add other than Stephen's unreasonable behaviour knocked my confidence thereafter. Stephen was someone I had greatly admired as a boy and up to the point of meeting him. They say never meet your heroes; this is particularly apt."

It is highly unusual in academia for such a powerful professor to marshal all of the forces at his disposal to try to destroy a mere graduate student. It's even more unusual for the student to prevail. Though Farley went on to complete his PhD, it was a Pyrrhic victory. He was never able to get a postdoctoral position, and that meant that any hopes of an academic career were dashed. Farley currently works as a compliance officer for a finance firm. "However, I never gave up doing my own theoretical physics research after Cambridge," he writes. "I'm currently doing my own research on gravitational lensing. . . . Ideally, I would now be doing theoretical physics research full time as this is my passion."[39]

Also dashed was Hawking's quarter-century-old friendship with D'Eath. The contentious "interrogation" about Farley's work in May 1999 was the last time that Hawking and D'Eath spoke to each other.

Ironically, Hawking was beginning to harbor doubts in his own mind about information loss; the string-theoretic work coming out at that time was powerful and convincing. Even though that sort of research wasn't quite in Hawking's wheelhouse, it shook his confidence.

In 2002, Hawking assigned his student Christophe Galfard the task of understanding one of the important Maldacena papers that had come out the previous year. "Stephen asked me to have a look at that paper so I took a little while to read it. A little while being about a year and a half," Galfard told a BBC film crew. The paper used some of the powerful new insights coming out of the theoretical community to imply that Hawking was wrong about the paradox. But the paper held firm despite their best efforts, with Hawking providing the big ideas via oracular guidance and Galfard working out the fine details. The documentary filmmaker told sociologist Hélène Mialet that, during the course of filming, "it became quite obvious early on that effectively, you know, his graduate students do all the actual work . . . in terms of hashing the numbers and working through the equations and all of that, and then they bring it to him, and he obviously assesses it and then points them in different directions." Working together in this way, Galfard and Hawking dissected the Maldacena paper. "He thought that there was something in there that was new and different," Galfard says. "I do strongly believe that it coincided with some deep ideas that he had for a long time, and this gave him the mathematical framework to actually verify whether or not the information paradox really was a paradox." And as they ventured deeper into the new territory, the two lay the groundwork for Hawking's Dublin announcement, his formal U-turn on the question of whether information is lost in a black hole.[40]

But before the two could complete their work, Hawking was struck down with a bad case of pneumonia; it was very grave, and Hawking was placed on life support. "We didn't know whether we would see him again. We were very, very concerned," Galfard said. It was three months before he was able to leave the hospital.[41]

. . .

Hawking never expected to survive long enough to see any of his theories falsified; he had already outlived his doctors' dire predictions by an almost unbelievable margin. But as his students and collaborators were keenly aware, Hawking spent most of his years teetering on a stark boundary. His emaciated frame seemed too fragile to contain a life within. Yet despite the way his condition consumed his being, he was always stoic about it—he even turned it into a source of fun. BBC reporter Pallab Ghosh wrote about one encounter he had with the physicist in 2004:

> The camera operator I was with wanted to make a last minute adjustment to his lighting and so he asked Prof Hawking's staff if he could pull out one of the plugs in the office so that he could use the socket for his equipment.
>
> Without waiting for a response he pulled the plug and the room was filled with a deafening siren.
>
> Prof Hawking then slouched forward and I feared that my colleague had inadvertently unplugged a vital piece of life-support equipment.
>
> Fortunately, it was the alarm to the uninterruptable power supply to his office computer and he was slouched forward with mirth at our incompetence.[42]

There was never a hint of self-pity, and seldom any vulnerability— which made it all the more striking when it appeared. Neil Turok describes one time when he was visiting Hawking in the hospital after a serious operation to repair the physicist's trachea. "One of the most moving moments I had with him was when his throat had collapsed. . . . I went to see him in hospital in London," Turok recalls. "And the first thing he said to me was, 'I nearly died.' And I just sort of sat there. Like what do you say to someone who just told you he nearly died? And so we sat there for a while and thought about that prospect. And then he said, 'Let's discuss physics.'"[43]

CHAPTER 7

INFORMATION (1995–1997)

There's a photograph from Hawking's wedding day in the early fall of 1995 in which the physicist and his new bride touch foreheads together. He sits in his wheelchair in a dapper gray suit, half-smiling at Elaine, who smiles back at him. The physicist's right hand sits in his lap, cradling the clicker which he uses to control his computer, and his wrist seems impossibly thin. The sleeve of his suit looks almost empty. His left hand is on the armrest of his wheelchair, cocked at an uncomfortable-looking angle, seemingly an attempt to touch his bride. But Hawking had almost no control over his arms, and the loving contact seems posed. His atrophied jaw and neck muscles can make it hard to tell the difference between a smile and a grimace. The only reliable hint at what is going on in Hawking's head comes from his eyes, which gaze at Elaine in a way that might indicate love, or anticipation, or even something else entirely.

Hawking jealously guarded his innermost thoughts, seldom letting information about his private life leak out to a public hungry for scandal. Only those who knew and loved him could readily interpret his emotions and gauge his well-being. Jane, of course, seemed of a mind that the physicist had been manipulated by Elaine. ("He is in the grip of forces that he can't control and which broke up our home," she insisted.) However, Jane had ceased being his confidante half a decade prior—and, due in part to the acrimony of their divorce, was not a reliable narrator of Stephen's needs or desires.[1]

Stephen had announced his intent to move out with Elaine in late 1989, but it was some time before he could get his affairs in order and find a place to stay. As was to be expected for a feuding couple (actually, two couples) living under the same roof, it got ugly—at one point, Jane writes, she put Stephen's suitcase outside the locked door in hopes that he would finally leave; the result was a brick hurled through a window. Finally, in February 1990, Stephen and Elaine moved out, but the troubles didn't end there. Tax collectors pursued both Jane and Stephen regarding profits from *A Brief History of Time*, and, as Jane put it, he was forced to pay a "monumental fine" after "having first protested that he had already paid enough tax to build a small hospital." ("I remember a story where he went to give some talks in Japan because it was tax time," former student Ray Laflamme recalls. "I don't know how true it was, but I remember the students having some rumors.")[2]

Making matters worse, Jane writes, Stephen "did not instigate divorce proceedings for a long time," leading to a five-year delay before the divorce became final in May 1995.* Jane was further hampered in her ability to move on with her life by an "unjust and whimsical quirk of English law": the moment she remarried, she would lose the ability to claim a financial settlement from Stephen.[3]†

However, had Jane consented to the divorce, Stephen would have been able to initiate the divorce after having lived apart from Jane for two years. Without that consent, English law required Stephen to wait until the fifth year post-separation before filing.‡ The timing of the divorce so soon after the five-year rule allows, coupled with the speed of Stephen's remarriage afterward, suggests that Stephen was just as eager to get on with his life as Jane was, if not more so.[4]

Jane had been closed out of Stephen's life and had almost no information about the physicist's state of mind. In fact, she only found out about the

---

* The petitioner on the divorce decree was technically Jane, not Stephen.

† Stephen, who apparently owned the intellectual property that formed most of the couple's assets, was free to marry his nurse; Jane, on the other hand, had to wait to marry her lover until after she officially had claim to some of that money. According to Jane, the settlement took eighteen months; she and Jonathan Hellyer Jones married in 1997.

‡ There are other grounds for divorce, but Stephen would not have been able to avail himself of them. He couldn't have claimed desertion or abuse. Even if it were perfectly clear that Jane's relationship with Jonathan was adulterous, the law said that excuse was only valid within six months of Stephen's finding out—and Stephen had known about the relationship since the late 1970s.

upcoming nuptials after the press did in July 1995. "I think he has been very ill-advised in what he is doing, if he is planning that," she told the papers when they asked her for comment. "His present relationship is nothing to do with me." As she later wrote, "With his marriage, Stephen effectively slammed the door on our remaining lines of communication. . . . I had no choice but to reconcile myself to the end of an era." Neither Jane nor their three children attended the physicist's wedding.[5]

The only person who could truly speak for Stephen was Stephen. Yet he was all but silent, even though a number of his friends and former students worried for his happiness and even his safety after his new marriage. Whatever information there was was safely locked inside his head.[6]

• • •

Stephen Hawking's work inhabited the very core of physics—and had even raised the hope of generating a theory of everything, a single mathematical framework that could reconcile the seemingly irreconcilable realms of the quantum and the relativistic. His analyses of black holes required elements from both of these incompatible worlds, and he had been able to find profound new truths—that black holes radiated—by looking at the region where those two theories clashed.

The most elegant expression of that clash was Hawking's black-hole information paradox: as Hawking interpreted it, the decay of black holes pointed directly to a flaw in quantum theory, a place where the existing framework in our understanding of the quantum realm needed to be patched. This was the very sort of region that could give birth to a theory of everything.

By the mid-1990s, however, physicists had trained new and powerful mathematical tools on the black-hole information paradox and were generating increasingly convincing evidence that Hawking's interpretation was wrong. The framework of quantum theory was withstanding his assault, and the answer to his paradox was turning out to be considerably more subtle than he had expected. It would be a number of years yet before he would concede this point—and his famous wager—but the mid-1990s marked the time that Hawking ceased being the leading expert on the paradox he himself had formulated. A new and younger generation of physicists was generating information about black holes that had evaded Hawking's powerful intuition.

• • •

Deep down, some of the biggest problems of twentieth-century physics revolved around the concept of information. In fact, both quantum theory and relativity can be thought of as rules about information: the laws governing how information can be gathered, exchanged, transmitted, and lost. In quantum mechanics, the uncertainty principle says that it's impossible to have perfect information about an object's position and momentum—or about how much energy it has and when it has that energy. In relativity, geometric rules dictate whether information can flow from one section of spacetime to the other. As abstract as the concept of information might seem, for a physicist, a battle over what happens to a chunk of information under a given set of circumstances is as consequential as it gets: it is an argument over the fundamental rules that govern how the universe works on both the smallest and the largest scales.[7]

Information is an abstract concept, but it's not so hard to understand in a concrete, everyday context, such as using information to describe, say, a marble sitting in a shoebox. Imagine that, for some reason, someone calls you up on the telephone wanting to know the precise position of the marble in the box. How do you answer the caller's question?

You might glance at the marble and say, "It's on the left side of the box." Or you might give a little more detail: "It's in the upper left-hand quadrant of the box." Or more: "It's a little more than an eighth of the way from the left side of the box, and nearly halfway down from the top of the box." Or more still: "It's 3.831 inches from the top and 2.632 inches from the left."

All of these responses are valid answers to the question, but they're not equal. As the answers become increasingly specific, they allow the caller to understand the marble's location with more and more confidence—with less and less uncertainty. And at its core, information is just this: a measure of how much a given transmission reduces uncertainty about a given question. The vague answers about the marble's position don't contain much information; detailed ones contain more information.

How much more? The information content in a message can be measured, just as the weight or position or speed of a marble can be. In the classical, ordinary world we're used to, the fundamental unit of information is the *bit*: a single item that can be a 1 or a 0, an on or an off, an up or a down. "It's on the left side of the box" is equivalent to one bit of information: 0 means left side of the box, 1 means right. "It's in the upper left-hand quadrant of the box" needs a second bit: a first 0/1 tells you if it's in the left/right side of the box, while a second 0/1 would tell you whether it's in the top/bottom half of the box. A two-bit message, 00, is enough to signify that the marble is in the top left quadrant. Something like "It's 3.831 inches from the top and 2.632 inches from the left" would require roughly forty bits to relay. The more you want to reduce the caller's uncertainty as to the position of the marble, the more bits it takes—the more information the message must contain.

This principle applies not just to the marble's position in the box but its color, its weight, its speed—anything that is interesting about the marble that you might choose to relay to a caller over the phone. You can describe everything about the marble, as perfectly as you could ever want, with a sequence of bits. Conversely, any sequence of bits that you wish to store can be represented by a marble in the box—if you position it just so, with the appropriate degree of precision, that marble "stores" a sequence of 0s and 1s that can be read out later by measuring the position of the marble.

If you allow specifying the properties of the marble with arbitrary precision, there would be no theoretical limit to the amount of information you could store in a marble/box system like this. Any message of 0s and 1s, no matter how long, could be encoded by the marble. It could be the opening line to *Moby Dick*: "Call me Ishmael," which would take 120 bits or so. It could be the King James Bible (35 million bits). It could even be the *Encyclopedia Britannica* (a few billion bits). In fact, one could theoretically store the sum total of human knowledge—everything that's ever been written, filmed, recorded, or otherwise produced by humans—in a marble positioned just so. There's a theoretically (if not practically) infinite storage capacity in a marble/box system . . . and that's even without troubling to add another marble to the box, which would (roughly) double capacity.

In the quantum world, however, there is no such thing as arbitrary precision. Even if we somehow were able to place an electron exactly at a given spot in the box, the uncertainty principle means that we would know *nothing* about its momentum or how fast it is moving; it won't stay in the place where we set it down, making it useless as an information-storage device, even for a tiny fraction of a second. In fact, the roughness and the uncertainty of the universe on the very smallest scale means that there is a fundamental limit on the amount of information that can reside on a quantum particle in a box. If you could capture that finite amount of information, and transmit it, say, over the phone, one could make a perfect copy of the system. There is no other information to be captured—unlike a classical marble in a box, which can theoretically store an infinite amount of information, a finite number of quantum "bits" would capture every possible thing that's meaningful to say about that particle in the box. The quantum information about the system is everything knowable about the system itself. That is, any lump of matter and energy in the universe sitting in a box—or any finite container, for that matter—could be 100 percent perfectly described by a finite amount of quantum information.*

It's a strict and somewhat bizarre law: in any given box, there's a finite amount of information, a maximum complexity to the stuff inside. What happens if you try to break the law? Well, you can always increase the complexity in the box by putting more stuff in it: throw in more particles, atoms of different kinds, tiny molecules vibrating and rotating in different ways, all the colors of the rainbow. Fill the box with more energy, heat waves rippling through the matter, light bouncing about, electrical currents swirling around. Yes, the complexity of the stuff inside the box is increasing, and you need more information to describe it. But as complex and varied as the stuff inside the box is, it still requires only a finite amount of information to be described. Keep putting more and more particles, more matter, more energy into the box, and the complexity goes up and up until . . .

Until something amazing happens.

The laws of general relativity dictate what happens next. When you cram too much matter and energy into a small enough space, spacetime curves more and more dramatically. The gravitational pull of the matter in the box increases and increases and increases some more until the curvature of spacetime is so huge that light from within the box can no longer escape. By stuffing all that matter and energy inside the box, we've caused a collapse into a singularity, a region of such intense curvature that the laws of physics no longer seem to apply. We've created a black hole.

All of a sudden, that hugely complex kaleidoscopic soup of matter and energy with its box has transformed into a black hole with its event horizon. And a black hole, far from being a complex jumble of matter and energy and information, is one of the simplest objects in the universe. For the rules of general relativity imply that a black hole is totally described by just three numbers: its mass, its charge, and how fast it's spinning. All other information about the system, all that information that allowed you to know every possible knowable thing about every single particle in the box, is gone.

---

* Quantum information is somewhat more complicated than classical information in a number of ways. Because a quantum particle, unlike a marble, can be in two (or more!) places at once, a quantum bit isn't just a 1 or a 0 like a classical bit, but can be both at the same time. Also, a quantum bit can't be copied without destroying the original—so while it's true we could make a perfect copy of the electron-in-box system described here, it's only possible to do so if the original electron-in-box system is ruined. So it's not really a "copy" as much as it is a "transfer" of quantum information.

It seems like the black hole is the ultimate eraser. All the quantum information about the stuff that becomes a black hole—or the stuff that falls in—is lost to an outside observer. We can't look at a black hole and tell what it's made of, what configuration of particles and energy has fallen behind the event horizon. Yet the rules of quantum mechanics seem to imply that quantum information can't just be lost. Information about matter and energy can't disappear without a trace, any more than the matter and energy itself can just up and vanish.

So what, then, happens to the information? There are two possibilities. First, the information is lost, seemingly violating the laws of quantum mechanics. Second, the black hole somehow stores the information. Now, this doesn't seem like such a bad option until you realize that a black hole can't store information indefinitely. One of Hawking's most important contributions to science is the proof that black holes radiate—they constantly emit energy—causing them to shrink, and eventually to disappear.* "The possibility of hiding information in a vault would hardly be a cause for alarm," writes physicist Leonard Susskind. "But what if when the door was shut, the vault evaporated right in front of your eyes?" A black hole couldn't hold onto that information forever. Worse, the energy black holes emit—known as Hawking radiation—can't carry information; it is as featureless as radiation can be. So a black hole is only a temporary storage site, yet the radiation it emits shouldn't be able to carry the information away, and the black hole itself disappears.[8]

This is the core of the black-hole information paradox that occupied so much of Hawking's brain for so long. When matter falls into a black hole, information is lost forever (which quantum theorists say is impossible), or it is stored indefinitely in the black hole (which general-relativity and black-hole experts, including Hawking, say is impossible). Two different sets of laws give two different answers to the question. This sort of contradiction is a sign that something new and important in physics is just waiting to be discovered. And it was a subject much on Hawking's mind in the mid-1990s.

• • •

Exactly twice in history has a physicist sold out the Royal Albert Hall. The first time was in 1933, when Albert Einstein was greeted with rousing

---

* Actually, they explode in an extremely violent eruption.

cheers.* The second was Stephen Hawking, in November 1995. Up on the stage, surrounded by five thousand of his fans, Hawking clicked through his lecture, titled "Does God throw dice in black holes?" And though probably only a small fraction of those in attendance truly knew what the talk was really about, it was about Hawking's thoughts about the black-hole information paradox—not the sort of stuff that one would expect to draw standing-room-only crowds at the Royal Albert. But people didn't buy tickets to a public Hawking lecture to learn physics; they were there to be in the presence of a great man.[9]

To those who understood the science, Hawking's Albert Hall lecture was a window into why the paradox had haunted the physicist for so long. It was about much more than what happens to particles falling into black holes. "This lecture is about whether we can predict the future, or whether it is arbitrary and random," the talk began.[10]

Classical physics, Hawking explained, presumes that the universe is deterministic. If you know the "state" of the universe—if you have perfect information about all the objects in the cosmos—then you could use the laws of physics to predict, with perfect accuracy, the state of the universe in the future. There would be no surprises, perhaps even no free will. This would no longer be true in a universe governed by quantum information, especially if that information, the very stuff that enables future predictions, is eradicated by black holes.

"What all this means is, that information will be lost from our region of the universe, when black holes are formed, and then evaporate. This loss of information will mean that we can predict even less than we thought, on the basis of quantum theory," Hawking told the audience. "Thus, the future of the universe is not completely determined by the laws of science, and its present state. . . . God still has a few tricks up his sleeve."

Hawking invoked the name of God over and over—at one point, accompanied by a crudely drawn image of a smiling, bearded deity throwing a pair of dice. (One die fell down the maw of a black hole.) "It seems Einstein was doubly wrong when he said, 'God does not play dice,'" Hawking declared. "Not only does God definitely play dice, but He sometimes confuses us by throwing them where they can't be seen." After a lull of more than sixty-two years, there was a new prophet in town.

---

* Einstein was such a celebrity that he didn't even have to show up in person to draw *Beatles*-sized audiences; three years earlier, a crowd of 4,500 eager viewers trying to see a film about Einstein and his theory of relativity stormed the American Museum of Natural History in New York. Wags dubbed it the first "science riot."

Hawking was at peak fame. For all of his bravado, however, the physicist was, at first, reluctant to take on Albert Hall, fearing that he wouldn't fill the venue.* He also betrayed no doubts whatsoever that he might be wrong about what happens to information that falls into a black hole. Nobody in the audience would have guessed that Hawking was becoming isolated, his viewpoint increasingly discounted by the scientific community.[11]

. . .

Einstein would never achieve his dream of finding a theory of everything, an overarching set of equations that explained all the interactions of matter and energy on all scales. But Einstein wouldn't even get a glimpse of the promised land before his death in 1955. Hawking, taking up the mantle of the prophet, suggested in 1980 that scientists would build such a theory of everything by the turn of the century.[12]

The Standard Model of particle physics does a wonderful job of explaining the behavior of matter and the forces that can affect matter, with the exception that it doesn't account for gravity properly. Scientists expect that a theory that explains all the forces of nature, including gravity, a theory of everything, should look something like the Standard Model but with more intricacy, with more subtle rules—with a different mathematical structure that extends the Standard Model and includes the laws about gravity as well. Back in the 1980s, a bit before Hawking started writing *A Brief History of Time*, he had seized upon a particular framework known as *N=8 supergravity* as the "only candidate in sight" for a theory of everything. And like other such mathematical frameworks, its fundamental simplicity is rather difficult to see behind its seeming complexity.[13]†

For one thing, N=8 supergravity is a *supersymmetric* theory, which means that for every particle we know of, there must be one (or more) partner particles that have yet to be discovered—not a deal breaker, but something that requires explanation. (After all, if such particles exist, why haven't we seen them yet?) Among the undiscovered particles were a graviton (a particle that carried the gravitational force), eight gravitinos, which mediate as-yet undiscovered forces related to gravity, and numerous

---

* When he eventually agreed to do the lecture, Hawking apparently donated all the proceeds to a charity for ALS.

† At core, these frameworks describe the symmetries of abstract objects in space. A bigger object looks more "complex" because it allows for (and generally requires!) more particles and ways they can interact. But it's "simpler" in the sense that you only need one object to fully explain everything rather than having to combine several different objects or theories together to examine the same phenomena.

others. On top of that, there were mathematical kinks that still needed to be worked out; it seemed that there might be some inconsistencies lurking in the dark corners of the theory. Despite these problems, in the late 1970s and early 1980s, N=8 supergravity was very popular as a candidate for the mathematical structure of a theory of everything. "It caught on with a fair number of people in the physics community quite soon," says Daniel Z. Freedman, a physicist at Stanford who helped develop the theory in the mid-1970s. "And Stephen fell in love with it for a couple of years."[14]

By the mid-1980s, a rival set of theories—known as string theories—began to emerge from the background. Like N=8 supergravity, the mathematical frameworks of string theory were complex-looking beasts that had relatively simple symmetries underneath. Instead of electrons and quarks and a whole host of fundamental particles, these theories assumed that everything was really composed of strings vibrating in different ways. While this seems like a strange assumption, it provided a mathematical structure that seemed to fit the requirements quite nicely. Also like supergravity, string theories had some intellectual baggage, predictions that didn't quite align with what we think we know about the real world. (For example, string theories tended to imply that our universe has ten dimensions rather than the four dimensions we're used to.) As crazy-sounding as this baggage seems to be, the mathematics was working out well. Very well. Theoretical physicists John Schwarz and Michael Green had proven that the underlying framework was well behaved; it didn't blow up or generate infinities that would make it a pointless endeavor to try to build a physical theory on top of it.* With that revelation, string theory looked like a suitable scaffolding for a theory of everything, and the physics community, particularly in the United States, was abuzz with excitement. "So overnight it became a major industry, at least in Princeton, and very soon the rest of the world," Schwarz later told a historian. "It was kind of strange, because for so many years we were publishing our results and nobody cared. Then all of a sudden everyone was extremely interested." This period came to be known as the first superstring revolution.[15]

The first superstring revolution happened just as Hawking was writing *A Brief History of Time*. In an early draft, Hawking wrote, "It is . . . my guess that these theories with extra spacetime dimensions are not the right

---

* In 1984, physicists John Schwarz and Michael Green figured out that with two particular types of symmetry, SO(32) and $E_8 \times E_8$, all the ugly elements in the theory that might give rise to anomalies—breakdowns in the theory—miraculously canceled. This gave theorists the green light to try to use string theories as a foundation for a theory of everything.

answer, but I could well be wrong." By the time the book came out, Hawking was not yet fully sold on string theories, but he was moving in that direction. Not only did the published version have several pages devoted to string theory, but Hawking seemed to admit that it had the edge over his favored N=8 supergravity. He ventured a prediction: "It is likely that . . . by the end of the century, we shall know whether string theory is indeed the long sought-after unified theory of physics."[16]

String theory started appearing in the popular press, mostly in caricature. It had become the archetype of an idea so difficult, so abstruse, that only a handful of people on the planet had the capacity to approach it. What relativity was to the 1920s, string theory was to those living at the turn of the millennium. And though Hawking had had a role in introducing string theory to the public with *A Brief History of Time*, he himself was not at the center of the revolution. If anything, he was behind the curve when it came to the latest scientific fashion, and perhaps for the first time in his career, he wasn't a trendsetter.

"[Hawking] had put his money on N=8 supergravity being the way to go, so he had always been a little bit cynical about string theories," says former Hawking student Marika Taylor. "One always wonders whether that was in part because he didn't really understand it or hadn't followed it. But certainly some of his friends were in the string theory camp . . . so he had an awareness, but he hadn't been following the details, that's for sure."[17]

"In the early days of string theory, [noted string theorist] Edward Witten came to Cambridge to give a talk, and I remember after the talk, he came to Stephen's office. We sat on the ground and listened to Ed and Stephen arguing with each other," recalls Ray Laflamme. "Cambridge had a bit of an anti-string attitude at the time—it was one of these new fashions which was not going to perturb the program in the Old World. But it was very interesting."[18]

Hawking had some good reasons to be cynical. There were lots of problems with string theory—not least of which was the fact that there were five different versions—but it had a staying power that N=8 supergravity didn't. Fad or not, string theory was gaining in importance over time rather than petering out. It was more than a passing fashion.

In 1995, Witten, then at the Princeton Institute for Advanced Study, argued that the five different versions of string theory were really different facets of a unique, overarching mathematical structure—a structure he dubbed *M-theory*.* It's something akin to Plato's allegory of the cave; the

---

* What the "M" stood for was never specified. Two possibilities are "matrix" or "membrane," but it may be something else entirely.

different versions of string theory were like the shadows of M-theory cast upon the walls. Each of them gives a hint about a hidden reality, but those shadows, not even in combination, allowed mathematicians a full under- standing of what the real, overarching theory is like.

It's a strange situation. Five different theories (actually, six, since an eleven-dimensional variant of supergravity was thrown in the mix) were different aspects of the same mathematical mother structure, M-theory. But knowing that M-theory exists is not the same thing as having it in hand; to this day, no mathematician or physicist has been able to character- ize a mathematical framework with the properties M-theory is supposed to have. But even without M-theory in hand, Witten's discovery showed that all string theories were fundamentally equivalent to one another; in math- ematical terms, they were *dual*. And this was enough to spark the second superstring revolution.

When two mathematical structures are dual to each other, theorists can take their separate knowledge of each structure and combine it into something more powerful. Even though M-theory itself was (and still is) not understood, physicists suddenly knew that all of their different inroads into each of the variants of string theory were reinforcing each other. For Hawking, though, the biggest implication of duality was that it suddenly made the whole field much more accessible to him. "I think the thing Stephen latched onto was bringing in supergravity as part of the whole picture of model equivalence," says Taylor. "And I think Stephen saw that as an entry point because he could do eleven-dimensional supergravity, right? Now you're telling him that this is equivalent to doing things in string theory."

Duality had taken a bunch of interesting, but niche, theories and bound them together into something more powerful and profound. Maybe they were even providing insight into what might become the framework for a theory of everything. And even though physicists couldn't prove any real connection between the string-theory models and the actual, physical real- ity of our cosmos, string theorists felt in their bones that their mathematics reflected something real.

This feeling got stronger in 1996, when physicists Andy Strominger and Cumrun Vafa used string theory's mathematical toolbox to build a simplified model of a black hole and to analyze its properties.* Through some intense calculations ("a mathematical tour de force," fellow phys- icist Lenny Susskind later called it), the pair counted up the possible

---

* In this case, "simplified" means "five-dimensional and carrying lots of electric charge." This made the calculations easier.

configurations of strings inside a black hole, which in turn led to a calculation of the black hole's temperature. Lo and behold, the calculations
matched exactly the very formula for black-hole temperature that Hawking had derived in the 1970s—the one that is inscribed on Hawking's tomb
in Westminster Abbey.[19]*

Hawking's formula was like a lighthouse, a beacon in the vast darkness of confusion surrounding black holes. And when Strominger and
Vafa saw it emerging from their string-theoretic calculations, they knew
that they were on course—that their model black hole behaved in fundamental ways like the ones in our real-life universe. "It's a little hard to
understand how amazing it was that you would get this thing on the nose,
all the details and everything," Strominger says. At almost the same time, a
hot young physicist at Princeton named Juan Maldacena performed a similar set of calculations; along with Curtis Callan, he, too, looked at certain
non-real-world versions of black holes using the toolbox of string theory,
and also discovered that Hawking's formula miraculously appeared. The
string-theoretic model was working; theorists were deriving fundamental
properties of black holes, but with a totally different set of mathematical
assumptions than Hawking himself had used. And those different mathematical assumptions led to some very different consequences.[20]

True, these two models dealt with hypothetical black holes rather than
the real-world ones, but there was a powerful reason to study such "toy"
black holes rather than astrophysical beasts. They were built up from first
principles in string theory, and as such, they were bound by the rules that
undergirded string theory's mathematical framework. Among these rules:
their model black hole cannot destroy information. If, in fact, these toy
models were getting at the fundamental properties of black holes, then it
made sense that the underlying assumption of no information loss would
be true, too. In other words, the success of these models made it much
harder to assert that information could be lost. "It was clear at that point
that [Hawking's] original thesis that information is destroyed couldn't be
correct," Strominger says.

It was a powerful mathematical argument against Hawking's idea of
black holes as information erasers. "String theory may or may not be the
right theory of nature, but it had shown that Stephen's arguments could not
be correct," Susskind writes. "The jig was up, but amazingly, Stephen and
many in the General Relativity community still would not let go."[21]

---

* Technically, they calculated the black hole's entropy, a concept intimately related to the
concepts of temperature and information in ways that will be made clearer in Chapter 13.

Hawking certainly didn't believe the results, but he couldn't ignore them. Unfortunately, he was ill equipped to attack them head on. Hawking, like his Cambridge colleagues, had kept string theory at arm's length for years, and he had an enormous amount of catching up to do. Even for a physicist or mathematician who could spend hours scrawling equations on the chalkboard or scratching figures into a notebook, it would have been a daunting task to get up to speed enough to fully understand the arguments, much less battle them head-on. Luckily, he had graduate students.

Every year, Hawking had his pick of the litter of the incoming class. "The way it works is, if you wanted to do a PhD with Stephen Hawking, you really have to do the 'Mathematics Tripos,' the famous part III," says Marika Taylor, referring to the legendarily difficult exam given to advanced mathematical physics students at Cambridge. "Stephen, in particular, needed his students to be extremely strong, because they had to be very independent."[22]

Taylor had been an undergraduate at Cambridge, and had often seen Hawking wheeling around campus. "When I was a second-year student, Stephen was living in the apartment block just behind the college house where I lived," she recalls. "And I had this perfect view from the window of his apartment block, and people would come to my room and watch Stephen going in and out. I would sort of come out of the window and sit on the roof and watch him from there." But what really inspired Taylor to pursue theoretical physics was a set of joint lectures that Hawking and mathematician Roger Penrose gave in 1994 about the nature of space and time. "It was looking at those debates that said, 'Okay, this is really where theoretical physics should be. This is the frontier of theoretical physics,'" says Taylor. "It really was the trigger for going to do part III, because that was the way to access that kind of physics."

When she aced the Tripos, Taylor was called to meet the master. "That, of course, was nerve-racking." Not only was she in the presence of an extremely distinguished physicist, but the conversation was slow. Nevertheless, the discussion was remarkably easy. "He just talked about what he was interested in, scientifically." And now Hawking's interests included M-theory. Hawking sent Taylor to speak to some of his more senior graduate students and to get some papers to read. "But then he made a joke at the end of the conversation: 'When you become my student, no more sitting on roofs, because it stresses out the nurses.'"

Taylor didn't yet know how difficult the task ahead of her was. Her thesis was going to be on M-theory, but Hawking was not an expert on the subject. Taylor would largely have to guide herself straight to the frontier of

an incredibly difficult branch of theoretical physics, digest all the important work of the past few years, and then teach Hawking what she had
learned before even being able to come up with a thesis idea. On top of
that, Hawking wasn't particularly enthusiastic about the string-theoretic
parts of the theory; he just cared about supergravity. "As I was starting to go
into those areas, I wouldn't say that he was skeptical," Taylor says. "He was
just not interested. . . . Actually, I think the real truth is that he didn't want
to engage with people on territory he was unfamiliar with."

And then, of course, there was the most fundamental problem with
studying M-theory under Stephen Hawking: M-theory says that information can't be lost in black holes—something that Hawking still firmly
argued in favor of. "You have to be stubborn in dealing with Stephen. He
and I had a long-standing sort of joke, because right from the beginning of
my PhD, I just didn't believe in information loss," Taylor says. When draft-
ing her thesis, she inserted "a sort of comment: 'Now that string theory has
shown that information is not lost in black holes . . . ' and Stephen just sort
of smiled when he read that section, and typed, 'Well, you can put this in,
but you may fail your PhD.'" (Taylor kept the phrase in, and she passed
with flying colors.)

Even so, Taylor and Hawking's other acolytes brought the master up
to speed on M-theory. It was just in time, because there was an even more
devastating attack on his beloved information-loss principle about to come.

. . .

"Like *A Brief History of Time*?" writer Dava Sobel told a reporter for the
Times of London in early 1997. "Everyone in publishing has been asking for years who will be the next Stephen Hawking? And I'll never forget
the day I woke up and realized: It's ME!"[23]

Sobel wasn't a professor; she wasn't even a scientist. Before she wrote
Longitude, the breakout best seller about the quest to solve a key naviga-
tional problem, she had been a reporter for the *New York Times*. The success
of *Longitude* was a shot in the arm for the pop-science book market.

The next Stephen Hawking could easily have been Stephen Hawking.
However, in the near-decade since *A Brief History of Time*, he had not proposed writing another book. Al Zuckerman, his agent, didn't pressure him.
Even though Zuckerman would make lots of money by inking a deal for
a new Hawking book, he was making plenty elsewhere; he was the agent
for thriller writer Ken Follett, who could be relied upon to churn out a
best-selling book every two or three years like clockwork.

For Hawking, writing *A Brief History of Time* had been utterly exhausting; even with the help of his graduate students, it had taken five years, start

to finish. Another book would take precious time away from physics and from his students. And, as Hawking must have known, the likelihood of matching the success of his blockbuster was exceedingly slim. *Brief History* had been named one of the top one hundred books of the century, according to British readers, along with works by James Joyce, George Orwell, and Franz Kafka. By comparison, Hawking's next book was almost guaranteed to be a disappointment.[24]

The Illustrated A Brief History of Time, which came out in 1996, was just what the title implied. By then, the original was eight years old, and sales were beginning to flag. Just a year before, Bantam had allowed the lucrative hardcover to be replaced by the cheaper paperback—a remarkable testament to sales, as most paperbacks come out a year after hardcover publication. "It has sold one copy for every 750 men, women, and children in the world, so there are 749 to go," he joked with a reporter in 1995.[25]

According to Zuckerman, the idea for the illustrated book wasn't originally Hawking's, but had come from a book packager named Philip Dunn. Book packagers are a sort of middleman in the publishing industry, a wholesaler who gets all the pieces of a book in place so that it's easier to sell to a publisher. "He came up with an idea to take Hawking's ideas and present them visually," Zuckerman says. "[It] is a book that tries to present *A Brief History of Time* as much as possible in a visual way, and this guy supplied the artists."[26]

By reissuing the book with oodles of beautiful illustrations, in hardcover, Bantam could squeeze even more money out of the Hawking phenomenon. But it did nothing to develop the physicist's reputation as a communicator of science; Hawking hadn't written anything substantial in a decade.

However, print was not the only medium out there. In a single week, a mediocre sitcom like *The Drew Carey Show* reached more households in the United States than even a publishing phenom like *A Brief History of Time* could reach over the course of a decade. And there was no shortage of people who wanted to help Hawking tap that market.

David Filkin, a veteran science popularizer with the BBC, approached Hawking to make a miniseries—six one-hour segments—intended for the highbrow television market. The result was *Stephen Hawking's Universe*, which aired on US public television in October 1997. It was a solid popsci offering, but Hawking was neither author, nor, for most of the series, a central character. He was more of a sales gimmick. A few of the episodes, such as one on how matter came to be and another on dark matter, were on subjects tangential (at best) to Hawking's research and interests. Nevertheless, there were places where the discerning could see Hawking's fingerprints: he flogged the no-boundary proposal, and, in a section on the rapid expansion of the very early universe, completely ignored the contributions of Paul Steinhardt, who had repeatedly been snubbed by Hawking. Hawking's ambivalence about string theory also shows through. Even as the narrator describes string theory as "the stuff of zealots" that "borders on mysticism," Hawking placed himself in the vanguard of the theorists: "By the end of the 80s I and a number of other physicists were beginning to wonder if string theory really was the ultimate theory of the universe," he declares. Nor could Hawking resist his signature appeal to the divine, suggesting that answering the ultimate question "would be to know the mind of God." Hawking had stamped his brand on the series.[27]

The Hawking brand was worth quite a bit of money, as advertisers had figured out. In 1997, U.S. Robotics, a company that manufactured modems, used the physicist to try to hawk its latest consumer model. "My body may be stuck in this chair, but with the Internet, my mind can go to the ends of the universe," he said in an advertisement. As it happened, Hawking had just received a spiffy new Internet-enabled computer for his wheelchair, courtesy of chip manufacturer Intel. "This computer makes me the most switched-on person alive," the physicist joked. "It's a bit slow, but I think slowly."* It took a few years—and a few upgraded models—but the back of Hawking's black monitor began to sport a prominent white Intel logo. Hawking's relationship with Intel continued through the years, culminating in a £7.5 million donation in 2001 to Cambridge by Intel founder Gordon Moore.[28]

Hawking was a fundraising superstar at Cambridge—one gift of £6 million for an endowed professorship in his name was nearly rejected on the grounds of its terms being "too generous." Yet Hawking never got the superstar salary from Cambridge that big grant-pullers tend to get in the United States. "Stephen was incredibly generous with his time for fundraising for the university, which they never properly gave him credit for, or respected him properly for, though they tended to use him," says Neil Turok, who was a professor at Cambridge in the mid-1990s.[29]

---

* Intel's press release put even more effusive advertising copy in the professor's (computerized) mouth. It read: "'Intel's newest Pentium processor technology keeps me connected to the world,' said Stephen Hawking, Lucasian Professor of Mathematics at the University of Cambridge. 'I have immediate access to the Internet and email wherever I am. I must be one of the most connected people in the world, and I can truly say, I'm Intel inside.'" Intel, "Professor Stephen Hawking Stays Connected to the World Through the Latest Intel Technology," press release, March 20, 1997, www.intel.com/pressroom/archive/releases/1997/CN032097.htm.

The Hawking name was indisputably glamorous. In 1996, the board of a small school in London for students with learning disabilities asked if they could name the facility after the physicist. He consented, and the Stephen Hawking School was born. Hawking himself never managed to work a visit to the school into his schedule, which is particularly odd, given Hawking's well-known advocacy for the disabled, particularly children. "[Hawking's] aide let it slip that whenever he traveled, he would ask his hosts to set up a meeting with local children with disabilities," a journalist wrote of his encounter with the physicist in 1993. "These visits were totally unpublicized, but I was lucky enough to go along and watch the pretty-great physicist answer questions from a half-dozen or so kids for an hour, their wheelchairs arranged around his in a semicircle. That's when I became convinced that even when you strip away the hype, Stephen Hawking may not be the world's greatest living physicist—but he's a pretty extraordinary human being."[30]

The year 1997 saw not one, but two theatrical productions about Hawking and his work. One, a play titled *A Brief History of Time: The Stage Show*, toured through England. It attempted to translate Hawking's work into "a theatrical language of movement, text, and metaphor," the producer told a journalist. "We present the history of cosmology as a fashion parade using six hats." The second, an American musical, *Falling Through a Hole in the Air*, focused on the physicist's triumph over ALS. "I have been fortunate," the tenor playing Hawking sang:

My sorrows bend in the arms of my love
And the laughter upraised
And when I close my eyes, my soul can wander,
For though my body's bound, my soul can wander free.
I rise like an eagle and fly.[31]

Hawking the symbol was transcendent, his place in the firmament all but assured, even as Hawking the physicist was engaged in an intellectual struggle—a battle that would bear directly on his scientific legacy.

•••

Hawking's grad students worked hard to get up to speed on the latest developments of string theory and M-theory. If Hawking were to have any chance of disarming the string-theoretic arguments that information was preserved in a black hole, his students would have to hack their way through a tangled thicket of ten- and eleven-dimensional mathematics to approach the battlefield. Yet Hawking was not able to help them at the blackboard, or with the details of the mathematics. His students were thrown right into the deep end. Christophe Galfard told sociologist Hélène Mialet about what it was like to begin working with the professor:

> The . . . first thing that he asked me to do concerned correlation functions in black hole [spacetime] decay to zero . . . that's all. I didn't know what a black hole space-time was. I'd had some lessons at university, vague notions, but nothing really practical, and correlation functions, I didn't really have a clue as to how to find that. And so all of those words, I had to find out one by one what they meant from a slightly deeper perspective than the partial way in which I'd learnt before.[32]

Even if Hawking had been a native string theorist, the pace of his communication gravely limited how much guidance he could offer his students. As he gradually lost control over the last muscles in his hand, using his computer by clicking the little pressure switch became more and more difficult. By the late 1990s and early 2000s, he was slowly losing the ability to talk even with the assistance of his computer. Galfard would attempt to speed up the pace by looking at Hawking's computer screen and attempting to complete the sentence, using the expressions on Hawking's face to guide whether he was on the right track.[33]*

Yet Hawking was confident that he and his students could point out flaws in the string-theoretic models of black holes. Somewhere, he was sure, the other physicists had made a mistake. Either their calculations or their assumptions were wrong. Information couldn't be preserved by black holes—it *had* to be lost. He would stake his reputation on it. Publicly.

Hawking was famous for his wagers. In early 1997, when he made his regular pilgrimage to Caltech in Pasadena, a wager was on the agenda. Two, actually.

The first bet didn't really involve information at all, but it did involve black holes, or, more precisely, singularities, which reside at the heart of every black hole.

---

* The hand-operated clicker would soon be replaced with a sensor that detected the twitching of Hawking's facial muscles. And Hawking could communicate to some extent, especially with people who knew him well, with his facial expressions; this became increasingly important as he grew older and the speed at which he could use his computer slowed.

The equations of general relativity assume that we all live in a nice, smooth manifold of spacetime; it curves and ripples, but is never discontinuous or pointy. It never misbehaves. Yet a singularity is a region in spacetime where that assumption no longer holds—and that tends to break the equations. The general-relativistic picture of black holes has a singularity at the center—the curvature caused by gravity is so great that it effectively punches a hole in the nice smooth surface of spacetime. But— and this is a very interesting but—we can never see the laws of relativity break down. Because the black hole has an event horizon around it—a surface beyond which no information can escape, not even if it travels at light speed—no information about the singularity leaks out. It's as if the event horizon were the boundary of a totally separate universe from the one we live in; nothing from inside the horizon can ever enter our universe at all.

But are all singularities similarly shielded from prying eyes? Are all places where general relativity's equations break down hidden behind an event horizon? In the late 1960s, Roger Penrose suggested that the answer was yes—that singularities could never be "naked," or directly visible.

Hawking, like Penrose, believed that there was some form of "cosmic censorship" that made it impossible to observe a singularity directly. But others weren't so sure—like Caltech's Kip Thorne and John Preskill. So in 1991, during one of Hawking's visits, the three drew up a bet:

> Whereas Stephen Hawking firmly believes that naked singularities are an anathema that should be prohibited by the laws of classical physics.
>
> And whereas Preskill and Thorne regard naked singularities as quantum gravitational objects that might exist unclothed by horizons for all the universe to see.
>
> Therefore Hawking offers, and Preskill/Thorne accept, a wager . . . [that there] can never be a naked singularity.
>
> The loser will reward the winner with clothing to cover the winner's nakedness. The clothing is to be embroidered with a suitable concessionary message.[34]

The wager was sealed with the two physicists' signatures and one's thumbprint.

In the intervening time, however, gravitational-wave theorists had come up with a weirdo scenario—one that wouldn't occur in a real universe—where a bunch of perfectly symmetrical gravitational waves collided at a point in spacetime. The sloshing of space and time would be so violent at that central point that if the energy were just right, it would create a singularity without generating a black hole. Hawking was forced to concede the bet—but he felt he had lost on a technicality, rather than a real counterexample to the cosmic censorship principle. Indeed, the fact that the energies had to be just so, and set up in just the right way to create even an infinitesimal singularity, convinced Hawking more firmly than ever that he was fundamentally right. But honor dictated that he concede. So he did.

At a public lecture, Hawking admitted defeat—but with a twist. Preskill was giving the lecture, and Hawking, who was in town, introduced him to the audience. "He said that I was the all-American boy because I like to drink Diet Coke and like baseball," Preskill remembers, "but then he wound up saying that he was conceding the bet to Kip and me and he had T-shirts for us to put on."[35]

Thorne later wrote:

It's not every day that Stephen gets proved wrong! With his concession, Stephen gave each of us the promised article of clothing: a T-shirt with his concessionary message. Sadly, I must tell you that Stephen's message . . . was not entirely gracious! He placed on the T-shirt a scantily clad woman. (My wife and Stephen's were aghast at this, but Stephen has never been politically correct. . . .) [T]he woman's towel says "Nature abhors a naked singularity." Stephen conceded, but he asserted that Nature abhors that which he concedes Nature can do.[36]

"[It] was not exactly concessionary-like language," says Preskill. "But he insisted that I put on this T-shirt and wear it during my lecture, which I can say I did. I was wearing a suit, so I put on the T-shirt and then put on my suit jacket." Though he tried his best to be a good sport, Preskill was quite uncomfortable wearing such a racy T-shirt at a public lecture. "Stephen thought that was hilarious. You know, I was such a fussbudget—so politically correct, the way he saw it—that I didn't think I should be parading around in that T-shirt. And he got a big kick out of that, that I was embarrassed by it."[37]

Though he had technically lost the bet, Hawking had turned his defeat into a taunt. The only possible response: another bet. So Preskill, Thorne, and Hawking inked a new wager, almost precisely the same as the last one, but with no wiggle room for technicalities. If Preskill and Thorne won, it would be because the cosmic censorship principle truly turned out to be wrong.

The wagering wasn't over, though. Naked singularities weren't foremost on Hawking's mind at the time. The black-hole information paradox was. And Thorne, like Hawking, still believed that the quantum theorists were wrong. Once something fell beyond an event horizon, its information was lost to the universe forever after. Perhaps it bubbled off into another universe (an idea that Hawking found attractive), but it could never reemerge into ours. Preskill, like most quantum theorists and string theorists, took the opposite view: information cannot be utterly destroyed, not even by a black hole. So the day after signing the naked singularity bet, the three—this time Thorne and Hawking versus Preskill—drew up the following:

> Whereas Stephen Hawking and Kip Thorne firmly believe that information swallowed by a black hole is forever hidden from the outside universe, and can never be revealed even as the black hole evaporates and completely disappears,
>
> And whereas John Preskill firmly believes that a mechanism for the information to be released by the evaporating black hole must and will be found in the correct theory of quantum gravity,
>
> Therefore Preskill offers, and Hawking/Thorne accept, a wager . . .
>
> The loser(s) will reward the winner(s) with an encyclopedia of the winner's choice, from which information can be recovered at will.[38]

Two signatures and a thumbprint, and the wager was on. And it was clear that Hawking thought he would win. Some of his students were busy learning M-theory and preparing to fight back against the string-theory assault. Unfortunately, they were about to get outflanked.

The second superstring revolution wasn't over yet, and hard-core string theorists like Ed Witten, Andy Strominger, and Juan Maldacena were filled with enthusiasm as they produced important result after important result. Just months after Hawking affixed his thumbprint to the bet, Maldacena, then only twenty-nine years old, found another duality—a place where two seemingly different mathematical models were actually equivalent—that landed the death blow on the idea of black holes destroying information.

As with any duality, there are two pieces, two separate mathematical objects, that wind up being equivalent in some fashion. In Maldacena's, the first piece is a well-studied form of spacetime known as anti–de Sitter space, or AdS. AdS is a strangely shaped kind of spacetime that doesn't look very much like the spacetime of our universe, but obeys all the physical rules of general relativity nonetheless. This hypothetical manifold serves as a mathematical toy; physicists can use it to play with the equations of general relativity (and string theory) in novel ways. For example, one can construct black holes in an AdS space, and because of the peculiarities of the manifold's shape, a large black hole never evaporates as it would in a more natural version of spacetime, whereas a small black hole does.

The second piece of Maldacena's duality is conformal field theory, or CFT. CFTs are mathematical frameworks that describe the behavior of particles without mass. And, as it turns out, we've already encountered something that's almost a CFT: the Standard Model. The Standard Model doesn't quite meet the definition of a CFT (many particles in the Standard Model have mass), but the underlying mathematical structures have a lot in common. Think of a CFT as a supercharged Standard Model that describes every particle's interaction in a universe without mass—and, consequently, without gravity.

Maldacena's 1997 paper suggested that these two very different-seeming mathematical structures were, in fact, dual to each other. Examining strings in an AdS universe with spacetime curvature and black holes and other effects of gravitational theory was essentially equivalent to studying particles in a CFT where gravity doesn't exist. Thus, according to Maldacena, one could build a black hole in an AdS space and then bop over to the tools of the corresponding CFT to analyze it.

This is a very abstract and counterintuitive concept. By studying the behavior of particles moving about in a universe without gravity, one can understand how black holes behave in the warped spacetime of a universe where gravity is of prime importance. That is, the AdS/CFT duality allows one to understand black holes without having to deal with the troublesome mathematics of singularities in gravitation. And within a couple of months of Maldacena's paper, Ed Witten published a follow-up paper where he did just that.*

Looking at a swarm of particles in the CFT space is exactly equivalent to observing a black hole in the AdS space—and just as particles have no special ability to wipe out information in the CFT space, black holes shouldn't be able to destroy information in the AdS space. You could build a tiny black hole in AdS, dump information into it, let it evaporate, and, voila! The information would still be there in the CFT side of things, so it had to still be there on the AdS side.

The AdS/CFT correspondence doesn't directly say anything about *real* black holes in our non-mathematical-toy universe. And it wasn't even a proven fact, but a conjecture. Yet the Maldacena paper rocked the world of physics, as did the Witten analysis. For those battling over information

---

* And even more. The Witten paper emphasized that the CFT theory lived on the *boundary* of the AdS space—think of CFT as a box and the AdS as living on the inside of the box. AdS/CFT showed that all the information that resides in the inside of the box is exactly the same as the information that is inscribed on the box itself. This is a profound concept known as *holography*.

loss in black holes, the burden of proof had shifted—there were convincing arguments that black holes couldn't erase information in various toy environments, and there was no reason to believe that a real black hole would behave any differently. If, for some reason, our universe behaved differently from all the other toy universes, that required explanation.

To some, the argument was over. "The moment I saw the Witten paper, I knew the Black Hole War was finished," physicist Leonard Susskind wrote. "Whatever else Witten and Maldacena had done, they had proved beyond a shadow of a doubt that information would never be lost behind a black hole horizon." Even Hawking was shaken, though he continued to insist that black holes destroyed information. He continued pressing his students to brush up on M-theory and the latest thinking about AdS/CFT, and to figure out convincing counterarguments that could preserve the hope of information loss. Yet they couldn't help the master out of his bind.[39]

"I think it became clear to him that he was wrong, and he didn't want to go down on the wrong side of history on this very important question," Strominger says. And so, in 2004, Hawking finally convinced himself—using his own mathematical constructs—that information could not be lost in a black hole. He realized that he was wrong, but in true Hawking form, he only came to that conclusion on his own terms.[40]

This meant he had to settle the bet. Preskill asked for a copy of *Total Baseball: The Ultimate Baseball Encyclopedia*, but in baseball-free Britain, that was a tall order. "At one point, we were having trouble getting it," one of Hawking's assistants told a newspaper, "and he tried to persuade John Preskill to take an encyclopedia of cricket, which of course we could find in England, but John Preskill is a baseball fan, being an American, so that wasn't good enough."[41]

Hawking took to the stage in Dublin to take his medicine and concede the wager. Preskill had, indeed, bet wisely; he had (probably) been correct about the answer to one of the most vexing physics questions of the late twentieth century. And Hawking was (probably) wrong on a subject that he had spent three decades of his life working on. Hawking's late "solution" to the information-loss paradox presented at Dublin convinced almost nobody—it was almost as if Hawking simply couldn't stand the idea of not answering the question himself.

"I think one has to put that in the context of a physicist wanting to be at the top of the game, wanting publicity," says Taylor. "For Stephen, being well appreciated, scientifically, was important to him, not just for his sort of science but also for his health—to give him motivation to carry on—and also in terms of his sponsors."[42]

Not getting the right answer to the black-hole information paradox should have been no source of shame. After all, Hawking was assured of his share of glory merely by asking such a profound and important question in the first place.

· · ·

By this time in his life, Stephen Hawking was used to losing bets, but this one was different. The bafflement and bemusement that surrounded the concession of the black-hole-information-paradox wager showed more clearly than ever before that Hawking was no longer at the intellectual center of physics. Even though he himself had formulated the paradox, Hawking's ruminations on the subject were not cutting edge.

Dublin was a graphic demonstration: there was no escaping the conclusion that Hawking the scientist was no longer of the first rank. Those days lay behind him. To see his once-great scientific mind in its full glory, one must turn back the clock even further.

CHAPTER 8

IMAGES (1990–1995)

The secret was out. "Wheelchair Physicist in Love Tangle with Nurse," blared the *Daily Mail*. Though Stephen had moved out of the household in February 1990, it wasn't until July that the tabloids got wind of the separation, thanks to an accident. Stephen's wheelchair was knocked over by a car—a surprisingly common occurrence for the physicist, who had a reputation for fearlessness when driving his wheelchair—and suffered a broken shoulder. As the press investigated the accident, they quickly realized that Stephen had changed addresses. Reporters, with their noses finely tuned to any scent of scandal, figured out that he had separated from Jane.[1]

Some newspapers looked for a cause. "The split partly came about through religious differences," an (anonymous) former student told a different paper. "She is a committed Christian and as he increasingly became interested in scientific rather than religious explanations, it became more difficult for them. They are still the best of friends but in those conditions you can't expect people to have a normal married life." Others looked for a victim. The *Daily Mail* chose David Mason, Elaine's husband and the father of her two children. David also happened to be the engineer who had adapted Stephen's computer speech system so it would fit on his wheelchair. "The whole situation is ludicrous and bizarre," he told the *Mail*.[2]*

The attention was unwelcome, and not just because of the embarrassment involved. It effectively notified their landlord, Cambridge University, that Stephen had moved out of the couple's house. "Once the separation had entered the public domain, the College lost no time in sending the Bursar across to enquire when we were going to move," writes Jane. "He was quite explicit: The College felt itself under no obligation to house the family if Stephen, with whom the College had signed the agreement, was no longer living there." Eventually, the university agreed to give Jane; her eleven-year-old son, Tim; her seventeen-year-old daughter, Lucy; and her partner, Jonathan, a year to find other accommodations.[3]

Stephen and Jane hadn't expected to reach their twenty-fifth wedding anniversary; it was almost miraculous that Stephen had survived to see their fifth. That didn't make it any less painful when their bond had finally sundered. Jane still clung to the hope that they could preserve the marriage—a hope that Stephen fed when he told the press there remained "a chance of reconciliation."[4]

Even so, Jane would have to move.

· · ·

In 1990, Stephen Hawking was trying to adjust to his newfound fame. He had been well known within the scientific community for decades, but the professor's sudden popularity was something entirely different—he had become the world's most recognized scientist. More, he had become a symbol to the public, a transcendent mind in a withered body. And as much as Hawking wanted to be known for the former, the latter was just as much a part of his public persona. And as a celebrity, he needed to cultivate and maintain that persona even when it didn't capture the complexity of the human underneath.

Luckily, Hawking had a genius, a natural talent, for nurturing his newfound celebrity. His wicked self-deprecating wit, coupled with an overweening certainty about his understanding of the natural world, made

---

* The paper didn't manage to get in touch with Jane, or catch wind of her relationship with Jonathan Hellyer Jones. When reporters came calling, "Jonathan, of whose existence they were unaware, managed to slip out the back door." The ruse was successful. "Jane, 48, is left to reflect alone," reported the *Mail*. Jane Hawking, *Music to Move the Stars: A Life with Stephen Hawking* (London: Pan, 2000), 575; Emma Wilkins, "Wheelchair Physicist in Love Tangle with Nurse," *Daily Mail*, August 1, 1990.

him into the perfect archetype of the scientific genius. He succeeded, even though for the first time antagonists were quite publicly trying to knock him down a peg or two.

...

By 1990, the year Stephen moved in with Elaine, *A Brief History of Time* had been on the best-seller lists for almost two years and was still selling strong. Not only had the book made the physicist a publishing phenom—and was beginning to earn him a tremendous amount of money in royalties—it had made him an international celebrity. Though fame was new to him, the limelight didn't feel uncomfortable.

"People are either drawn to it or they aren't," says Peter Guzzardi, the editor of *Brief History*. "What is that intangible something . . . that unqualified impulse to step under the big Klieg lights and do your thing? I think Stephen definitely had that. He loved the media spotlight, and it just—he came alive. He's one of those people that just came alive when he was the center of attention. As well he might, you know; it must have been tough, just being Stephen."[5]

Hawking's celebrity, or, more precisely, the wealth that accompanied it, ameliorated the physicist's difficulties finding and affording sufficient care to keep him alive and able to work. Since his tracheostomy operation in 1985, he needed 24-hour care, 365 days a year, and, according to Jane, "only a tiny fraction of this expense would be borne by the National Health Service." Typically, ALS sufferers would require home care, institutionalization, or hospice care only for a relatively short time; the system was simply not set up to deal with the singular presentation of the professor's affliction. His longevity—a surprise to everyone, including himself—turned what was usually a short-term situation into something that required a long-term solution.[6]

In the late 1980s, the Hawkings had received help from the John D. and Catherine T. MacArthur Foundation and other charitable sources—administered through Cambridge University—to help pay for the nurses. Jane estimated that the £36,000 per year they received—worth about $170,000 in today's US dollars—"just covered the bills."* As Stephen's condition worsened, the cost steadily climbed. However, now that Hawking's book was a success (and he was married to his chief nurse), the financial pressure that he and his family had been under was greatly reduced. And

---

* The numbers have been converted to 2020 US dollars using late 1980s exchange rates and forty years of US inflation.

he was able to start turning his attention to other people in worse straits than himself.[7]

Hawking had particular empathy for children who suffered physical disabilities; unlike them, he had had a relatively normal, able-bodied youth, only to be afflicted in early adulthood. "It is very important that disabled children should be helped to blend in with others of the same age," he told an audience in 1990. "How can one feel a member of the human race, if one is set apart from an early age? It is a form of apartheid." Hawking certainly chafed at being set apart from his peers because of his disability instead of his brain; he saw nothing heroic in his struggle against his disease. "I find it a bit embarrassing in that people think I have great courage," Hawking told a reporter in mid-1990. "But it is not as if I had a choice and deliberately chose a difficult path."[8]

Hawking typically made light of his disability, and he denied that it hindered him in his work. "I was lucky to have chosen to work in theoretical physics because that was one of the few areas in which my condition would not be a serious handicap," he said at an ALS conference in 1987. Not only didn't his disability hinder him, he would tell his audience, but it actually helped him. "In fact, in some ways, I guess it had been an asset: I haven't had to lecture or teach undergraduates, and I haven't had to sit on tedious and time-consuming committees," he wrote. "So I have been able to devote myself completely to research." And that's on top of compensation for not being able to write: his enhanced visual sense.* "If you can't write it, you become better at holding diagrams in your brain," cosmologist Alan Guth told sociologist Hélène Mialet in 2005. It was part of the mystique; Hawking had transcended his body. What would be a crushing blow to mere mortals had not only failed to defeat Hawking, it had strengthened him, given him power.[9]

In public, Hawking's stoicism almost never wavered. Almost. In 1988, a journalist asked the scientist whether, given the option, he would want to "[regain] the power to walk and feed himself at the expense of being intellectually mediocre." Hawking's surprising reply: "Yes." Then Hawking backtracked. "I don't want to be anyone else. People should be who they are."[10]

. . .

---

* However, an enhanced visual sense only extends so far. Hawking wrote, "It is impossible to imagine a four-dimensional space. I personally find it hard enough to visualize three-dimensional space!" Stephen Hawking, *A Brief History of Time* (New York: Bantam Books, 1998), 24.

A fter Hawking lost the use of his hands, he had to do calculations entirely in his mind. And his was a geometric mind.

In mathematics and physics, there are often multiple ways to solve the same problem. Actually, "solve" isn't quite the right word. Mathematicians grasp for something deeper than a solution: a profound understanding, a weaving of the answer into one's intuition, an incorporation of the essence of the problem directly into one's brain so that it becomes a part of you. You'll often hear mathematical types try to describe this feeling with the term *grok*—a transcendent form of understanding described in Robert A. Heinlein's science fiction novel *Stranger in a Strange Land*.

Pure mathematicians or physicists don't seek solutions as much as they seek this grokking of something beyond themselves, an extension of their own brains into new domains—domains that become brand-new mental playgrounds, new sources of things to explore.

Different mathematicians have different means of grokking a problem. Some might seek understanding in patterns and symbols, and the relationships among them and the rules for manipulating them: this is an *algebraic* viewpoint. It is a world of formulas and numbers and functions and operations. Others might try to understand problems by drawing mental pictures: seeing shapes in space and how those shapes move and interact with and relate to each other. This is the *geometric* viewpoint. For example, an algebraic thinker might envision a square number as a number that can be expressed as $n \times n$, an integer multiplied by itself. A geometric thinker might envision a square number as dots that can be perfectly arranged into a square pattern; only numbers that can be so arranged are square numbers (hence the name).

These two viewpoints aren't mutually exclusive—they bleed into and reinforce each other. And even if a mathematician tends to gravitate toward one viewpoint or the other when trying to grok a new area, he or she can use algebraic techniques to survey the landscape and build up a deep geometric intuition later, or vice versa. That's one of the beautiful things about mathematics: one can approach the same idea in many different ways, each of which might give a different perspective and understanding.

To understand how a body moves through general relativistic spacetime, an algebraic thinker would examine the equations of general relativity. By manipulating the symbols in the equations—representing mathematical objects called *tensors* that encode the warp and weft of spacetime—he could figure out the object's path. A geometric thinker who has properly trained her mind's eye might simply be able to "see" the object moving and conclude how it rolls around on the abstract four-dimensional surface.

These are both valid ways to try to understand general relativity. Sometimes (but not always), geometric thinking is more intuitive and algebraic reasoning more precise, but both approaches are powerful—and natural—ways to try to understand the cosmos.

The rules of spacetime, in particular, are naturally suited to a geometric thinker. After all, the algebraic formalism of general relativity, the equations describing how gravity works, describes curvature—a geometric concept. It might be a funny sort of curvature in four dimensions (or even more dimensions in toy models of the universe!), but it is still something fundamentally geometric, something that could be understood reasonably well by drawing mental pictures. If you think in the right way, you can "see" the rules of Einstein's spacetime. And a diagram of spacetime is a good example: If you can imagine your path through spacetime as a little dot moving upward (through time) and to the right (through space) on the diagram, the dictum that you can't move faster than the speed of light is exactly equivalent to saying that you can't take a path on the diagram steeper than 45 degrees from the vertical. Thinking geometrically, you can immediately see that there are regions in spacetime that you're not allowed to reach by virtue of the light-speed limit, and you can immediately divide the universe into two: the parts you can reach, and the parts you can't. If you grasp these kinds of geometric rules, you can train your mind to grok Einstein's theory. You might even be able to make discoveries.

Even before ALS began to take its toll, Hawking seemed to be attracted to geometric modes of thinking, which may be part of the reason general relativity appealed to him in the first place. As his disability became more profound, he shifted radically to the geometric side of the spectrum, avoiding algebraic symbol-manipulation whenever he could. As he said in 1998, he typically tried to "avoid problems with a lot of equations or translate them into problems in geometry. I can then picture them in my mind." Without the ability to write things on paper, without an easy way to keep track of complex manipulations of symbols, he was largely foreclosed from making new discoveries using heavy-duty algebraic reasoning. At best, he could show his students the instruments of calculation and let them do the work. But when it came to geometric reasoning, he could more than hold his own. And as he developed his mind's eye to handle the geometric intricacies of general relativity, he was able to hone his intuition—and be led by it. As his friend Kip Thorne wrote, "Hawking is a bold thinker. He is far more willing than most physicists to take off in radical new directions, if those directions 'smell' right." Or, as Thorne knew very well, to throw cold water on an idea if it smelled wrong.[11]

In the summer of 1985, the astronomer Carl Sagan sent Thorne a draft of his novel *Contact*—in which Earth travelers had to use an alien craft to travel almost instantly to the center of the Milky Way and back—and asked for Thorne's help coming up with a way to do that. The light-speed limit would seem to prevent that; there was no way the travelers could make a round trip to the center of the galaxy (about thirty thousand light-years away) in fewer than sixty thousand years (as measured by a clock on Earth). It was equivalent to asking to find a path into the "forbidden" region of a spacetime diagram; there was no way to do it without, at some point, having a line extend more than 45 degrees from the vertical. However, that assumes that spacetime is a nice, smooth, unbroken sheet. What if the sheet had a hole in it? A black hole with its singularity?[12]

"I have my crew going through a black hole, and I know you're not going to like that," Sagan told Thorne. "Can you help me rework it?" Inspired by Sagan's question, Thorne and his graduate students started calculating.* And Thorne quickly concluded that a black hole couldn't be used as a means of transport. But there was another possibility: instead of a singularity, a point of infinite curvature, what about a smooth, non-singular tunnel? A *wormhole*?[13]

If a black hole is a chaotic gash in spacetime, a wormhole is its mild cousin: a tunnel that connects two distant points in spacetime. If the opening and exit to the tunnel are nice and smooth, as is the spacetime inside the tunnel itself, then such a tunnel isn't expressly forbidden by the laws of general relativity. Indeed, the question of whether wormholes exist is a *topological* problem. Just as relativity doesn't tell us whether the universe is flat or curved like a sphere or curved like a saddle, it doesn't tell us whether the universe has holes like a donut or a pretzel or a colander. So if there were some way to create such a tunnel, it could be like a shortcut into a "forbidden" region of spacetime—even though a traveler never moved along the sheet of spacetime faster than the speed of light, it would effectively allow faster-than-light-speed travel.

They found, surprisingly, that spaceship-sized, traversable wormholes could exist under certain conditions. Those conditions are pretty outlandish, though: scientists would have to find some form of exotic matter or spacetime-stretching field that would keep the wormhole from collapsing. As a consequence, Thorne concluded in 1988, wormholes are "an intriguing possibility for actual construction by advanced civilizations." And suitable for a Carl Sagan novel.[14]

---

* Sometimes involuntarily. Thorne put a question about wormholes on the final exam of the fall 1985 session of his introductory course to general relativity.

Hawking's intuition, however, took him in a completely different direction. Unlike Thorne, he quickly came to the conclusion that using a wormhole like the one in *Contact* would forever remain the stuff of fiction; wormholes, he wrote in 1992, "are no good for space or time travel." He couldn't prove it conclusively, but he made a powerful argument that's still debated today.[15]

Hawking realized (as did Thorne) that once you bend spacetime enough to get a wormhole, strange things begin to happen: not only do you manage to travel great distances instantly, seemingly violating the speed-of-light limit, but you also wind up being able to travel backward in time. "This would mean that at some point in the wormhole's history it would be possible to go down one mouth and come out of the other mouth in the past of when you went down." This, in turn, sets up a Groundhog Day–like paradox—you enter the wormhole at noon and emerge before you came out—at, say, 11:59, which allows you to enter the wormhole again at 12:00. You then emerge at 11:59, enter the wormhole again at 12:00, emerge at 11:59, and so forth. You can do this hundreds and hundreds and hundreds of times, looping around and around on the same track through spacetime. In general-relativistic terms, you're following what's called a "closed time-like curve," and on a spacetime diagram it looks like a little closed loop.[16]

A person trapped on a closed timelike curve is looping forward and backward and forward in time, again and again, hundreds and thousands and millions of times. But to an outside observer, all those hundreds and thousands and millions of loops take place in the span of a minute—the minute between 11:59 and 12:00. During that minute, hundreds and thousands and millions of copies of you emerge from the wormhole all at once, and, all in tandem, reenter it and disappear. Hawking realized that this would break spacetime. All those millions of copies of you add mass and energy to a small region of spacetime—enormous amounts of it as the number of copies of you build up without bounds. And this was happening on a subatomic scale all the time, with particles circulating in and out and in the wormhole again: an incalculably huge amount of mass and energy crammed into a small space like that would itself alter the curvature of space. Hawking reasoned that this was enough to defeat any attempts to construct a wormhole: if you were to try to bend spacetime in such a way to create one, you'd start getting paths through spacetime that began to look like closed timelike curves, building up a lot of energy. In such a scenario, "spacetime will resist being warped so that closed timelike curves appear [or] spacetime would develop a singularity which would prevent one reaching a region of closed timelike curves," Hawking concluded. Thus, Hawking had convinced himself that timelike curves were impossible.[17]

In other words, it is impossible to travel into one's own past—or to build a wormhole like the one Thorne had pondered. Hawking called this the *chronology protection conjecture*; it seemed like there was some fundamental principle of nature that prevented anything from looping around and around through time.

It wasn't an airtight argument, but it was pretty convincing. For any doubters, Hawking ended his paper with a half-serious joke: "There is also string experimental evidence in favor of the conjecture from the fact that we have not been invaded by hordes of tourists from the future."[18]

Thorne's own calculations seemed to suggest that the instability caused by these subatomic particle loops would quickly stabilize, and wouldn't destroy the wormhole. But Thorne trusted Hawking's intuition almost more than he trusted his own. As he wrote a number of years later:

> Hawking has a firm opinion on time machines. He thinks that nature abhors them, and he has embodied that abhorrence in a conjecture, the *chronology protection* conjecture, which says that *the laws of physics do not allow time machines*. (Hawking, in his characteristic off-the-wall humor, describes this as a conjecture that will "keep the world safe for historians.")
>
> . . . Hawking seems ready to bet heavily on this outcome.
>
> I am *not* willing to take the other side in such a bet. I *do* enjoy making bets with Hawking, but only bets that I have a reasonable chance of winning. My strong gut feeling is that I would lose this one.[19]

Contact notwithstanding, Hawking's mental map of how spacetime behaved all but demolished Thorne's idea of traversable, spaceship-sized wormholes.

• • •

"I rely on intuition a great deal," Hawking told a BBC radio host. "I try to guess a result, but then I have to prove it. And at this stage, I quite often find that what I had thought of is not true or that something else is the case that I had never thought of."[20]

In the early 1990s, *A Brief History of Time* was still setting sales records, and the press couldn't get enough of the new scientific superstar. Everyone was trying to get inside Hawking's head, to try to figure out how his mind worked. Not since Einstein had the public been treated to such an archetype of pure intellect, a superlative mind that appeared to have such a tenuous connection with the worldly plane of everyday existence. Hawking had been appreciated—and highly honored—within his field for years. But with the publication of *A Brief History of Time*, he had become a household name. A celebrity. And, to some, a hero. The demands on his time were unceasing: there were talks, interviews, invitations, and honors.

Other schools, including his beloved Caltech, tried to lure him away from Cambridge. "Trying to pry him away from the Lucasian Professorship was certainly a long shot, but Kip and I went on a secret mission in 1991," says John Preskill. They offered him a newly minted professorship, the Richard P. Feynman Chair, but in the end Hawking decided to stay in England. "The outcome of the discussions was that we agreed he could come and visit every year. These visits were quite expensive, because he would bring a whole entourage—his medical team, and he would bring students—but we got a foundation, the Sherman Fairchild Foundation, to pay all those expenses and to agree to do so on an ongoing basis." Hawking was in such demand that people would break the bank to get a little piece of him.[21]

Stephen loved the attention—as did Elaine, who almost literally bounced off the walls with excitement at her soon-to-be husband's appearances. (She would reportedly do cartwheels in odd places, such as at a reception where Stephen received a degree at Harvard University, or on a film soundstage where Hawking was being interviewed for a movie.) But the professor steadfastly denied that the worldly comforts of fame and fortune had changed him—if anything, he said, his newfound celebrity was an inconvenience and a distraction.[22]

"It has not made much difference," Hawking told an interviewer for *Playboy* in 1990. "Even before the book, a certain number of people, mainly Americans, would come up to me in the street, but it has made that sort of encounter more frequent. And other things like interviews and public lectures have taken up the limited time I have to do research. However, I'm now cutting down on such things and getting back to research."[23]

Even in the early 1990s, when Hawking's international fame was still new, his interviews had the practiced polish of a man who was comfortable with media attention. Part of this was because journalists simply couldn't get unfiltered, spontaneous thoughts out of the physicist. At best, Hawking's interviewers couldn't get more than a question or two in in real time, and by virtue of the labor and time that Hawking had to expend in crafting those responses, they were more fussed over and deliberate than most off-the-cuff statements by other celebrities. Most of the time, even that wasn't an option; because the physicist communicated so slowly, the interviewers had to submit their questions in advance. Hawking and his assistants had the luxury of time to craft the perfect answer to each question—a response that would burnish the image of the physicist that he himself was building.

The image the public had fallen in love with was not an easy one to create or maintain. Hawking had to be humble at the same time that he was being portrayed as the intellectual heir of Newton and Galileo and Einstein. He couldn't be defined by his disability even though the way he so graciously coped with ALS was a major reason for his immense popularity. And he had to be a success—a wealthy celebrity—while playing the role of physics ascetic, eschewing worldly things in favor of cosmic knowledge.

In the *Playboy* interview, which appeared in April 1990, Hawking couldn't seem to settle on whether his origin story should emphasize how he coped with his disability ("I chose my field because I knew I had ALS. Cosmology, unlike many other disciplines, does not require lecturing.") or not ("From the age of twelve, I had wanted to be a scientist. And cosmology seemed to be the most fundamental science.")* And even as he insisted to the *Playboy* interviewer that the publication of *Brief History* had made almost no difference in his life, the repercussions were already immense; after all, he was in the brief interregnum between having left his wife of a quarter century and the press finding out about it.[24]

Jane, for her part, was crafting a counternarrative distinct from Stephen's. Jane, too, was trying to walk a fine line with her image of her husband. In it, she had recoiled from a Stephen who was arrogant, selfish, and spiritually crippled; the single vision of his hyper-logical worldview had robbed him of some of his humanity. Jane emphasized his brilliance—after all, his fame, and, by reflection, hers, was founded on the quality of Stephen's brain. At the same time, Jane implied that Stephen was precisely the opposite of transcendent. Within Stephen's brain spun the motes of suns and worlds and spaces, yet his logic confined him to the dust, ciphering things beyond his ken.

And Jane knew it even if Stephen didn't. Not long before the separation, she presented this image to a sympathetic reporter. "In the beginning, I felt Stephen was a scientist and he shouldn't involve himself in areas that

---

* The latter is closer to the truth; Hawking's decision to study cosmology couldn't have been influenced by his disease because he wasn't aware of his illness at the time he chose his field. Hawking enrolled in Cambridge to study cosmology and started his studies in October 1962; he was diagnosed with ALS in January 1963. In Hawking's words, "The doctors told me to go back to Cambridge and carry on with the research I had just started in general relativity and cosmology." Given that the diagnosis meant that Hawking had a two- or three-year life expectancy, he couldn't have thought at the time that he would survive long enough to advance far enough in his career to have to lecture. Hawking, *My Brief History*, 47.

didn't really concern him. Now he is coming up with such astounding theories . . . that can have a very disturbing effect on people . . . and he's not competent," she said in 1988, shortly after *Brief History* first came out. "But I pronounce my view that there are different ways of approaching it and his mathematical way is only one way—and he just smiles." After the divorce settlement in 1995, she dialed up the volume. "A spiritual home as well as a dependable family home is essential for every person born into this world," she would write in the first version of her memoir. "It is at our peril that we neglect those deep-seated needs in favour of materialism, egoism, science or the extremes of rationality."[25]

It was a battle of well-worn tropes. Stephen's tended to carry the day, for they were much more compelling. And Stephen was a master at crafting his image—and subtly helping others do it for him.

In 1992, Hawking was the guest on an unusually lengthy episode of the BBC radio show *Desert Island Discs*. Sue Lawley, the host, would chat with celebrities and talk about what music they would want to have if they were stranded on a desert island. "In many ways, of course, Stephen, you are already familiar with the isolation of a desert island, cut off from normal physical life and deprived of any natural means of communication," Lawley said to him, straight off the bat. "How lonely is it for you?"[26]

"I don't regard myself as cut off from normal life, and I don't think people around me would say I was," he responded. "I don't feel a disabled person—just someone with certain malfunctions of my motor neurons, rather as if I were color blind."

As for Hawking's choice of music, he couldn't resist a selection from *Die Walküre*, because Wagner "suited the dark and apocalyptic mood [he] was in" when diagnosed with ALS. A Puccini aria, some Brahms, some Poulenc, and a light Beatles song accounted for four more of his eight choices. His remaining three were just as on-message as Wagner. A Beethoven string quartet played by a character in a novel who knows he's about to die. The beginning of Mozart's *Requiem*, the composer's last, unfinished, work before his death at the age of thirty-five. And Edith Piaf's "Je ne regrette rien." "That just about sums up my life," Hawking said. It was a powerful message.

So powerful was it that Hawking could make people choke up even with television commercials. In 1993, Hawking was the star of a minuteand-a-half-long television advertisement for British Telecom—the UK equivalent of Ma Bell. In his robotic voice, Hawking begins, "For millions of years, mankind lived just like the animals. Then something happened which unleashed the power of our imagination. We learned to talk."

Hawking's wheelchair slowly drives under Mayan vaults and into a Greek amphitheater as the physicist extols the virtues of communication. As a giant BT-branded radio telescope fills the screen, Hawking concludes with a signature touch. "With the technology at our disposal, the possibilities are unbounded. All we need to do is make sure we keep talking."[27]

"I saw an advert on the television in England, for a telephone company, and [Hawking's] voice was on this advertisement. And this advertisement nearly made me weep," David Gilmour, guitarist and vocalist for Pink Floyd, told a radio interviewer a year later. "I've never had that with a television advertisement before, or with a commercial on the television. . . . I just found it so moving that I felt I had to try to do something with it." So Gilmour took Hawking's voice from the BT advertisement and made it into a track on the newest Pink Floyd album, *The Division Bell*.[28]

As Gilmour plainly saw, any performance that can make people teary-eyed during a telephone commercial—that's something unusual. That's art.

• • •

At times, the line between science and art is hard to see. Theoretical physicists, like artists, are often guided by a sense of aesthetics, a desire to grasp something beautiful that feels just out of reach. A discovery, a paper, a result, will often reveal but one aspect of a deeper truth that the physicist feels, but cannot yet express fully. But there is, of course, a difference between art and science: nature. Nature is the final arbiter of not just what is true, but what is beautiful. The most aesthetically pleasing models wither and rot if they are contradicted by experiment. Conversely, ideas considered ugly, even revolting, come to be considered beautiful if they explain the way the natural world works. Quantum mechanics, relativity, even atomic theory were all rejected by eminent scientists who simply found the theories distasteful. Over time, new experiments shape scientists' ideas of how the universe works, bending their aesthetic sense into alignment with natural law, turning once-unthinkable ideas into paragons of beauty and elegance. Even when that beauty is gathering dust.

By the late 1980s and early 1990s, the fields of general cosmology and relativity were in something of a rut. There hadn't been any real breakthrough experiments or observations since the late 1960s and early 1970s. As a consequence, both fields were ripe for experimental discovery; cosmologists and gravitational physicists were hankering for some new observation that would test their intuition. And prospects for something new in cosmology, at least, were extremely high. The year 1990 saw the launch of the Hubble telescope, the most capable orbiting telescope the world had yet seen. Hubble would allow astronomers to look deep into the cosmos to see very faint and distant objects, and to set better limits than ever before on the rate of the expansion of the universe, as well as the conditions in the young cosmos. And 1989 saw the launch of COBE, the Cosmic Background Explorer. At long last, physicists would begin to see the hot and cold regions in the cosmic microwave background that theorists had predicted, but that no one had yet spotted.

General relativity was having an even tougher time. There was a bright spot, however, with the 1991 launch of a NASA telescope designed to detect gamma rays—light with even higher energy than X-rays. While gamma-ray astronomy didn't directly test theories of gravity, gamma rays, like X-rays, might give glimpses of events so violent that they could only be the handiwork of black holes. Indeed, a certain mysterious kind of gamma-ray emission discovered in the 1970s seemed to hold the key to one of Hawking's early predictions, one that carried Hawking's best chance of ever winning a Nobel Prize.

The gamma-ray mystery first emerged in the early 1970s, thanks to a set of secret satellites launched by the US Air Force. The idea was that if someone detonated a nuclear bomb in the atmosphere or in outer space, these instruments—pairs of bizarre-looking, black, twenty-sided things studded with sensors—would detect the flash of gamma rays in the bomb's fireball. But unexpectedly, even though there were no nuclear explosions, the satellites were spotting brief bursts of gamma rays from deep space. Such bursts would have to be born in an extremely violent event—something like a supernova in which a star collapses into a neutron star or a black hole—but searches for supernovae that might be sources of the mysterious gamma-ray bursts came up empty. If the bursts weren't caused by the collapse of a star, what other event could be responsible?

Stephen Hawking thought he had an answer. In 1976, along with one of his students at the time, Don Page, Hawking argued that the solution to the mystery might lie not in supernovae, not in the birth of black holes, but in their death. For several years, he had been arguing that the universe was populated with mini black holes—only as massive as a mountain or an asteroid, rather than a star—created in the aftermath of the Big Bang. If they existed, these primordial black holes should be ending their lives in massive explosions. All around us, there should be mini black holes exploding, releasing a blast of gamma-ray radiation every time one gives up the ghost.

Page and Hawking calculated the very rough properties that these gamma-ray bursts should have if they in fact came from primordial mini black holes. A definite observation of gamma-rays from a primordial black hole would be a tremendous vindication of general relativity and quantum theory and would give us important information about the early universe and strong interactions at high energy that could not be obtained in any other way," the pair wrote. Unfortunately, the equipment of the day wasn't good enough to make the necessary measurements. The physicists realized that sort of observation would have to wait until engineers could build a gamma-ray telescope with high resolution, and fly it above the gamma-ray-absorbing atmosphere.[29]

In 1991, NASA launched the Compton Gamma Ray Observatory, an enormous (and extremely expensive) orbiting gamma-ray satellite with instruments that could map out where these mysterious gamma-ray bursts were coming from and what their properties were. If the telescope had found gamma-ray bursts that were the signature of mini black holes, it would have very likely led to a ceremony in Stockholm for Hawking. "People have searched for mini black holes of this mass, but have so far, not found any," Hawking told audiences years later. "This is a pity, because if they had, I would have got a Nobel prize."[30]

Unfortunately for Hawking, they hadn't, and he didn't. The Compton satellite and its successors, coupled with LIGO, implied that some gamma-ray bursts (the short ones) came from neutron stars slamming into each other to create a black hole; others (the longer ones) do, indeed, seem to come from distant, extremely powerful supernovae. Primordial black holes were not the answer.

Hawking would begin to sour on the possibility of primordial black holes. By 1993, along with a Cambridge colleague, John Stewart, Hawking used a toy model of black holes—black holes in a two-dimensional universe rather than our own four-dimensional one—to investigate what happened when a black hole evaporated. In the model, one of two things happened, either of which was bad news. The first was that it would create a naked singularity, an open wound in the fabric of spacetime. Hawking was absolutely convinced that this was impossible. (More on this shortly.) The other possible outcome was what they dubbed a "thunderbolt": a breakdown in the equation that they interpreted as a violent burst of high-energy particles. What's more, Hawking and Stewart reasoned that the same sorts of energetic explosions would happen in our four-dimensional universe. But, they admitted—possibly with a tinge of sadness—these gouts of particles couldn't explain the mysterious gamma-ray bursts. "It would be tempting to try to connect such events with the gamma ray bursts, but there is a problem with the energies involved," they wrote. Such events "would have to be extremely mild and could not account for the observed gamma ray bursts. If the universe does contain black holes that are reaching the end points of their evaporation, it seems they will do it without much display."[31]

Within a few years, Hawking seemed to be moving toward giving up on the concept of primordial mini black holes altogether. One reason might be the work that he did with one of his grad students, Raphael Bousso. In the mid-1990s, Hawking tasked Bousso with calculating how many such mini black holes would have been created immediately after the Big Bang.

"The project he actually gave me, he tried to explain in terms of some kind of argument he was having with Roger Penrose," Bousso explains. "He basically wanted me to work on something that would show that the universe doesn't have to begin the way that Penrose says it should. . . . He basically wanted me to calculate the probability that black holes would be created by some kind of quantum process during a very early phase of the universe called inflation, when space was expanding at an exponential rate."[32]

"It was a new student, just starting with Stephen. And Stephen was suggesting some problem for him to work on. And this problem was to show that something that I had done was wrong. So the student had to do this," Penrose told sociologist Hélène Mialet in 1998. "Then I went to talk to Stephen. And . . . I came out of the room, and I found the student waiting there, rather nervously. And he came up to me, and he asked me, 'What is it I'm supposed to do?' Because he . . . it was quite funny." As Penrose later wrote, "Being a student of [Hawking's] was not easy. . . . Hawking might ask the student to pursue some obscure route, the reason for which could seem deeply mysterious. Clarification was not available, and the student would be presented with what seemed indeed to be like the revelation of an oracle—something whose truth was not to be questioned, but which if correctly interpreted and developed would surely lead onwards to a profound truth."[33]

Bousso soon got over his initial confusion, tackled the problem, and began calculating the probability—which, in turn, would give an estimate of how many primordial black holes would have been created. The answer was pretty stark: none. Once the two plowed through the detailed calculations, they realized that while there could possibly be lots of subatomic-sized black holes forming, getting any black holes larger than that was exquisitely difficult. The subatomic black holes would have evaporated away almost instantly and wouldn't have survived to the present day. There would therefore be "no significant number" of primordial mini black holes left over from the very early universe.[34]

The argument wasn't perfect, and Bousso and Hawking made a number of assumptions, particularly about the evolution of the universe in the first moments after the Big Bang. Even without those assumptions, Bousso's work wouldn't kill the idea of primordial black holes entirely. "There are different kind of questions you could be interested in—Are you producing just some kind of black hole in the universe, or specifically those kinds of black holes that would finish evaporating now?" Bousso says. "If you just want some kind of black hole, you have much more freedom." To this day, scientists look for signatures of primordial black holes (admittedly without really expecting to find them).[35]

However, the idea of primordial mini black holes, which featured prominently in 1988's *A Brief History of Time*, didn't make an appearance in Hawking's later works *The Universe in a Nutshell* or *The Grand Design*. Even though Hawking was deeply attached to the idea—it was one of his youthful theories and one that got him a lot of attention over the years—he was willing to abandon his intuition in the face of greater evidence. Up to a point.

Even if there were no *primordial* mini black holes from the first moments after the Big Bang, that said nothing about mini black holes created on Earth. As ludicrous as that sounds, it's not out of the question: if engineers figure out how to pour enough matter and energy into a small enough space, they might be able to create a tiny black hole that appears and suddenly evaporates away in a burst of energy. And pouring lots of matter and energy into a tiny space is precisely what particle colliders like the Large Hadron Collider at CERN are designed to do. "Some of the collisions might create micro black holes," Hawking said. "These would radiate particles in a pattern that would be easy to recognize. So, I might get a Nobel prize after all."[36]

. . .

When a possible naked singularity appeared in one of his calculations, Hawking knew in his heart that it couldn't be; the alternative, no matter how distasteful, had to be true. (The so-called thunderbolts were, indeed, distasteful; they showed that the equations were failing.) This intuition went back to his years as a student, and was directly inspired by the work of a mathematician who was very influential on the young Hawking: Roger Penrose.

Penrose was a decade older than Hawking, a young professor at the University of London when Hawking was a student at Cambridge. Trained as a pure mathematician, he got interested in cosmology thanks in part to Cambridge physicist Dennis Sciama, and quickly made a splash by bringing novel mathematical techniques to bear on problems in cosmology and general relativity—including the problem of singularities. Along with Brandon Carter, a student of Sciama's at Cambridge, Penrose came up with a clever method of visualizing what happens even in the seemingly unvisualizable infinities that appear in the middle of a black hole.

The so-called Penrose (or Penrose-Carter) diagram* begins with an ordinary spacetime diagram like the ones in Chapter 4. Such a spacetime diagram has infinities in it—you can imagine a path, say, that moves through time without stopping, going into the infinite future, but it's impossible actually to plot such a path on the diagram. You'd need an infinite piece of paper. But Penrose realized that you could use a mathematical trick in some ways analogous to what painters use to portray unbounded distances in their artwork: they create a vanishing point, a little dot in the middle of their canvas, that is supposed to represent a place infinitely far away. Parallel lines that all stretch off to infinity are all pinched together into that vanishing point, and our brains automatically understand that the flat, finite image represents something with infinite depth. Similarly, a Penrose diagram pinches off all the parallel paths that stretch out through time, causing them to converge at a point that represents the infinitely distant future.

Where the Penrose diagram gets more complicated than your average perspective drawing is that there's more than one vanishing point. Just as you can imagine paths stretching to the infinite future, there are paths coming from the infinite past—and the diagram helpfully includes a vanishing point for that, too. But a spacetime diagram has infinities not just in time, but also in space; you can travel in an infinite direction to the right or to the left, so there are two more vanishing points for those two infinities. So, essentially, a Penrose diagram is a spacetime diagram where the top, bottom, left, and right are each pinched off into a vanishing-point infinity, making it look like a diamond. And just as the corners of the diamond represent infinities, so, too, do the lines that connect them; they can only be reached (if at all) by traveling infinitely in one direction or another.

---

* Technically, Carter diagrams, named after physicist Brandon Carter, are a class of Penrose diagrams for spacetimes that are rotationally symmetric—but even experts in the field don't tend to make the distinction between the two.

Just as in a traditional spacetime diagram, the light-speed-limit rule applies in a Penrose diagram; 45 degrees is the path of a light beam, and it's impossible to travel at a sharper angle than that without breaking the light barrier. But unlike a traditional spacetime diagram, the Penrose diagram captures all the infinities in an entire Einsteinian universe. Every infinity has a place, an explicit representation, on the chart—even the infinity of a singularity at the heart of a black hole.

That's the real strength of the Penrose diagram: how it's able to create a visual representation of black holes and other cosmological singularities. For example, to construct a Penrose diagram for an idealized version of the simplest kind of black hole—one that's not spinning or electrically charged—you start with a diamond-shaped Penrose diagram that represents all the spacetime in our universe. Grafted on the diamond, sharing one of its sides, is an upside-down triangle. That's the black hole. It takes a little unpacking to figure out what the diagram means, but it contains some profound insights.

If your path starts in the diamond, you can cross over into the triangle without any difficulty. However, if you start in the triangle, you can't cross into the diamond without going faster than the speed of light. No matter

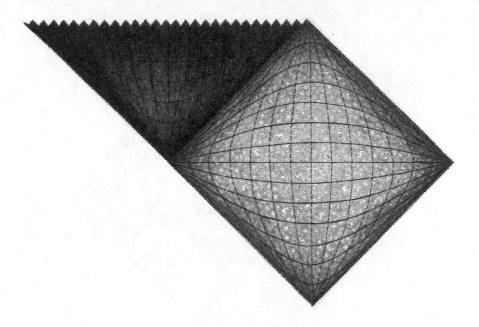

what you do, so long as you don't move on a path steeper than 45 degrees, your path eventually winds up hitting the top (zigzaggy) line of the triangle. Well, that zigzaggy line is the singularity of the black hole, and the border between the triangle and the diamond is the event horizon. It's a one-way barrier: you can cross the event horizon from our universe into the black hole, but you aren't coming back; you're doomed to hit the black hole's singularity. This means that if you ever manage to take a snapshot of the singularity itself, you're already past that one-way barrier, and your photo will never make it back into the universe outside the black hole. The event horizon effectively shields the singularity from outside observers— it totally and utterly blocks the singularity from view, cutting it completely away from our cosmos. And that's just one of the details that's hidden in the Penrose diagram; if you look at the image in the right way you can figure out a number of other fascinating and unexpected properties of black holes.*

Penrose diagrams quickly became a key tool for gravitational and cosmology researchers. "[Penrose] diagrams are very valuable. I can't think

___

* A quick example for the brave: Notice that the event horizon is at a 45-degree angle, just like a light beam. This means that in some sense, the event horizon is moving at the speed of light—even though it doesn't get larger when you observe it from the outside, from the point of view of someone trying to escape from the black hole, the horizon is essentially receding at light speed; it's impossible to overtake it and escape.

without them," physicist Leonard Susskind said in a lecture a number of years later. All gravitational physicists need such tools for visualization, Brandon Carter, who co-invented the diagrams, explained to sociologist Mialet. "To communicate with others, but also to communicate with ourselves, to think about our own ideas, we need a visual medium . . . but Stephen is the extreme case, he's absolutely, he's even more dependent on this."[37]

The Penrose diagram was an integral part of Hawking's way of thinking about general relativity—an indispensable component of his mechanisms for trying to understand the workings of the universe. "This technique of representing an entire space-time in a little diagram like this, this is something that Stephen uses all the time," Hawking's student Christophe Galfard told Mialet. "As soon as one speaks to him about a certain universe, a certain space, a certain solution to Einstein's equations, he asks . . . for the Penrose diagram every time, because this, this is visual." The geometric insight the diagram gave Hawking allowed him to avoid the algebraic difficulties of solving equations. "You don't need to do forty thousand calculations; you can just see it," Galfard explained. And, he added, other people's ideas, even if they were algebraic, rather than geometric, had to be translated into visual language.[38]

Penrose's visual aid works not just for black holes, but for pretty much any sort of spacetime pathology. No matter what weirdo general-relativistic structure you can think of—singularities of different flavors, wormholes, whatever—chances are good that you can capture its essence on a Penrose diagram, just as you can pin down its properties in algebraic manipulations of the equations of gravity.

As Penrose played around with various scenarios in his diagrams and equations, he noticed that he was unable to come up with a configuration where a singularity was born without an event horizon blocking access to it in some fashion. In other words, it seemed like there was no way that you or I or anyone in our universe could ever be able to see a singularity directly; such singularities all seemed to be shrouded by event horizons that blocked any attempt to view them. In the late 1960s, after trying for years, without success, to imagine an event-horizonless, "naked" singularity, Penrose began to suspect it might be impossible. But it was just a hunch. Thus, the cosmic censorship conjecture was born: all singularities must be hidden behind event horizons rather than "naked" and observable from our universe.* Penrose floated the idea in a conference and then wrote it

---

* With one glaring exception: the Big Bang.

up in a paper, only half believing it.* "I was rather presenting the case that perhaps it's false," Penrose says. "That would disprove cosmic censorship."[39]

Hawking, on the other hand, showed no uncertainty. "Stephen took the argument in a rather curious way," Penrose says, "as a demonstration that cosmic censorship was true, because I had failed to disprove it." Hawking strongly believed in the censorship principle, in part because if it were untrue, he would have to drastically rework his intuition about the way spacetime works. It was on this foundation that Hawking bet John Preskill and Kip Thorne that nature abhors a naked singularity.

Three decades later, the question is still undecided.

. . .

In 1992, Stephen Hawking turned fifty. This was a ripe old age for a theoretical physicist or a mathematician. Maybe even overripe.

"No mathematician should ever allow himself to forget that mathematics, more than any art or science, is a young man's game," mathematician G. H. Hardy wrote in 1940. "Galois died at twenty-one, Abel at twenty-seven, Ramanujan at thirty-three, Riemann at forty. There have been men who have done great work a good deal later; Gauss' great memoir on differential geometry was published when he was fifty (though he had had the fundamental ideas ten years before). I do not know an instance of a major mathematical advance initiated by a man past fifty. If a man of mature age loses interest in and abandons mathematics, the loss is not likely to be very serious either for mathematics or for himself."[40]

There are oodles of counterexamples. (In fact, arguably the biggest mathematical advance of recent years is Zhang Yitang's 2013 proof of an important theorem regarding primes; in 2013, Zhang was fifty-eight.[†]) Yet Hardy was just articulating an undercurrent of thought that has permeated the fields of mathematics and theoretical physics for decades: you're washed up by forty, if not earlier. Afterward, you're fit only for helping the next generation. Not everybody buys into this prejudice, to be sure, but that undercurrent is real, and it's pervasive. "There are people who do really

---

* Technically, Penrose presented a related conjecture about a set of inequalities about mass and spacetime—if one found a counterexample to the inequalities, it would automatically mean that the cosmic censorship principle was wrong.

† There are plenty more that Hardy should have been aware of: Karl Weierstrass, Joseph Fourier, Leonhard Euler, Pierre-Simon Laplace . . . and that's without even taking issue with his assumption of gender. Emmy Noether was making major advances non-stop through her late forties and early fifties and would no doubt have continued had she not died of cancer at the age of fifty-three.

wonderful things later in life," cosmologist Andreas Albrecht says. "I've never trusted that narrative. There [are] people who really placed boundaries on themselves with that narrative, and sort of have a kind of resignation."[41]

Though Hawking was adamant about not setting boundaries, the issue weighed on him as he entered his forties and fifties. When a reporter asked the physicist in 1995 (at the age of fifty-three) whether physics was a discipline for the young, his answer was unequivocal: "It is. I'm definitely over the hill." However, he added, his mind in its senescence was still a match for anyone else's: "But I still feel I'm doing good work, at least as good as my younger colleagues, but in a rather different direction. I've always gone in a somewhat different direction."[42]

Hawking's position as the Lucasian Professor at Cambridge no doubt made him particularly sensitive to the passage of time. The university rules were that he would have to relinquish the position at the age of sixty-seven—a rule first enforced with his predecessor's predecessor: the physicist Paul Dirac.

Dirac was responsible for one of the most important breakthroughs of twentieth-century science: in 1928 (merely twenty-six years old) he combined some of the theoretical underpinnings of special relativity with the rules of quantum mechanics, and derived an equation that rocked the world of theoretical physics. The Dirac equation, as it came to be called, was not only a masterpiece of mathematical physics, but its symbols also carried the seeds of an important discovery: antimatter. When Dirac used his equation to describe the behavior of the electron, he was forced to conclude that there was another version of the particle with equal and opposite charge—and that it wasn't merely a mathematical formalism, but actually had to exist as a particle distinct from the electron. Four years later, experimentalists found this particle: the positron. That same year, Dirac accepted the position of Lucasian Professor of Mathematics at Cambridge. In 1933, at the age of thirty-one, Dirac won the Nobel Prize. And that marked the end of his contributions to physics. In 1937, shortly after he got married, he submitted a short paper on cosmology to *Nature*. Upon reading the paper, Niels Bohr, one of the founding fathers of quantum theory, reportedly snorted, "Look what happens to people when they get married." After decades of inconclusive work, Dirac was—unwillingly—booted from his Cambridge chair at a forced retirement age of sixty-seven. (Dirac's predecessor, Joseph Larmor, departed the position at the age of seventy-five.)[43]

Just like Dirac, when Hawking turned fifty, the insights that had made him a physics star were roughly two decades old. While he was far from done as a physicist, Hawking suspected that he was not as intellectually nimble as he once was. "One is more agile and more open to ideas when one is young," he explained to a journalist. "Later on, one has . . . intellectual capital to protect. I now have a lot of capital."[44]

A few weeks after *A Brief History of Time* came out in 1988, Hawking told a reporter, "I have no enemies. I think it helps being disabled. People are not jealous of me." He probably wouldn't have made that comment in the early 1990s, after the success of his book turned him into a major celebrity and a wealthy author. He did indeed have a lot of capital to defend—and for the first time, he would have to defend it while in the public eye.[45]

. . .

A Brief History of Time rubbed a number of critics the wrong way. Despite the famous last line about knowing the mind of God, to the discriminating reader Hawking's resolute atheism shone through—Hawking's universe had no need for a deity, and he was unashamed of trumpeting that fact even at the gates of the Vatican. In *Brief History*, he tells the story of when he and his colleagues had an audience with the pope in the early 1980s after attending a cosmology conference organized by the Catholic Church:

> He told us that it was all right to study the evolution of the universe after the big bang, but we should not inquire into the big bang itself because that was the moment of Creation and therefore the work of God. I was glad then that he did not know the subject of the talk I had just given at the conference—the possibility that space-time was finite but had no boundary, which means that it had no beginning, no moment of Creation. I had no desire to share the fate of Galileo, with whom I feel a strong sense of identity, partly because of the coincidence of having been born exactly 300 years after his death![46]

If that weren't enough to rile up theologians and philosophers, Hawking displayed his scorn for twentieth-century philosophers, who he thought "[had] not been able to keep up with the advance of scientific theories. In the eighteenth century, philosophers considered the whole of human knowledge, including science, to be their field and discussed questions such as: did the universe have a beginning? However, in the nineteenth and twentieth centuries, science became too technical and mathematical for the philosophers, or anyone else except a few specialists. . . . What a comedown from the great tradition of philosophy from Aristotle to Kant!"[47]

The counterattacks not only went after Hawking's theology and his "scientism"—elevation of scientific inquiry above all other means of gathering knowledge—but also tried to tear down his science. William Lane Craig, a philosopher and Christian theologian, pooh-poohed many of Hawking's ideas, such as imaginary time (see Chapter 10), as "metaphysically absurd":

> For what possible physical meaning can we give to imaginary time? Having the opposite sign of ordinary "real" time, would imaginary time be a sort of negative time? But what intelligible sense can be given, for example, to a physical object's enduring for, say, two negative hours, or an event's having occurred negative years ago or going to occur in two negative years? . . . Could anything be more obvious than that imaginary time is a mathematical fiction?[48]

Hawking's response was to double down. In a 1992 speech, he reiterated his view that philosophers simply were too mathematically inept to be able to comment intelligently on modern physics. Worse still than the run-of-the-mill philosopher was the "subspecies called philosophers of science," he complained. "Many of them are failed physicists who found it too hard to invent new theories and so took to writing about the philosophy of physics instead. . . . Maybe I'm being a bit harsh on philosophers, but they have not been very kind to me." Hawking never softened his position; if anything, he got harsher over time, declaring, for example, that "philosophy is dead" in *The Grand Design* two decades later.[49]

Stephen's newly estranged wife, Jane, was among the detractors. "The truth was that the supercilious enigma of the smile which Stephen wore whenever the subject of religious faith and scientific research came up was driving me to my wits' end," she later wrote. Since the publication of the book, she had found an audience for her grievances, including Bryan Appleyard, a journalist for *The Times* of London. (Jane later described Appleyard as one of "just two professional journalists whose writings about our situation I respected.")[50]

In 1993, trying to capitalize on Hawking mania, Bantam released a collection of the physicist's lectures as a book titled *Black Holes and Baby Universes*. Appleyard used the opportunity to take a whack at Hawking, arguing that he was "a product of hype rather than reality," haughty, contemptuous, and wrong:

> There are in this new book dozens of examples of confused, contradictory and imperfectly grasped philosophy, which I have not the space to list and disentangle. It need only be said that a certain abrupt, careless impatience combines with an alarming belief in his own publicity

to produce some appalling nonsense. . . . Hawking remains in his life a figure of great courage and heroism. Sadly, his books are arrogant and narrow-minded. The truth of this contradiction may be that the demands of marketing have created a false picture; that the great cosmic speculator has nothing to do with the impatient, patronising hack philosopher.[51]

"He has a real chip on his shoulder," Hawking told a reporter when asked about Appleyard's critiques. "I don't know that I've seen him write approvingly of anyone. I feel that he is a failed intellectual and so he has to decry everyone else."[52]

Hawking could easily dismiss attacks from people outside of his field; after all, they didn't have the training to fully understand his work. However, Hawking had antagonists even within his own scientific community. John Barrow, a scientific sibling—he, like Hawking, had had Dennis Sciama as his PhD adviser—was a prominent voice in a brutal takedown of Hawking in *The Spectator*. One passage read:

Hawking is now in his 50s, and it is a general rule that most theoretical physicists will have shot their bolts by their mid-30s. Einstein was 26 when he published his major paper on relativity, and by 34, with his first paper on general relativity, his life's work was effectively done. Barrow says: "To compare Hawking to Newton or Einstein is just nonsense. There is no physicist alive who compares to Einstein or Bohr in ability. But those rather grottily researched little biographies of Galileo and Newton in *A Brief History* do rather invite you to put Hawking in the same sequence. In a list of the 12 best theoretical physicists this century Steve would be nowhere near."[53]

That wasn't all. Barrow also implied that Hawking's work, even the "interesting" parts, would wind up being of no lasting value.

For a man struggling to transcend his mortality by seeking eternal knowledge, there could have been no insult more keenly felt.

CHAPTER 9

FLASH (1987–1990)

"**S**o he asked me to find a picture of Cygnus X-1," recalls Ray Laflamme. Laflamme, a French Canadian, was in the last stages of finishing up his PhD thesis at the time. But his adviser's book was a higher priority; it was less than a month away from being sent to the printers, and the graphics hadn't been completed. Laflamme's task was to find a picture of the first black hole ever discovered—and the subject of the famous Hawking bet: Cygnus X-1. "Okay. I was a student in quantum gravity, so a picture of Cygnus X-1 is kind of . . ." Laflamme shrugs.[1]

Black holes are by their nature not very photogenic, even if one could image them directly, which was well beyond the reach of science until very recently. Back in 1987, the best one could do for an illustration was to show a picture of a field of stars—little seemingly random dots against a black background—with an arrow pointing to where a black hole was known to lurk. But as a theorist, Laflamme's experience with observational astronomy was just about sufficient to tell him which end of a telescope to look through. Even getting a picture of the correct field of stars, in this case, from a patch of sky inside the constellation Cygnus (the swan), was beyond what he himself could do. "So I go and talk to a few people who were kind of astrophysics people and I phone them and say, 'I want to have a picture of Cygnus X-1,' so they ended up sending me to the Greenwich Observatory," Laflamme remembers. "And they told me, 'This is a lot of work to go and find this in some plate somewhere. Maybe you could go to the Institute of Astronomy in Cambridge.'" Laflamme goes to the Institute of Astronomy and they manage to give him a photographic plate with the right patch of sky. Somewhere in that photograph lurked a black hole. But where?

Cygnus X-1 was totally invisible in the photographic plate, which wasn't able to capture any X-rays. The location of the black hole had to come from a satellite that was able to see X-rays, and Laflamme had to put an arrow at the proper coordinates. However, the photographic plate that he got from the Institute of Astronomy didn't have a proper coordinate system attached to it—there was no indication of precisely where in the sky the telescope had been pointed when the photograph was taken, or even how much of the sky was captured in the plate. So Laflamme turned back to the institute for help. "They said to me, 'What's the scale?' And I said, 'I don't know. Somebody gave me this picture without the arrow, and I want to know where it is in there,'" he recounts. "And they said, 'We have to find our records. And this will take many months.'"

Not acceptable. "I go back, and Stephen says, 'I need it for two weeks from now, because they are waiting to print the book. Get your ass moving, and tell me where [Cygnus X-1] is.'" Laflamme calls around, and nobody is able to help him, not even some of the most esteemed astronomers in Britain. "So, I said, I went to the Greenwich Royal Observatory, and they had no clue where it was in this picture. I went to Martin Rees, not the astronomer royal yet, but he had no clue where it was. I went to [Donald] Lynden-Bell, who was one of the top astronomers, and he had no clue where it was. The guy in the street isn't going to know, so I took up an arrow and I said, 'It's gonna be . . . '" Laflamme moves his finger in random circles and suddenly jabs at an imaginary point in space. "'There.'" He grins. "'Stephen,' I said, 'this is where Cygnus X-1 is.'"

. . .

The years 1987 through 1990 marked one of the greatest transitions in Stephen Hawking's life. At the start of this period, he was a well-respected, if not terribly well-known, physicist who had tried his hand a few times—not so successfully—at science popularization. At the end, he was one of the world's most recognizable international celebrities, a scientist superstar the likes of which hadn't been seen since Einstein.

A Brief History of Time, published in 1988, was the spark for his rise to fame, even if it had little to do with his appeal to the public. There were better-written, even better-selling popular science books out there, many of which were written by first-rate scientists. But once *Brief History* hit the shelves, the publishing world reeled. So did Hawking. His comfortable life in Cambridge with Jane and the children, his career as a theoretical physicist laboring in semi-obscurity, were both roiled by the sudden apparition of fame and fortune and attention.

But for Stephen, the recognition couldn't come soon enough. He had worked for more than half a decade—nearly succumbing to his disease in the middle—to make his dream of a physics best seller come to fruition. Now nothing could stand between him and his desire.

• • •

There would be no more delays. It was five years since Hawking first drafted the little one-hundred-page manuscript that would become the core of *A Brief History of Time*. He had signed his contract with Bantam Books and had begun discussing strategy with his editor, Peter Guzzardi, in mid-1984. A year later, Hawking was in a coma in a hospital in Geneva. He would have to learn how to communicate again, first without, and then with, a computer embedded in his wheelchair. That work on the book began again in 1986 was a testament to Hawking's force of will. For there was a lot of work to be done to turn that dense little draft into a book that people would want to read. Or at the very least, buy.

"The material itself was not that promising, but the story of Hawking was promising," admits Guzzardi. "So the hope was we can take this material which is kind of dry and uneven and sometimes written with all kinds of assumptions about the reader already understanding a lot of physics and astrophysics, and in some cases, written a bit like a middle-school level. So it was very uneven."[2]

The early draft of Hawking's manuscript was uneven, perhaps, because parts of it were derived from his prior lectures, most of which were intended for physicists, or at least scientists. Hawking made an attempt to smooth over the language and make it more accessible to the general audience, but he was not always successful. For example, in the draft manuscript, he writes,

> There are only a limited number of extended supergravity theories. In particular, there is one called the N=8 theory that contains the largest number of particles. They are shown in table 11a. Large though this number is, it is not large enough to contain all the particles that are thought to occur in the theories of the electromagnetic, weak and strong forces.[3]

In 1980, when he was appointed the Lucasian Professor, Hawking—or, more precisely, one of his students—gave a speech to the gathered academics at Cambridge University:

There appear to be only a finite number of such theories. In particular, there is a largest such theory, the so-called N=8 extended supergravity. This contains one graviton, eight spin-3/2 particles called gravit[i] nos, twenty-eight spin-1 particles, fifty-six spin-1/2 particles, and seventy particles of spin 0. Large as these numbers are, they are not large enough to account for all the particles that we observe in the strong and weak interactions.[4]

Other than eliminating some details and bumping others to a table, the language is only cosmetically different, and is not the sort of material that would appeal to a typical lay reader. Nor did the draft display any real interest in the rich intellectual history of the field of cosmology. There were few historical details, and none of the kind of humanistic information that could put Hawking's work in context, lighten the prose, and give the reader some space to breathe in between difficult technical passages about astrophysics and quantum mechanics.*

At one point, a frustrated Al Zuckerman suggested using a ghostwriter. Guzzardi refused, but he did hire astronomer and science journalist John Gribbin, who had known Hawking for a number of years (and later co-wrote a biography of the physicist), to take an early crack at reshaping the manuscript. "What I did with Peter—and Peter didn't tell anybody else at the time—was to rearrange it so it made more sense logically," says Gribbin. "I don't want to say 'disappointed,' but I would have liked to have done more. You know, if Peter had felt able, if Peter had actually asked him, Stephen would have let me do it if he had known it was me, because we had a good relationship. But Peter said that he promised that he wouldn't do this, that only he would look at it."[5]

Worst of all, at least from a sales point of view, was that there wasn't a hint about his personal life—nothing at all about his day-to-day existence, his struggles with ALS, his wife or his children, his youth. Hawking wanted the book to be all about physics, not about him. "The material was kind of underwhelming. You had to see the potential in it," Guzzardi says. "I mean, everybody could see *Hawking's* potential." But Hawking the person wasn't what Hawking wanted to write about, and that wasn't going to change. Even so, Guzzardi was excited about the possibilities.[6]

The first step was to figure out the audience. "The question is, you know, where's the reader? How do we smooth this material out and aim at a reader at a certain level, and where is that level? Who is that reader?" Guzzardi asks. "That kind of decision had to be made, and then moving the

---

* This is an unavoidable hazard of writing about Hawking's research.

material in that direction, . . . to leaven it with examples and illustrations and whatever else we might be able to use to lighten and open the material." Space wasn't a problem; the one-hundred-page manuscript needed to be expanded somewhat to make it to book size. However, even after Gribbin restructured the book draft, splitting up some chapters, combining others and reordering the material to extract a coherent narrative, it still needed a lot of work. "It's iterative, iterative," he explains. "So you got a draft and you . . . go through it, and you make all kinds of comments and notes. You say, 'What about this?' And 'I lost you here,' and 'I really love this part over here.' And then the author will address that. Another draft and you go through that process again. And he typically goes through maybe three drafts before it gets to the place where you think, 'Okay, this is close enough where I can really bear down on it in a sentence by sentence.'"

That's the most labor-intensive part: going through it line by line, making sure every idea is understandable and every sentence is clear. "You push and push and push for explanations for clarity. Until, you know, until I get it basically, or mostly get it," Guzzardi says. "Probably, there are places where I finally would have had to pull up short when it came to a particle spin, gluons, and quarks. But I just figured this gives you enough of a sense of how that plays, so that if you really want to know more, you can dig into some other books, but we're close enough."

Despite Hawking's famous stubbornness, the physicist didn't resist the changes his editor was suggesting. "You have to understand how eager he was to make this work," says Guzzardi. "He was highly, highly motivated. . . . And if this was going to be part of that process, well, by God, he was gonna buckle down and do it." Hawking—and his students—were all in.

Of Hawking's students, Brian Whitt, not Ray Laflamme, was the one who helped Hawking the most with the writing of the book. "Yeah, it was Brian," Laflamme said. "Stephen probably thought that my French Canadian writing, that people would be upset that, say, pages 92 through 96 were written by a French Canadian guy."

"Stephen looked at his students for help because he couldn't do it on his own," Daksh Lohiya, another student of Hawking's, who had graduated a few years earlier, told a reporter. "Brian Whitt, a little bit of Bruce Allen (now, the director of the Max Planck Institute for Gravitational Physics in Hanover), a little bit of me—everybody made some contribution."[7]

Hawking, Whitt, and the other students wrestled with the structure and the prose, responding to round after round of Guzzardi's edits. The book slowly began to take form, and it was improving markedly at each step. The beginning of Hawking's one-hundred-page draft had sounded like the opening to a high-school essay:

From the start of civilization, Man has asked questions such as "When did the universe begin?" "What happened before the beginning?" "Will it have an End?" "Is space finite or infinite?" "What is the nature of time?" "What is the difference between the future and the past?" The aim of this book is to explain some of the answers to these long-standing questions that are suggested by modern developments in physics and cosmology.[8]

By late 1986, Hawking's opening to the book was much more engaging:

There is a story about a well-known scientist who was giving a public lecture on astronomy. He described how the Earth orbits around the Sun and how the Sun, in turn, orbits around the center of a vast collection of stars called our galaxy. At the end of the lecture, a little old lady at the back of the room got up and said: "What you have told us is all wrong. The World is really a flat plate supported on the back of a giant tortoise." The scientist thought he could deal with the old lady quite easily. "What is the tortoise standing on?" he asked. "Oh, no, you don't catch me out like that," said the old lady. "The tortoise is standing on the back of another tortoise, and before you ask me what that tortoise is standing on, let me tell you that it is standing on the back of another tortoise, and so on."

Most people nowadays would find the picture of an infinite tower of tortoises rather ridiculous, but why do we think differently? Do we know whether the universe had a beginning, and if so, what happened before then? What is the nature of time? Will it ever come to an end? Recent developments in physics suggest answers to some of these longstanding questions. . . . [9]

The final version of the opening of *A Brief History of Time*, circa 1988, was striking even by the standards of professional writers:

A well-known scientist (some say it was Bertrand Russell) once gave a public lecture on astronomy. He described how the earth orbits around the sun and how the sun, in turn, orbits around the center of a vast collection of stars called our galaxy. At the end of the lecture, a little old lady at the back of the room got up and said: "What you have told us is rubbish. The world is really a flat plate supported on the back of a giant tortoise." The scientist gave a superior smile before replying, "What is the tortoise standing on." "You're very clever, young man, very clever," said the old lady. "But it's turtles all the way down!"

Most people would find the picture of our universe as an infinite tower of tortoises rather ridiculous, but why do we think we know better? What do we know about the universe, and how do we know it? Where did the universe come from, and where is it going? Did the universe have a beginning, and if so, what happened before then? What is the nature of time? Will it ever come to an end? Can we go back in time? Recent breakthroughs in physics, made possible in part by fantastic new technologies, suggest answers to some of these long-standing questions.[10]

That was prose to be proud of.*

Better still was the ending. It was a long way from Hawking's original draft:

When we combine Quantum Mechanics with General Relativity, there seems to be a new possibility that did not arise before: that space and time together might form a finite four dimensional space without singularities or boundaries like the surface of the earth but with two extra dimensions. It seems that this could account for the observed features of the universe such as its large scale uniformity and also for the smaller scale departures from homogeneity, like galaxies, stars and even human beings. It might also account for the arrow of time that we observe.[11]

The final version, of course, ended with one of Hawking's most memorable quotations:

However, if we do discover a complete theory, it should in time be understandable in broad principle by everyone, not just a few scientists. Then we shall all, philosophers, scientists, and just ordinary people, be able to take part in the discussion of the question of why it is that we and the universe exist. If we find the answer to that, it would

---

* There are a number of variants of the turtle anecdote. Hawking's retelling—with an old lady antagonist saying "very clever" and "turtles all the way down"—is most similar to a version published in a 1975 issue of *Reader's Digest*, which took it from a *Natural History* magazine article about turtles in 1974, which apparently got it from a 1967 PhD thesis. Interestingly, Hawking appears to have substituted Bertrand Russell for philosopher William James, who starred in the other accounts. Bernard Nietschmann, "When the Turtle Collapses, the World Ends," *Natural History* 83, no. 6 (July 1974): 34; John Robert Ross, "Constraints on Variables in Syntax" (PhD diss., Massachusetts Institute of Technology, 1967).

be the ultimate triumph of human reason—for then we would know the mind of God.[12]

All that hard work had transformed the book. Guzzardi had brought out the best in Hawking. Even the title was significantly improved; originally, Hawking had suggested *From the Big Bang to Black Holes: A Short History of Time*. Guzzardi countered with the much snappier *A Brief History of Time: From the Big Bang to Black Holes*. Initially, Hawking resisted the change. "What I came up with by way of a defense was in a kind of intuitive moment, I knew how much Stephen loved humor. And I just said to him, 'You know, the difference in my mind is that *A Brief History of Time* makes me smile, and *A Short History of Time* does not.' That argument carried the day."* At one point, Guzzardi even had to keep Hawking from cutting that famous last line. ("Had I done so, the sales might have been halved," Hawking later wrote.[13])

By early 1988, the manuscript (title and "mind of God" line intact) was finalized, all the illustrations were done, the cover was selected, and it was sent out to the printers. According to Lohiya, in that period of waiting for the books to be produced, Hawking told Whitt and the other students that the book might actually turn a profit: "'I think I should share some of the profit with you,' he said. 'So would you like a percentage of the proceeds or a downpayment?'" After some initial reluctance, Whitt agreed to the sum of £500: "500 pounds was a hell of a lot of money because our scholarship was 125 pounds a month," Lohiya recalled. "Judy Fella, the secretary, gave us the cheques. We got them cashed, went to the pub, and thought we had gypped the poor man."[14]†

Nobody had any inkling what would happen next. "You try to do everything right. You work your ass off, you work on the manuscript, you work on the cover, you work on the marketing plan, and then you put it out there," says Guzzardi. "Then lightning has to strike. You can have all the elements, but without the lightning strike, it's just going to disappoint you. And yet lightning doesn't strike all that often."[15]

· · ·

---

* According to Guzzardi, this exchange occurred when the *Brief History* title was in place and Hawking wanted to change it to *Short History*. However, archival drafts demonstrate that the latter was the working title up until Hawking's final submission of the draft, suggesting that it more likely happened when the change to *Brief History* was first broached.

† After publication, Lohiya told the reporter, lawyers approached Whitt in hopes of convincing him to claim a share of the royalties, but Whitt refused.

When the book came out in April 1988, Peter Guzzardi had just left Bantam Books for another publisher, Harmony. So someone else had to organize the recall and destruction of all forty thousand copies of *A Brief History of Time*, not to mention the pacification of a pissed-off author.

A few weeks to months before a book goes to market, publishers send out "galleys" of the manuscript to publications and influential people in hopes of getting the book reviewed in prominent outlets. The esteemed peer-reviewed journal *Nature*, which had published one of Hawking's most important papers (and a number of his father's), received a copy, and the editor there assigned the review to Don Page, one of Hawking's former students, who had since become a physics professor at Penn State. Page, an evangelical Christian, took issue with Hawking's theology and had some quibbles about controversial elements of the book, but told *Nature* readers that *Brief History* "should be acquired by anyone seeking to learn about some of the latest ideas and speculation in cosmology, and to gain familiarity with the insights that have sprung from the extraordinary mind of Stephen Hawking." Behind the scenes, though, Page had found the book riddled with minor errors—errors that were obvious to those with rudimentary physics knowledge. For example, an illustration depicting tracks of subatomic particles was labeled as a field of stars and vice versa. Page called Bantam to tell them about the problem.[16*]

As Hawking put it, "the changeover at Bantam led to such confusion that the first printing contained a large number of errors, like photographs and diagrams being in the wrong places or wrongly labeled. They therefore had to recall it before it got to the shops and do another printing." However, Hawking's memory of the incident was apparently faulty; the US edition had already reached shops, as it had shipped for a sale date of April 1. (The recall notice—which told readers that the corrected version would have a blue cover instead of the original silver one—hit the newspapers on April 3. The UK edition was published several weeks later, perhaps accounting for Hawking's error.) When Bantam sales reps started calling bookstores to

---

* Hawking gives the details in *My Brief History* (London: Bantam Books, 2018), 96–97. Interestingly, the passage is worded almost identically to a passage in Kitty Ferguson's biography, *Stephen Hawking: An Unfettered Mind* (New York: St. Martin's Press, 2017), 142, which was written a number of years earlier. It is unclear whether Hawking borrowed the text originally written by Ferguson for his own autobiography, or Ferguson based her text on an unpublished manuscript or speech written by Hawking. Or perhaps it's even some combination of the two. This is a good example of what sociologist Mialet called the "hall of mirrors" that one finds when trying to trace the "origin" of a passage attributed to Hawking. Hélène Mialet, *Hawking Incorporated: Stephen Hawking and the Anthropology of the Knowing Subject* (Chicago: University of Chicago Press, 2012), 154.

try to get the books shipped back, as Hawking biographers John Gribbin and Michael White put it, "to their amazement, there were no unsold copies left. . . . According to executives at Bantam, this was the first sign that they were on to something really big." Lightning had struck.[17]

. . .

Stephen wasn't the only one who misremembered details about the recall. Jane wrote, "The first edition had to be pulped at the last minute because of the fear of legal action on account of certain aspersions cast in the text on the integrity of a couple of American scientists." (Jane found a silver lining in the recall, because "this misfortune allowed a minor omission to be rectified. Stephen had dedicated *A Brief History of Time* to me, a gesture which came as a much appreciated public acknowledgement, but the dedication had been left out of the American edition.") Jane was correct about the dedication, but the aspersions cast at the integrity of two scientists were not the cause of the recall. In fact, the offending paragraphs sat unchanged for several months, much to the chagrin of his insulted colleagues.[18]

Hawking's attack had come as a complete surprise to its targets. "I think it was Paul [Steinhardt] who got in touch with me and said, 'You should know about this thing in *A Brief History of Time*,'" says Andy Albrecht, presently a cosmologist at the University of California at Davis. Nestled in the middle of the book was a paragraph that contained serious, possibly career-destroying, allegations against both Albrecht and Steinhardt. In fact, it had already caused significant damage to Steinhardt. "[Steinhardt] really tried to protect me," Albrecht recalls. "I remember him saying, 'I don't want you to be worried about this. You're very young, you should be pursuing your own career and not worrying about this.' And he said, 'I'm going to take care of this.'"[19]

The allegations stemmed from some work that Albrecht, then a PhD student at the University of Pennsylvania, had done with Steinhardt, his thesis adviser, in the early 1980s. It was a time of rapid advances in cosmology, and Albrecht and Steinhardt had devised a new—important—theory called *slow-roll inflation* (more in Chapter 11). But they hadn't been the only ones to develop the theory. Hawking knew that Andrei Linde, a physicist then at the Lebedev Physical Institute in Moscow, had come up with a similar idea a few months earlier—indeed, Hawking and Linde had bonded while discussing the idea at a Moscow conference in 1981.

It's common in physics for two sets of physicists to come up with the same idea in parallel. Perhaps the most famous example involved the most celebrated occupant of Hawking's academic chair, Isaac Newton, who invented the mathematical machinery of calculus at the same time that continental mathematician Gottfried Leibniz did. It's also common for these incidents to lead to accusations of plagiarism and to cause rifts in the scientific community; this happened with the Newton/Leibniz fight, which divided British scientists from those on the European mainland for more than a generation. When it happened with slow-roll inflation, Hawking was the catalyst.

After Steinhardt and Albrecht published their slow-roll paper in 1982, Hawking began murmuring to his colleagues that the two had stolen Linde's idea—that Steinhardt and Albrecht had heard Hawking mention it in a Philadelphia seminar in late 1981, and had quickly written a paper to claim credit. When Steinhardt found out, he was livid; this was the sort of allegation that could wreck his reputation. Even though Steinhardt, a newly minted assistant professor, was greatly outranked by Hawking, who occupied one of the most prestigious offices in all of academia, he had to fight back.

Steinhardt wrote a letter to Hawking stating that he didn't recall Hawking saying anything about slow-roll inflation at the Philadelphia meeting—and besides, the work had already been underway before Hawking came to town. Steinhardt even sent copies of some of his correspondence to prove it. Hawking's conciliatory response seemed to put the issue to bed; he apparently accepted that Steinhardt and Albrecht had come up with the idea on their own. And that was that. Until *A Brief History of Time*.[20]

Sometime early on in the writing process—as Hawking added padding to his first draft—Hawking had inserted the following passage:

> I spent most of the seminar talking about the problems of the inflationary model, just as in Moscow, but at the end I mentioned Linde's idea of slow symmetry breaking and my corrections to it. In the audience was a young assistant professor from the University of Pennsylvania, Paul Steinhardt. He talked to me afterward about inflation. The following February, he sent me a paper by himself and a student, Andreas Albrecht, in which they proposed something very similar to Linde's idea of slow symmetry breaking. He later told me he didn't remember me describing Linde's ideas and he had seen Linde's paper only when they had nearly finished their own. In the West, he and Albrecht are given joint credit with Linde for what is called "the new inflationary model," based on the idea of a slow breaking of symmetry.[21]

The implication was clear, and it wasn't pretty. Hawking had dredged up a nasty accusation and given it a second life in a very, very public manner. And it was already causing damage; Steinhardt reportedly found out about the paragraph when it disrupted his attempt to get a grant from the National Science Foundation. Steinhardt had to figure out what to do, and fast—not just to protect his own reputation, but Albrecht's.[22]

Luckily for Steinhardt, his protégé had some vivid memories of that fateful seminar at Philadelphia's Drexel University in 1981. "That was the first time I had met Hawking," Albrecht says. "And one of the things that was happening at that talk which Hawking gave was that some of the Drexel students were videotaping it. And I remember having this fierce reaction, like, they should just come and listen to the science, but they're fiddling around with technology instead." Albrecht's annoyance with the Drexel students in 1981 turned to a sense of gratitude in 1988 when Hawking's attack was published. "When Paul got nuts and said, 'There's this thing in *A Brief History of Time*,' I said, 'Well, you know, it was videotaped.'" Steinhart managed to dig up a copy the videotape, and there was no mention whatsoever of Linde's version of slow-roll inflation. Steinhardt sent a copy to Hawking and to Bantam Books—and told the story to *Newsweek*, which was doing a cover story on Hawking and his new book. When the *Newsweek* reporter asked for a response from Bantam, the publisher announced that the offending passage would be "expunged from future editions."[23]

Hawking himself didn't apologize until, trying to appease a cosmologist friend who was close to Steinhardt, he offered to publicly make amends by writing a letter to the editor of a physics magazine.[24] The letter, which appeared in the February 1989 issue of *Physics Today*, read, in part:

> Some people have interpreted this passage as implying that I am suggesting Paul Steinhardt and Andreas Albrecht plagiarized Linde's proposal. This is definitely not the case. I have always been quite sure that Steinhardt and Albrecht came to the idea of a slow rolldown completely independently of Linde. And I have now seen a videotape of the seminar that, although not quite complete, does not show me mentioning Linde's idea. . . . I am very sorry if some people have gotten the wrong impression from what I wrote.[25]

That is, any perceived insult was merely a result of misinterpreting what Hawking had written. "It was very maneuvering. And then he mentions a gap in the tape from Drexel," says Albrecht. "It's almost in the league of how Rosemary Woods might have accidentally erased the [Watergate] tape. . . . It's just this tiny, tiny gap in the flow."* He shakes his head. "Face. Trying to save face."[26]

Over the years, Hawking occasionally, slyly, alluded to the incident in a way that made clear precisely where he stood. "[I] will just say that I first encountered the idea when I visited Andrei Linde in Moscow in October 1981," he told a group of cosmologists at a birthday celebration-cum-physics-conference in 2008. "Again I'm not going to stir up a hornet's nest by trying to assign credit for this. I leave that to the Nobel committee." Steinhardt's reputation never recovered in certain circles, such as Hawking's beloved Caltech (where Linde landed after the fall of the Berlin Wall). "I don't have a very kind view of Steinhardt as a scientist," says Daniel Z. Freedman, a supergravity theorist at Caltech close to Hawking, Linde, and Linde's wife, Renata Kallosh, also a top-flight physicist at Caltech. However, Freedman quickly adds, "But he's done some very, very good things in other fields."[27]

"I think the fact that [Hawking] had been pretty mean made me a little bit wary around him. Who knows what he's going to do next. So it did have that kind of impact, but I sort of sought to transcend that as best I could," Albrecht says. "The attitude I chose to take was that human beings do this kind of thing."

Albrecht was probably unaware that his words echoed his mentor's statement to *Newsweek* more than thirty years earlier, in 1988. "Hawking is an outstanding physicist," Steinhardt had said. "But he's not a god. He's a human being."[28]

...

Day after day, *A Brief History of Time* sold copy after copy. Not since Einstein had a theoretical physicist been the sort of person who would merit a profile in *Time* and the cover of *Newsweek* (back in the days when those magazines were extremely influential). Nobody in the publishing world had seen lightning like this before. It was getting easier to mistake Hawking for a god, or at least a rock star.

In Chicago, two superfans, Susan Anderson and Bill Allen, printed five hundred T-shirts bearing the words "Stephen Hawking Fan Club"— and they instantly sold out. So the pair printed more. Within two months, the number of shirts on the street had skyrocketed to eight thousand, and they were appearing all over the city. Anderson and Allen started getting

---

* This refers, of course, to Richard Nixon's secretary, who was supposedly responsible for a mysterious gap of eighteen and a half minutes in the Watergate tapes.

requests for shirts from all around the world—including from a certain physicist in Cambridge (who wanted them in medium and large sizes).[29]

At the peak of the craze, one Chicago high-school senior admitted that his T-shirt confused his classmates. "My friends look at the shirt and ask, 'What rock group is this Hawking in?'" he told *People* magazine. "Worse, I have friends who claim they have his latest album."[30]

At first, Hawking couldn't quite fathom how famous his book had made him—or what his fame would mean. According to Laflamme, the winter after *A Brief History of Time* first came out, Stephen decided to give a series of eight lectures to undergraduates at Cambridge based on the book, one lecture for each chapter.

On the day of the first lecture, Laflamme picked Hawking up at the house, but Hawking was in a rotten mood. "Stephen was very unhappy, and he was grumpy. Everything I was doing was wrong, so I stopped and I said to Stephen, 'I don't put you the right way in the chair, or it's too late to go to the bathroom, or the tea is too hot or too cold. What is wrong?'" Laflamme recalls. "Peevishly, he looked at me and says, 'I'm worried about my lectures.' He says, 'I'm worried that nobody will show up.'"[31]

Soon, it was time to go. The two physicists made their way across campus to the lecture hall. "You roll from the back door into the guts of the building, which doesn't have too many stairs, and we arrived in the room and it was packed. Packed with people. People sitting on the stairs, probably breaking all the rules for safety," says Laflamme. "And suddenly Stephen has this big grin—that smile. That tells you that even he didn't expect to catch that fire."

For better or for worse, Hawking was a celebrity, recognized not just by the Cambridge locals who regularly came within a hair's breadth of running him over as he wheeled himself heedlessly into the street. He was suddenly in demand—at galas, for lectures, at festivals. By day, he was feted with champagne and salmon, which his nurses would carefully feed him, and he would spend the night spinning maniacally around the dance floor in his chair. ("He was reckless, driving his wheelchair around and dancing at parties with it," says former student Raphael Bousso.) He famously used his chair as a weapon, allegedly crashing into people who walk too slowly in front of him, ramming cars that block his ramp, and running over the toes of people he didn't particularly like, including Prince Charles. ("That's a malicious rumor," Hawking told a reporter in 2000. "I'll run over anyone who repeats it.")[32]

Now, after the publication of *A Brief History of Time*, he had access to the toes of the richest and most famous people in the world. Laflamme remembers one infamous visit by a film star: "Someone comes into my office, and said, 'Shirley MacLaine is coming, and Stephen is worried,'" he says. "'He doesn't want to be alone with her. Can you come to lunch with us?' So, as a student, free lunch, OK? I had no idea who she was, and that was probably a good thing." Laflamme rounded up a small handful of students and went to a little restaurant near the department to join Hawking and await MacLaine's arrival, and waited, and waited. "She finally arrives, and she goes on about how lucky she is to meet this great man, and touching his hand, and I can see Stephen is trying to stay away from all of this," Laflamme continues. "And then she's there talking about the mind, energy, and all of this, and all the students, we're rolling our eyes." One of the students got the bright idea to take a spoon, and quietly bend it under the table and pass it to his fellow students. "So the students suddenly bend spoons. I remember she was totally, totally over the top." Jane's biggest complaint was that MacLaine was rude: "She spoke only to Stephen and largely ignored the rest of us."[33]

Just as fame brings fans, it brings detractors, too. Even as some reviewers praised the clarity of Hawking's prose, others declared it unreadable. Even the UK publisher of the book, Mark Barty-King, admitted at the time that the book was "one which I personally found quite difficult to read because of the subject matter, but one which I considered to have enormous appeal." In literary circles, Hawking's magnum opus quickly became known as the "most unread bestseller of all time."[34]

Hawking was well aware of the claims that purchasers of *Brief History* "don't read it: They just have it in the bookcase or on the coffee table, thereby getting the credit for having it without taking the effort of having to understand it. I am sure this happens, but I don't know that it is any more so than for most other serious books, including the Bible and Shakespeare," he wrote. "On the other hand, I know that some people at least must have read it because each day I get a pile of letters about my book, many asking questions or making detailed comments that indicate that they have read the book, even if they don't understand all of it. I also get stopped by strangers in the street who tell me how much they enjoyed it." That line of criticism didn't really sting at all. But another one cut deep.[35]

In October 1998, *New York Magazine* ran a long story about the publishing industry that turned into a vicious attack on Bantam:

> How did a book as complex and incomprehensible as *A Brief History of Time* end up as the likely best-selling non-fiction book of the year? The simple answer, I believe, is that Bantam Books is perhaps the top

student of best-seller strategy in American publishing. And it knew that the only way to guarantee a bestseller in this case was to exploit the illness of Stephen Hawking to promote his book—in a way that is at best irrelevant and at worst shameful. . . .

Nearly a third of the cover is taken up by a striking photograph of Hawking in his wheelchair. Then there's the dust-jacket copy: "From the vantage point of the wheelchair where he has spent the last 20 years trapped by Lou Gehrig's disease," Bantam's promoters have written, "Professor Hawking has transformed our view of the universe."

. . . This is an almost unprecedented exploitation of a nonfiction author. I defy Bantam to name another nonfiction book in America— *any* nonfiction book, other than autobiography or biography—with a picture of its author on the front cover. Even Carl Sagan, whose books on cosmology and the universe and whose wide TV exposure have made him a widely recognized face, has never had his own photograph on the front of a book.[36]

This went straight to the core of Hawking's identity, of his struggle to make his mark as a physicist, as a communicator, and as a human being despite his disability rather than because of it. And it hit home, even as Hawking denied any such exploitation of his disease. "Undoubtedly, the human-interest story of how I have managed to be a theoretical physicist despite my disability has helped," Hawking wrote. "But those who bought the book from the human-interest angle may have been disappointed because it contains only a couple of references to my condition: The book was intended as a history of the universe, not of me. This has not prevented accusations that Bantam shamefully exploited my illness and that I cooperated with this by allowing my picture to appear on the cover. In fact, under my contract I had no control over the cover."[37]

"Jeez! Exploitation? Doesn't that mean somebody has to suffer?" Guzzardi bristles. "And Stephen was a terrific advocate for people who had his ailment and other handicaps. And he certainly didn't suffer. In fact he loved the limelight and loved the cover . . . but I may just be being insensitive in some way." In fact, Guzzardi felt an unusual degree of empathy with Hawking's plight. "It turns out that I had polio when I was a kid," he says. "So I would think I would be attuned to somebody kind of dissing me in some way, or exploiting people like me."[38]

Simon Mitton, a friend of Hawking's—and a publisher who tried to purchase the rights to *Brief History* but was outbid by Bantam—sees it in a slightly different light. Mitton thinks that Bantam's intent from the start was, indeed, to exploit Hawking's disability to boost sales, but that Hawking made the deal with his eyes fully open. "He accepted that that was what was going to happen with Bantam," Mitton says. "He needed the money. And he accepted that the publishers have got to be able to get their money back."[39]

• • •

The media had begun to discover Stephen Hawking a number of years before *A Brief History* came out; he was the subject of several major profiles, including one in *Vanity Fair* and another in *The New York Times Magazine*, and had even been the subject of a short biography titled *Stephen Hawking's Universe*. But it was *A Brief History of Time* that catapulted him into the stratosphere, and the media couldn't get enough. And as he discovered, Hawking was really good at interviews, and got better and better with each one.

Hawking had the perfect personality to be a media star; his quick wit and self-effacing humor would have endeared him to even an antagonistic journalist under any circumstances. On top of that, his disability afforded him leeway right away that only A-list celebrities typically get: a list of questions in advance. After a while, though, even that advantage didn't matter much, as most journalists' questions were depressingly predictable. Hawking was able to select from his computerized library of previous responses, or grab a section of a lecture he had given, give it a little polish, and he had a spontaneous-sounding on-message bon mot every time—as in this 1990 *Playboy* interview:

> PLAYBOY: . . . Can you tell us a little about your early life, before the secrets of the universe caught your interest?

> HAWKING: Yes. I was born on January eighth, 1942, three hundred years to the day after the death of Galileo. I was born in Oxford—even though my parents' home was in London—because Oxford was a good place to be during the war.

> PLAYBOY: Galileo was tried and imprisoned for heresy by the Catholic Church for his theories of the universe. Did he have something in common with you?

> HAWKING: Yes. However, I estimate that about two hundred thousand other babies were also born on that date. [Smiles] And I don't know if any of them were later interested in astronomy.[40]

Though it seems like a quick, off-the cuff exchange, it was anything but. It was an extremely subtle way of crafting his image.

The birthday connection with Galileo seems silly and inconsequential—and it is, as Hawking himself suggests. Yet biographers of Newton are quick to point out that the discoverer of gravity was born in the same year that Galileo died. You don't have to believe in metempsychosis to understand the symbolism: Newton was Galileo's intellectual heir. With a deft touch, Hawking taps into that imagery and, simultaneously, pokes fun at the idea that there's any significance to the coincidence. On the conscious level, readers are struck by Hawking's humor and humility; on the subconscious, they elevate him to the stature of Galileo.

This was no accident. Hawking used the same phrasing, slightly reordered, in a speech he gave in 1987:

> I was born on January 8, 1942, exactly three hundred years after the death of Galileo. However, I estimate that about two hundred thousand other babies were also born that day. I don't know whether any of them were later interested in astronomy. I was born in Oxford, even though my parents were living in London. This was because Oxford was a good place to be born during World War II.[41]

He told this story numerous times over the years in speeches and interviews as well as in movies about him. It appeared almost word for word in his 2013 autobiography, and in the book he published posthumously in late 2018. This little anecdote was the public relations equivalent of a shark—it was so uncannily efficient at its task that it no longer needed to evolve.[42]

Hawking had been adamant about *A Brief History of Time* being about physics rather than the physicist, but he realized that the media—and by extension, the public—wanted to learn more about the person. So when a producer picked up the movie option for his book, it caused a bit of a dilemma, not just for Hawking, but for the moviemakers.

"It's like any producer, you have a property and you want to turn it into something. I don't think anybody had any idea of what to do with it," says director Errol Morris. "And I don't think I had any idea of what to do with it at first. It was only later, not so long after, but later that I decided that I could see the movie, that I can see what can I do to make it."[43]

Steven Spielberg signed on as the executive producer—the person holding the purse strings—and soon settled on a director. Errol Morris had directed a handful of critically acclaimed documentaries, including *The Thin Blue Line*, which led to the release of a man who had been wrongly convicted of murder and put on death row. But what made Morris an inspired choice was that he had trained to be a historian and philosopher of science. He was not just fascinated with how humans acquired knowledge and sought truth, but also, like Hawking, was repelled by certain elements of modern philosophy of science. (At Princeton, Morris had so frustrated Thomas Kuhn, the eminent philosopher of science, that Kuhn had chucked an ashtray at him.) Yet Morris well knew that the real story of Hawking couldn't be solely about physics. As he recounted in 2019:

> One of the hallmarks of genius is that [one] is doing something so unexpected and so utterly different. And so utterly original. In Hawking's case, I would say that this is not an argument about his science versus his biography. I think it's not the correct way to look at it. It's that he created this persona that was a hybrid of all of this. Not to denigrate the science in any way; you know the discovery of Hawking radiation or any one of a host of things was incredible mathematical physics in its own right. But what brought him beyond all of that was this strange mixture of science and biography that made it unlike anything. Unlike anything. It made it unique.[44]

This was not the approach that Hawking initially wanted. "I mean, Hawking was annoyed with me at times, because he kept saying that it should be really about the science," Morris says. "But I said it's an adaptation of the book. And in the end, he liked the movie."[45]

Jane was harder to convince. "I tried very, very hard to get Jane Hawking to be in the movie and failed," Morris recalls. He went to dinner at the Hawkings' house, and happened to mention that he played the cello.* And they promptly brought out a cello and some sheet music—the Fauré *Elegy*—and handed it to the surprised director. "The Fauré *Elegy* isn't impossibly difficult. It's not easy, but it's not impossibly difficult. And if I had been practicing, that would have been okay, but I hadn't been practicing," he says. "I said this to Stephen one time, that if only I had played a little better, she would have agreed to be in the movie."[46]†

The movie could survive without Jane, especially since many of Hawking's friends and other family members decided to participate, including

---

* Morris had studied at the Fontainebleu School under Nadia Boulanger, which is how he met the composer Philip Glass, who scored many of his films, including *Brief History of Time*.

† The timing of the filming for *Brief History* couldn't have been worse for Morris; it was happening at the very moment that the Hawkings' marriage was breaking up in late 1989 and early 1990. Jane's refusal to be in the film was her first act of defiance after Stephen moved out of the house.

his mother, Isobel. But the show could not go on without Steven Spielberg, who was also extremely unhappy with Morris' approach. According to Morris, Spielberg apparently envisioned the film as being more like Ray and Charles Eames' *Powers of Ten*, which was intended to instill in a viewer an awe for the natural on various scales, from the subatomic to the supergalactic. But for the intervention of Spielberg's partner, Kathleen Kennedy, Spielberg would have fired Morris. In the end, Spielberg merely wound up pulling his name from the movie credits.

It was an expensive movie for the time—$3 million—and especially expensive for a documentary. Morris did many of his interviews at the sound stage in Elstree Studios in England so that he could capture audio of sufficient quality. But that meant building a replica of Hawking's office on the set. Morris' production designer, Ted Bafaloukos, constructed a perfect facsimile, down to the smallest details, including the Marilyn Monroe poster on Hawking's wall.

During the filming, Morris wound up building a "dictionary" of different shots of Hawking that could be edited with the voiceover added, thanks to a duplicate of Hawking's speech synthesizer, which Hawking allowed the director to use. Morris' shots—a close-up of Hawking's eye, of his hand squeezing the clicker switch that controlled his computer, of his visage reflected in the screen of his computer monitor as it flickers with words of incipient speech—extracted the greatest possible dynamism out of a mostly immobile human being. And in the end, Hawking was thrilled with the outcome (he thanked the director for making his mother a movie star), and so was Morris. "An interview is good if somehow it captured the complexity of a person that you were interviewing, that on some level—not totally—but on some level, you created a complex portrait," he says. "And I do believe *A Brief History* is a complex portrait and does capture what I think is truly interesting about the book." Even thirty years later, Morris looks back fondly at the movie. "I don't always like my work. Often I don't like it. But I do like [*Brief History*]," Morris says. "And I adored Stephen Hawking. I don't like most people, but he was something else."[47]

• • •

Supernova. This was the moment when the star ignited into a blaze of light that outshone the rest of the galaxy. Hawking's runaway best seller transformed the scientist into a public figure—one with a backstory that turned him into a symbol beloved by billions across the globe. Before, there had been only an extremely clever scientist working on profound and arcane questions. After, there was a scientific celebrity the likes of which hadn't been seen since Einstein. Before this moment, one needed to understand Hawking through his physics; after, his physics was almost totally obscured by his celebrity.

And what a celebrity it was. By the fall of 1988, Hawking was a superstar, not just in the English-speaking world, but internationally. Crowds surrounded him wherever he went. When he flew to Spain in October for the launch of the Spanish-language version of *Brief History*, he was engulfed in a sea of people who spontaneously broke into applause as he wheeled by. "He was good at staying at the center of attention," says Raphael Bousso, his student a number of years later. "Everything he said was recorded, and was a headline in the paper. And you know, he enjoyed that. He enjoyed that very much."[48]

Jane, who accompanied Stephen to Spain along with their youngest son, Tim, was having a harder time coping with celebrity. "So much sudden attention was gratifying and disturbing at one and the same time," she later wrote. "I felt ill-at-ease in the public eye, conscious of the way I walked or I held my head or even at the way I smiled. However, the crowds and the cameramen were not very interested in Tim or me, often pushing us out of the way in their eagerness to film Stephen, so it was not difficult for us to blend unnoticed into the throng." Even at home, Jane felt she had nowhere to escape to. Since the tracheostomy, there had been team of six or more nurses attending to Stephen around the clock, and that alone was a big disruption to the peace of the household. Now that the media wanted access—and Stephen wanted to grant it—this, too, had become a source of tension. "It was bad enough having nurses in the house all the time: with television cameras and reporters as well there would be no privacy for anyone anywhere," Jane wrote. "My arguments cut no ice. Quite the contrary, they were represented as yet further evidence of my disloyalty to the man of genius." By the middle of November, Jane had picked out a property in France where she and Stephen and the children and Jonathan could, "once again, find the unity and harmony, the modus vivendi" out of the harsh glare of the media spotlight.[49]

It was not a normal modus vivendi. In the late 1970s, Jane struck up a friendship with a choirmaster, Jonathan Hellyer Jones, whose wife had recently died of leukemia. The relationship rapidly grew—from giving the young Lucy piano lessons to staying after dinner and helping around the house to something much more.

Stephen had consented. "I would have objected," he later wrote, "but I too was expecting an early death and felt I needed someone to support the children after I was gone." Jane put it in slightly different terms: "He almost seemed glad that there was someone else who could relieve him of the burden of my emotional insecurities so that he could get on with the more important business of physics." Regardless, Jonathan had become a key support for Jane, indeed, a regular part of the family, and someone who would almost certainly formalize his role when Stephen finally succumbed to his disease. It was a delicate and sensitive arrangement that required a great deal of discretion not to embarrass Stephen. Even so, a number of people who knew the Hawkings at that time relate how awkward it was to have dinner not just with Stephen and Jane but also with Jane's lover at the table.[50]*

When Stephen had his tracheostomy in 1985, this modus vivendi was seven years old, having lasted longer, almost certainly, than anyone expected. But the tube newly inserted into in his throat meant that Stephen needed twenty-four-hour nursing, and the cadre of nurses—and, to Jane, their "prying eyes, listening ears, and gossiping tongues"—upset the fragile balance of the household. Jane perceived that she and her children Lucy and Tim had become "second-class citizens as if we, the family, were the lowest of the low, crouching on the bottom rung of a ladder, at the top of which the angels—the Florence Nightingales—administered to the deity—the master of the universe," she wrote. "In between there were the several echelons of students, scientists, computer engineers, all of whom were obviously more important than we were." Jane soon had recurrent nightmares of being buried alive.[51]

The nurse wielding the biggest shovel was Elaine Mason. Mason had met the Hawkings through her husband, David Mason, who had helped engineer the computer system and speech synthesizer on Stephen's wheelchair. A professional nurse, she joined the nursing rotation, and Stephen soon preferred her company over the other nurses. The attachment was mutual. "Elaine volunteered to travel with Stephen at every opportunity," Jane wrote, marveling that Mason didn't feel a "tremendous wrench" when she left her husband and children at home for such long periods. Even though—perhaps because—Elaine's personality was so markedly different from Jane's, flamboyant and volatile rather than steady and reserved, Stephen kindled a romance with the nurse. Jane later wrote that

---

* Jane doesn't explicitly reveal when her relationship with Jonathan turned physical. Reading between the lines of her memoir, it seems that she was primarily concerned not with staying faithful to her husband, but with maintaining discretion—in part out of kindness to Stephen, but also in part for fear of embarrassment and being called a hypocrite. Discretion notwithstanding, a number of visitors to the Hawking household assumed that Jonathan was Jane's lover.

she wouldn't have begrudged the two a physical relationship, especially given her own situation with Jonathan, so long as the pair were discreet—and so long as Elaine posed no threat to her relationship with Stephen. But, over time, Jane had come to believe that Elaine was undermining the marriage. "In public and at home, she seemed to be busily usurping my place at every opportunity, sometimes apeing me, sometimes undermining me, often flaunting her influence over Stephen," she wrote. "She had an unassailable stranglehold over the nursing rota and had so successfully ingratiated herself that all remonstrance was useless: any comments would be reported back to Stephen and I would be castigated for my interference." And, she continued, her complaints to the Royal College of Nursing were of no avail.[52]

Even before the thunderbolt of fame struck, Jane and Stephen's relationship was under great strain—strain that even began to affect the staff at Cambridge University. After the book came out in April 1988, though, vast fissures began to appear in public.

These cracks are apparent in the 1989 BBC documentary *Master of the Universe*. On the surface, the film is upbeat, and Stephen hits all the notes that so endeared him to the public: his genius, his perseverance, and his humility. The film ends with a voiceover from his speech synthesizer as he and Jane tuck Tim into bed. "I have received a lot of help. This has meant that my disability has not affected my outlook," he said. "I have a beautiful family, I have been successful in my work, and I have written a best seller. One really can't ask for more."

Off script, the camera captured a marriage nearing its end. Some of the details were subtle; before departing the house in the morning, Jane doesn't lean down to give Stephen a kiss, but instead tousles his hair and gives him two pats on the head, almost as if he were an Irish setter rather than a spouse. Some were blindingly obvious. Jane says that she felt excluded by the awards and honors Stephen was receiving—that they created distance between Stephen and his family. "I'm not an appendage of Stephen's as I very much feel I am when we go to some of these official gatherings," she explained, seemingly on the brink of tears. "I mean, sometimes I'm not even introduced to people, I come along behind." Perhaps the most striking moment in the documentary was when Lucy—then a lively young girl nearing the end of high school—addressed the cameras: "I'm not as stubborn as him. I don't think I'd want to be that stubborn," she said. "And I don't think I've got quite his strength of mind, which means he will do what he wants to do at any cost to anybody else, which I suppose you have to have, really, in his position."

The very moments of Stephen's greatest honors had become dangerous stressors for his family, and midsummer of 1989—a bit more than a year after *A Brief History of Time* hit the market—delivered a one-two-three combination of honors in quick succession. The last would set into motion a series of events that led to the end of his marriage.

On the 15th of June, Stephen was to receive an honorary degree from Cambridge University. It was extremely rare for Cambridge to honor one of its own in such a way, and it was a major event. Prince Philip, the Queen Consort, was in attendance in his role as chancellor of the university. (As Jane writes in her memoir, the prince asked Hawking about his wheelchair as he rolled by. "Self-propelled, is it?" the prince asked. "Yes," Jane replied. "Watch out for your toes!")[53]

An even bigger honor came the next day. Every year in June, the queen announced a list of people who had been granted knighthoods and various other honors that she saw fit to bestow upon her citizens. On June 16, the world learned that Stephen Hawking was the newest "Companion of Honour," one of some threescore luminaries in the British Empire who had made distinctive contributions to the arts, sciences, or government. Jane and Stephen would have a royal audience at Buckingham Palace a few weeks later. When the couple presented a thumb-printed copy of *A Brief History of Time* to the queen, Jane later wrote, the queen asked a baffling question: "Was it a popular account of his work such that a lawyer might give?"[54*]

The third honor came on the 17th of June with a benefit concert in Stephen's honor. Jonathan was the conductor of a music ensemble, the Cambridge Baroque Camerata, which had recently lost its sponsorship. In May, Jane realized that if Jonathan put on a concert, he could simultaneously pay tribute to Stephen, raise money for charities such as the Motor Neurone Disease Association, and raise the profile of the Camerata in hopes of finding a new sponsor. Stephen agreed, but on the day of, something about the concert conducted by his wife's intimate companion had stuck in his craw. Stephen, Jane writes, "was edgy and disgruntled. His perceptions of the event seemed to be coloured by the grudging view that Jonathan and the orchestra had obscured his share of the limelight."

---

* Queen Elizabeth II was never much of a reader. "She is a woman more read about than reading, and no book about her suggests that reading plays an important role in her life," wrote author Valerie Grove, who, in the late 1980s, sat on a panel to suggest books for the royals to read. But hope always springs eternal—and in the summer of 1988, *Brief History* topped Prince Charles' list. Valerie Grove, "What the Royal Family Will Be Reading This Summer," *Sunday Times* (London), August 28, 1988.

In the argument that followed, Jane asserted, Stephen pointedly reminded her that since the "Companion of Honour" was not a hereditary title, she "had no part in it."[55]

As soon as she could, Jane retreated, along with Tim, to the new home in France, where she could "hide [her]self away from the tyranny of the outside world." A few weeks later, hoping to defuse the situation, she invited not just Jonathan and Stephen, but the whole Mason family—David and Elaine and their children—to France. As Jane put it:

> While I had no intention of interfering in any fond attachment that might have developed between Elaine and Stephen, I thought that she might just be persuaded to see that the success of our task depended on finely balanced teamwork. . . . Naively I trusted too that if she realized that Jonathan and I did not, as a matter of course, sleep together in the same room or indulge in riotous orgies, she would learn to respect the *modus vivendi* which enabled us to go on caring for Stephen and the children indefinitely, come what may.[56]

But the modus vivendi was broken. Stephen and Elaine gazed with increasing antagonism at Jane and Jonathan. The only thing everyone could agree on seemed to be their universal lack of empathy regarding David Mason, who was finding himself the fifth wheel in his own marriage. (David was still, to some extent, in thrall of the physicist. "If he raised an eyebrow, you would run a mile," David later told *People* magazine. "He uses people." Even after Elaine left him for Stephen, he would continue to service Stephen's wheelchair.)[57]

Stephen made it quite clear that he wasn't enjoying his stay in the French countryside, and tensions rose. Within days, an argument over how Stephen had treated one of the nursing staff escalated and blew the lid off of all the resentments that had been building up over the years. "Flames of vituperation, hatred, desire for revenge leapt up at me from all sides," Jane later wrote. Stephen and the Masons returned to England, leaving a shell-shocked Jane hunkering down in France, waiting for her "lost child" to come to his senses. However, when Jane finally returned to England in September, Stephen announced that he was going to move in with Elaine.[58]

For so long, Jane and Stephen had kept their marriage together. It had survived as the spousal affection between wife and husband had slowly transformed into something more maternal. It had even survived as Stephen's replacement waited in the wings and began to subsume some of his spousal and fatherly duties.

It had also survived their disagreement over religion. Though Jane and Stephen had deep differences in their philosophy and theology, that situation was little changed since the beginning—with, perhaps, the sole exception that suddenly people were taking his thoughts on God much more seriously. Jane herself seemed to view Stephen's atheism as a primary reason for the breakdown of her marriage, as she was more than happy to make known. Journalist Bryan Appleyard later wrote that he was shocked by Jane's willingness to air her grievances during a 1989 interview, when she "started fiercely criticising her husband before I could even turn on my recorder. She was religious and he was not. In fact, he was aggressively anti-religious, and his anger with her faith was becoming intolerable. He would not be in the same room as her devout friends." Even so, it was Stephen, not Jane, who sought to dissolve the marriage, and there's little evidence that, to Stephen, Jane's religious beliefs warranted anything more than slight amusement coupled with a mild dose of condescension. Jane, who felt the religious schism much more deeply, would have been content keeping the union intact. What's more, the woman Stephen next married, Elaine, was just as religious—and even more demonstrative about her faith—than Jane ever was.[59] *

No, it was not religion that killed their marriage, but the fact that the Hawking household became increasingly Stephen-centered, first as a seemingly endless parade of nurses devoted themselves solely to the welfare of their patient, and then as journalists, celebrities, and adoring fans prostrated themselves before the newly consecrated god of best-seller lightning. It's no coincidence that the bond between Stephen and Jane broke at precisely the moment that Stephen achieved his biggest triumphs.

The term "triumph" comes from a singular honor bestowed upon a victorious general by the Senate of ancient Rome. There would be an enormous parade through the streets of the city, with the conquering general—the imperator—taking pride of place in a special chariot drawn by four horses. In the chariot with the general, standing slightly behind him, rode a slave

---

* Elaine tended toward evangelism much more than Jane did with her more reserved, if no less devout, faith. In fact, Elaine was convinced that God had spoken to her during a fit of despair on a "cold, isolated" beach in England, when from deep within her came a loud voice that exclaimed: "I am here, and I died on that cross for you because I want to surround you with my love." Stephen didn't seem to mind Elaine's religiosity. "I was indeed able to share my faith with Stephen. He would often grin and say he didn't mind coming second place to God." Elaine Hawking, "Baptism Testimony of Elaine Hawking," Internet Archive, April 2018, https://archive.org/details/chipping-campden-baptist-church-513231/2018 -04-01-Baptism-Testimony-Elaine-Hawking-Elaine-Hawking-46453311.mp3.

who held a golden crown above the general's head. But that slave had another task as well. As the triumphal procession wound its way through the cheering crowds, he would periodically whisper in the imperator's ear: "Look behind thee; thou art mortal."[60]

As Jane told Appleyard in 1989, her role was no longer to be a caretaker for Stephen in his illness, but "simply to tell him that he's not God." Yet of all people on the planet, the one most conscious of his own mortality was none other than Stephen Hawking.[61]

CHAPTER 10

IGNITION (1981–1988)

———————

Behind St. Peter's Basilica, where tourists aren't allowed, the chatter of crowds gives way to the squawking of parrots. The grounds of the Vatican are green, overwhelmingly green, with immaculate gardens and fountains and trees of all varieties. Perched atop a small hill is a squat little building that looks almost like a pagan temple—complete with a statue of the goddess Cybele. Inside, the walls are adorned with quotations from Seneca and Cicero. It is almost gaudy in its opulence, with bright frescoes covering the walls and ceiling and almost every unpainted surface bedecked with marble of every color known to humanity.

This is the building where luminaries from around the world gather every other year to advise the pope on matters scientific. Like a number of other countries, the city-state of the Vatican has a scientific advisory board—the Pontifical Academy of Sciences—and in 1981 Stephen Hawking was a guest of the board (and would be appointed a member a few years later).*

This time, the meeting was about cosmology, and the subject of Hawking's talk was a wee bit touchy: "the possibility that space-time was finite

———————

* Hawking's atheism was not a barrier to his appointment to the Pontifical Academy; members are only required to have "acknowledged moral personality." When Charlie Townes, the co-inventor of the laser, was becoming a member, a church official told him that members needed to have good moral character. "How do you know if I do?" Townes asked. The official's response: "We have our ways." Charles Seife, "Science and Religion Advance Together at Pontifical Academy," *Science* 291, no. 5508 (February 23, 2001): 1472–1474.

but had no boundary, which means that it had no beginning, no moment of creation," the physicist recalled in *A Brief History of Time*. "I had no desire to share the fate of Galileo, with whom I share a strong sense of identity, partly because of the coincidence of being born exactly 300 years after his death!"[1]

Hawking was joking, but at the Vatican meeting, the physicist proposed something so bold and radical—and controversial—that it might have made the pope rather uncomfortable. For the so-called no-boundary proposal, Hawking argued, dispensed with the need for a creator, for a prime mover who set the laws of the universe in motion. There was no need for a watchmaker to fashion the clockwork of the cosmos. There was no need for God.

Hawking thought the no-boundary proposal to be one of his greatest contributions to physics. Other scientists aren't quite so certain; even after decades of work on it, neither Hawking nor his collaborators managed to convince the wider community that it was worthy of much attention. But the no-boundary proposal had a certain elegance, a beauty that was itself convincing to those inclined to believe that the solutions to fundamental problems must themselves be beautiful—and radical.

• • •

In the early 1980s, Hawking was bursting with optimism about the field of cosmology. There were a bunch of brand-new ideas about the very early universe—about the first few fractions of a second after the Big Bang—and Hawking had been instrumental in giving them life. He was most excited about his own pet theory, the no-boundary proposal, even though it was viewed skeptically by most of his peers and almost impossible to explain for a lay audience. Yet Hawking decided to try. For Hawking was deciding to write a book about physics. He was bubbling with excitement about his cosmology and work on black holes, and he hoped that his enthusiasm would translate into a best seller.

Hawking had his eye on a much bigger audience than his Cambridge University connections could give him: he wanted to be widely read. More, he wanted to be understood. It was a drive almost as great as his drive to understand.

However, as Hawking knew all too well, time was not on his side; the mere effort of putting words down on paper was excruciatingly difficult for him. And in 1985, as he lingered in a coma, it seemed for a time that he would never get a chance to acquire the wider audience that he always wanted.

...

Of all of Hawking's work, the no-boundary proposal is perhaps the hardest for a layperson to grasp, even in thumbnail, which makes it all the more remarkable that it was so central to *A Brief History of Time*. It's wrapped up with the baffling idea of imaginary time and the counterintuitive mathematical formalism that physicist Richard Feynman used to understand the behavior of quantum-mechanical objects. It brushes up against philosophical problems that are swept under the rug in the ordinary thinking about quantum theory. Even the idea of the "boundary" in the no-boundary proposal is not quite what a non-physicist would think.

To a cosmologist trying to gain a deep understanding of the universe, it's not enough to know the laws of physics. An abstract collection of rules, important though those rules might be, doesn't necessarily tell us much about the universe we live in, any more than knowing the rules of chess tells you what a particular game is going to look like, or whether the person playing black or white is more likely to win. The rules of the game are only one part of a full mathematical model of the universe.

The other part is what physicists refer to as the *boundary conditions*, a description of a physical system at a given point in time. The laws of nature take the conditions at the boundary and transform them over time; the system evolves according to the rules of physics into something new. It's the combination of laws and boundary conditions that give physicists a mathematical model of our specific universe rather than an abstract Platonic set of equations that might or might not mean anything special.

For example, the laws of gravity apply to all sorts of different configurations of matter in a given solar system. It doesn't matter whether the system has one big planet, or seven thousand little ones, or is composed of nothing but dust; the fundamental laws are the same in each case. But if you want to ask a question about our solar system—something as simple as "Where am I now?"—there's no way for any mathematical model to yield an answer without boundary conditions: the positions of the Earth and the other bodies in the solar system at a given time. It could be the positions of those bodies a few seconds ago, in which case, the question would be really easy to answer. It could be the positions of those bodies a million years ago; it would take some effort to calculate the motions of all those bodies over the course of a million years, but with precise enough information and a careful enough calculation of the motions dictated by the laws of nature, physicists could derive the positions of the bodies in the solar system with high accuracy. The boundary conditions could even dictate the positions of those bodies a billion years in the future; the calculations work backward just as they do forward, so physicists could figure out where we are now based upon the boundary conditions a billion years hence. Once you have a mathematical model—laws plus boundary conditions—you can roll it forward or backward in time as much as you want, answering any sorts of questions you can ask of such a model. But the boundary conditions are as important to the model as the underlying physical laws are.

This poses a particular problem for cosmologists who are trying to come up with models of the very early universe. It's not just that they don't know the underlying physical laws that were operating in the very dense, hot conditions shortly after the Big Bang—that's the realm of quantum gravity, which hasn't been worked out yet. But even with a grand unified theory in hand (and, at the time, Hawking was fairly confident that physicists would build one by the turn of the century), that wouldn't be enough to build a mathematical model of the cosmos that, as Hawking put it, "would account for the whole universe at one go":

> It now seems possible that we might find a fully unified field theory within the not-too-distant future. However, we shall not have a complete model of the universe until we can say more about the boundary conditions than that they must be whatever would produce what we observe.[2]

Because cosmologists are looking at the universe as a whole, the boundary conditions are a description of the entire universe at a given moment, which is problematic, to say the least. Worse, if you want to grasp the ultimate prize—to model the very birth of the universe—you have to run your model backward in time to the Big Bang. But as Hawking himself proved early in his career, the Big Bang must have a singularity, a point where mathematics breaks down. You simply can't roll back time in your model in this way; to understand the Big Bang, you need the boundary conditions at the very beginning of the universe, which is precisely what you want the model to tell you in the first place! You're stuck in a logical loop, a chicken-and-egg problem that you can't get out of. Unless . . .

That "unless" is the reason for all the complicated mathematical and intellectual backflips that Hawking's no-boundary proposal requires—it's a high cost for a high payoff. If you can break the deadlock and somehow get a hold of the boundary conditions at the very start of the universe, you could conceivably come up with a mathematically complete model of the cosmos. At its core, that's what Hawking was attempting to achieve with the no-boundary proposal. "If you have a universe which lasts forever, you have the chicken-and-the-egg problem," says Neil Turok. "His no-boundary proposal had a chance of literally solving the problem and in a completely conclusive way. So it was very, very appealing." But the proposal came with a lot of baggage, to say the least.[3]

. . .

Perhaps the best place to start the story of the no-boundary proposal is in the late 1940s, when Stephen Hawking was only six years old. Richard Feynman was then a young physicist, having recently left bomb work at Los Alamos for more academic pursuits. And he was hard at work wrestling with the fundamental puzzles of quantum mechanics.

According to the rules of quantum theory, subatomic particles behave in ways that violate the laws of motion that we're used to—worse, they violate common sense. It's impossible for a brick or a cat to be in two or three or four places at the same time. However, protons and neutrons and electrons do this all the time. Indeed, the laws of physics describe such particles as being in a state of "superposition"—doing several mutually contradictory things simultaneously—as a matter of course. If you throw a Ping-Pong ball at a screen with two holes in it, the ball must go through the left hole or the right hole (or neither); it can't go through both at the same time. Shoot an electron at a barrier with two holes, and unless you explicitly prevent it from doing so, it will automatically zoom through both the left and the right at the same time.

As strange as this behavior might seem, the mathematics of quantum mechanics makes perfect sense. If you want to find out what happens to an electron being shot at a two-slit barrier, all you have to do is set up the equations of quantum theory in the right manner, and out pops the answer, encoded in a mathematical object known as a *wavefunction*. That wavefunction—so named because the quantum-theoretic equations treat objects almost like they're waves—contains all anyone could possibly want to know about the electron. (In fact, it contains *more* than anyone could possibly know; thanks to uncertainty, an observer won't be able to obtain all the information from the wavefunction, just some of it.)

That's all very nice and good for a physicist who needs to do calculations. The answer always comes out correct. But to Feynman, getting the correct answer wasn't sufficient. He wanted a deeper understanding of what was happening to the electron, and he didn't feel that he really understood something until he could visualize it. And he couldn't really visualize a wavefunction, whatever that wavelike object might be, whether it was passing through two slits or doing something more complicated. So he had to come up with another way of describing quantum objects, a way that could be seen in his mind's eye. As he told an interviewer,

What I am really trying to do is bring birth to clarity, which is really a half-assedly thought-out pictorial semi-vision thing. I would see the jiggle-jiggle-jiggle or the wiggle of the path. . . . It's all visual. It's hard to explain. . . . An inspired method of picturing, I guess.[4]

In 1948, Feynman created a new approach to describing quantum particles moving through force fields—one that gave the same answers as the standard equations of quantum theory, but used actual objects moving in classical ways under the hood rather than mysterious, unvisualizable wavefunctions.

To use Feynman's method to calculate the behavior of, say, an electron going through a two-slit barrier, one starts by considering every possible path that an electron can take. It might go through the left slit. It might go through the right slit. It might hit the barrier in the middle. But those are just the obvious paths; the Feynman method needs to consider *all* paths. It's possible that the electron will go in the right slit, make a U-turn, and come back out the left. Or it might go through the right, out through the left, do a loop-de-loop, and zoom through the left slit once more. Every single possible path, no matter how absurd, has to be considered; as physicists are wont to say, "everything that isn't forbidden is compulsory." Each of those possible paths makes a contribution to the final answer, lesser or greater depending on the likelihood that an electron will take a given route. When summed up, the total result is the same thing as what one would get by treating the electron as a wavefunction. As Feynman put it:

The electron does anything it likes. It just goes in any direction at any speed, forward or backward in time, however it likes, and then you add up the amplitudes and it gives you the wave function.[5]

But unlike in the standard way of solving quantum-theoretical equations, in Feynman's so-called *path integral* method one doesn't have to try to imagine the electron as some smeary semi-particle semi-wave. Instead, the electron behaves as a classical-ish particle ought, and it's easier to envision a large ensemble of these little electrons all taking their own paths than it is to imagine some kind of semi-wave semi-particle with counterintuitive properties trying to move through a barrier. Feynman's path integral method—and the closely related Feynman diagrams, which he used to visualize how particles interacted with one another—quickly became important tools to the study of subatomic particles and fields. And, thirty years later, more than that.

"The idea of the Feynman path integral, I know, was Stephen's driving intuition," says Turok. "You know, the one physicist who he really respected was Feynman. He was a hero to Stephen, as for many of us, but they resonated, particularly at Caltech. . . . So Stephen loved the Feynman path integral, and he thought it was the way to do gravity."[6]

That is, Hawking so loved the Feynman path integral framework that he sought to use it not just on the subatomic particles and fields that it was typically used for, but on the problems of cosmology that most interested him. To say that this was an ambitious plan is rather an understatement. Usually, the Feynman path integral method was used to calculate the wavefunction of a particle like an electron. Now Hawking was proposing to use it to calculate the wavefunction of the entire universe, the whole cosmos, in one go. And instead of trying to sum up every single path that, say, an electron might travel, applying the path integral method to the cosmos meant trying to envision every single possible evolution of the universe—*all* possible universes all at once—and somehow adding them together. But by considering the wavefunction of the entire universe, Hawking thought he could get new insight into the early universe, and even potentially break the chicken-and-egg problem of trying to figure out the boundary conditions of the early universe.

Plus, there were some good reasons for thinking that treating the early universe like a quantum-mechanical object with a wavefunction could yield some interesting insights. A wavefunction is a complicated object with a lot of components, but it's not so difficult to tease apart these components and study them separately to get a better understanding of the whole object. The simplest component of such a wavefunction is known as the *ground state*, and it was possible that understanding the ground state of the wavefunction of the universe could yield some profound insights. "The way I was thinking about it in those days, was that the evidence, the observations, were that the universe is a simpler place earlier than it is now," says James Hartle, a physicist at the University of California at Santa Barbara, who helped Hawking develop the no-boundary proposal in the early 1980s. "It was more homogeneous, more isotropic, more nearly thermal equilibrium. And it was, in some sense, completely smooth. Those are properties that characterize, typically, ordinary quantum mechanical systems when they're in their ground state, the lowest possible state in ordinary quantum mechanics." In other words, it's possible that the nature of the early universe could be understood by treating the universe as a wavefunction and calculating its ground state—and it might not be a coincidence that the early universe (at least as we understand it) looks remarkably like what the ground state of a wavefunction ought to look like. To Hawking, this sparked a great sense of excitement. But there was a major problem.[7]

Feynman's path integral method works great when it comes to particles and fields, but it blows up when it encounters spacetime.

The geometric properties of Einsteinian, Lorentzian spacetime are very different from those of the ordinary, Euclidean space that we're used to. The formula to measure distance in spacetime—

$$s = \sqrt{(x^2 + y^2 + z^2) - t^2}$$

—has a minus sign next to the time coordinate, t, and this means that the manifold of spacetime behaves in a way that defies ordinary geometric expectations. This sign difference forces a departure from the ordinary, Euclidean geometry that we're used to; spacetime is not a manifold whose geometry is Euclidean, but Lorentzian instead.

The Feynman path integral method works in a manifold that has a Euclidean geometry, but not in a Lorentzian one. "You can't do it for technical reasons," says Hartle. "But you could do it by Euclidean construction." That is, there's a mathematical trick to make the Lorentzian spacetime manifold of relativity look much more like an ordinary Euclidean geometric object, and when you use that trick, the Feynman path integral method suddenly begins to work.[8]

This mathematical trick is to introduce the concept of imaginary time. For a non-physicist, this idea is deeply, deeply weird to the point of incomprehensibility—and almost certainly it was the cause of numerous copies of A Brief History of Time being tossed across the room. "Imaginary time is a difficult concept to grasp," wrote Hawking, "and it is probably the one that has caused the greatest problems for readers of my book." To a physicist, however, there's nothing particularly strange about it, at least mathematically speaking.[9]

As those who took high school algebra might remember, there's a number known as "i"—the "imaginary" unit—that's defined as the square root of negative one. There's an interesting little trick in relativity that replaces the ordinary variable of time, t, with one that's closely related: $i \times \tau$—that is, a time variable τ multiplied by the imaginary unit, i. If you substitute that into the equations, the new distance formula looks like this:

$$s' = \sqrt{(x^2 + y^2 + z^2) - i^2\tau^2}$$

And since i is the square root of negative one, the i^2 kills the negative sign in front of the τ^2. That is, all of a sudden, the distance formula looks like this:

$$s' = \sqrt{x^2 + y^2 + z^2 + \tau^2}$$

The troublesome minus sign has disappeared—there's no hint of the strange Lorentzian geometry that plagues ordinary spacetime. It looks Euclidean, so long as we use imaginary time in our equations rather than time.

Mathematically, this is no big deal. In some sense, going from time to imaginary time is roughly equivalent to turning your head sideways—rotating your view by 90 degrees—in a complicated higher-dimensional space. (In fact, scientists call it a *Wick rotation* after the person who first used the technique, Italian physicist Gian-Carlo Wick.) This substitution-cum-rotation does, in fact, make the mathematics much, much easier. But philosophically, it means that the equations no longer describe the motions of objects in the universe in the ordinary way. Watching something moving "forward" in imaginary time doesn't have the same obvious interpretation as observing something evolving as ordinary time passes. "So let's give up on time being real. Let's pretend time is imaginary. And now everything's fine," says Turok. "The problem with it is you can never calculate anything in real time from this Wick-rotated picture, or it's very difficult to do that." Physicists are okay with this loss of interpretability if they gain more in understanding than they lose. For any non-physicist who wants to follow what's really going on, though, it's a huge barrier to comprehensibility.[10]

The Wick rotation takes a Lorentzian manifold of spacetime, with all its strange geometry and bizarre relativistic effects, and transforms it into a manifold of space-imaginary-time with a super-simple Euclidean geometry that's much easier to handle. For all the disadvantages that this device creates, it has a number of big advantages. For Hawking, the decisive one was that the Feynman path integral method—which simply didn't work in spacetime—was just fine in space-imaginary-time. Hawking could start calculating the wavefunction of the universe, so long as he could come up with a consistent way of describing all possible cosmoses to feed into the Feynman path integral machinery.

In 1981, when Hawking visited the Vatican, his thinking on the subject was still relatively new, and Hawking was operating mostly on intuition and hope. But when he was looking for a consistent description of cosmoses, his intuition was telling him something that was extraordinarily profound, if true. Hawking's visualizations of the cosmoses were telling him that *all* possible universes, viewed through the lens of imaginary time, were smooth, closed, finite, and "compact." They didn't have singularities, and didn't extend forever in any direction.* And one can prove, mathematically, that a universe with such properties would have no boundaries—neither in space nor in (imaginary) time. It would have no beginning and no end.

From a physicist's point of view, this notion destroys the problem of boundary conditions at the Big Bang entirely. There can't be boundary conditions to the universe because the universe has no boundaries! The Big Bang isn't really a boundary, an end, an "edge," of our cosmos any more than the South Pole is an "edge" of the Earth. And so in Hawking's no-boundary model, the laws of physics are enough to determine everything about the cosmos. As Hawking put it, "if the universe is in a no-boundary state, we could, in principle, determine completely how the universe should behave, up to the limits of the uncertainty principle."[11]

Hawking liked to put a theological spin on his idea of a universe with no boundaries—the idea of a universe with no beginning and no end, to him, seemed to obviate the need for a creator. In some sense, God was necessary to set the boundary conditions, to give the universe its beginning or bring it to an end. But a universe without a beginning or end would just be: it wouldn't need to be created or destroyed. Needless to say, Hawking's interpretation wasn't terribly popular with theologians.

The no-boundary proposal wasn't born as an attempt to banish God; it was a complicated way of applying a thirty-year-old technique for handling quantum systems to the universe as a whole. And, as one might expect, given that it's an attempt to sum up all possible universes, it rests on a whole bunch of assumptions and intuitions. Even after he worked with Hartle in the early 1980s to work out some of the kinks in the theory, and labored on the topic on and off in the decades after that, the no-boundary proposal wasn't embraced by most of the scientific community. "The idea has never really been accepted," Turok says. "I would say 90 percent of cosmologists or theoretical physicists who have an opinion . . . most people don't even form an opinion; of those who do, 90 percent of them would say they probably didn't agree with it, or they thought there was some problem with it."[12]

---

* This might seem like it's a direct contradiction to what we know—indeed, what Hawking himself proved—in that the universe has to begin with a singularity. But a no-boundary universe could still be singularity-free: what would appear as a singularity to a person living in spacetime would not be singular when looked at with the mathematics of space-imaginary-time.

To Hawking, however, it was a central pillar of his legacy. And it was a potent metaphor. Quite naturally, the last chapter of his last book—his autobiography—was entitled "No Boundaries."[13]

...

In the early 1980s, Stephen Hawking's voice is distinctly baritone, but it's almost impossible to understand what he is saying. His larynx and tongue and lips still have enough control to make sounds that come out with the pacing and the structure of ordinary speech, but it doesn't sound like language at all. He makes a creaky, keening sound with a guttural rumble behind it. It sounds almost as if a great cat were trying—with infinite patience—to explain something in English to an uncomprehending audience. So, every few seconds, he has to break off as one of his inner circle, such as his graduate assistant or someone else who can interpret his utterances, translates his words.

Even so, his lectures and chats are frequently punctuated by peals of laughter. Hawking was somehow able to display his wicked sense of humor in his talks, just as he could in his writing, even though he had to speak through a translator, or later, a voicebox, which nearly obliterated any exercise of comedic timing. "Somehow it worked for him," explains John Preskill. "Often, he'd make some sort of comeback, and it wouldn't be immediate; you'd know it was coming up . . . somehow, for him, the delay was part of what made it funny. Sometimes, he'd immediately say 'Rubbish!' That was one of the ways he would disagree with you. Very British, you know. You'd say something, and he'd say 'Rubbish!'" Even though people tended to treat Hawking with deference, Preskill, by his own account, treated him in a somewhat snide way. "I'd make fun of him, and he'd like this," Preskill recalls. "Whenever we'd get together, I'd be irreverent around him, and when he would make a comment, I'd say, 'What makes you so sure about that, Mr. Big Shot?' and that kind of thing. And he really liked that."

And by the early 1980s, Hawking was a big shot, and to a greater and greater extent, not just in the physics circles he moved in. The press was beginning to notice the young scientist, just entering his forties, who was so highly regarded by his peers, and who had such a unique and heartrending backstory.

Even in the early 1980s, Hawking was no stranger to the media: science journalists had quoted him in newspapers and magazines quite a few times over the years as an expert on black holes. But around that time, reporters' interest in Hawking was beginning to shift from Hawking as a subject-matter expert to Hawking the human being—he was no longer being used to comment on a story, but instead becoming the story himself.

In 1981, the *New York Times* ran a piece by journalist Malcolm Browne titled "Does Sickness Have Its Virtues?" The article clearly articulated one major leitmotif in the coverage to come:

> Even the gravest physical illness seems sometimes to be highly compatible with scientific creativity. Since the early 1960's, the British physicist Stephen Hawking, now 39 years old, has been progressively crippled and otherwise incapacitated by amyotrophic lateral sclerosis. But during the same period, Hawking's mathematical analyses of space and time in the vicinity of black holes have earned acclaim by colleagues as the work of rare and authentic genius.
>
> In fact, sickness of one form or another seems to be fairly common at the pinnacle of scientific creativity, as it has been for such artistic luminaries as Dostoyevsky, Proust, Van Gogh, and Berlioz.[14]

When Hawking's face with its "mild cetacean smile" graced the cover of the *New York Times Magazine* in early 1983, readers were treated to a heavy dose of this physical-infirmity-begets-genius theme:

> The art [of theoretical physics] demands an exceptional ability to concentrate, to remember, to make connections between ideas. It is perhaps significant, then, that Stephen W. Hawking, a physicist whose insights about gravity and matter are changing the way we look at the universe, should have attained his intellectual stature while his body was failing him, atrophying, shaping him increasingly into a cerebral being.[15]

And this was after Hawking objected, after seeing a draft of the magazine article, that the reporter had "made too much of his physical condition." Yet as much as Hawking wanted journalists to focus on his work and not his disease, the disease always would be front and center to any reporter trying to tell a compelling story. In 1983, when the BBC science show *Horizon* ran a forty-five-minute profile entitled "Professor Hawking's Universe," it begins with Hawking's death sentence ("Twenty years ago, Stephen Hawking, a young research student at Cambridge University, began to show the first symptoms of an incurable disease and he was told it might kill him within a few years") and ends with a *memento mori* (Hawking, gazing upon an empty space where Cambridge will eventually hang his portrait: "It's like seeing your own tombstone").[16]

As galling as it was to Hawking to be treated not as a human but as a metaphor for transcending humanity, he welcomed—and enjoyed—the press attention. His stock was rising higher and higher not just within the cloister of physicists, but with the public at large. It was nearly time to sell.

· · ·

Lucy Hawking was just turning twelve, and was graduating from Newnham Croft, the local elementary school. She had gotten in to Perse Girls, a well-regarded school—but it was one the Hawkings would have to pay for. Thus was born the idea for *A Brief History of Time*. "I first had the idea in 1982 of writing a popular book about the universe," Stephen later wrote. "The intention was partly to earn money to pay my daughter's school fees."[17]

While this might have been the main impetus for Hawking to attempt to write a popular book, it was not his entire motivation. For Hawking wanted to be well known outside of his community. He was not one of those scientists who eschewed (or actively scorned) popularizers. In fact, he wanted to be one himself. Over the years, he had tried his hand at writing a few articles for popular consumption in magazines such as *Scientific American* and *New Scientist*. The time seemed ripe for a more ambitious attempt. Carl Sagan's *Cosmos* had shown that the market was there, and Hawking thought he might be able to leverage his increasing popularity in the press to reach a wider audience than before. Hawking later wrote that the impulse for writing a book came from the message he wanted to convey: "The main reason was that I wanted to explain how far we had come in our understanding of the universe. How we might be near finding a complete theory that would describe the universe and everything in it."[18]

The year 1982 was a high point in Hawking's excitement about cosmology. There was a brand-new theory, inflation, that seemed to explain what had happened in the first few moments after the Big Bang; a year prior, Hawking was at the center of an attempt to work out the kinks in the idea. (More about this in the next chapter.) And, of course, after a number of years of research, he had just unveiled his no-boundary theorem; in 1982, Hawking was busy hammering out the details with Jim Hartle. And in his boundless enthusiasm, Hawking—echoing the gospels—was telling his audiences that some of those present would likely live to see a theory of everything.* It was almost inevitable that Hawking would start writing a book to spread this excitement—and these ideas—to a wider audience.[19]

Hawking had published a few books before; they were academic works intended for a technical audience published with Cambridge University Press. Like most books published by academic presses, they were never expected to make much money for the publisher, much less for the author. However, there are always exceptions, and in the mid-1970s, Hawking and his friend George Ellis had written a very technical treatise explaining some of the latest developments in relativity theory from base principles, and it had been very well received by the community. "Now that had done incredibly well," says Simon Mitton, who was then the director of science publishing at Cambridge University Press. "And it meant my colleagues at the press both in Cambridge and New York were very anxious. 'Simon, can you get Stephen to do something for us?'"[20]

Mitton, himself an astronomer, had been the departmental secretary of Cambridge's Institute of Theoretical Astronomy when Hawking worked there half time around a decade prior; he had gotten to know Hawking quite well, as his departmental duties had included looking after Hawking's needs. The two remained in contact after Mitton accepted his new position at Cambridge University Press. "I got on with him very well when I had responsibilities for him at the Institute of Astronomy. And I got on with him well when he consulted me about publishing matters," Mitton says. "Stephen's secretary, Judy Fella, contacted me. And she said, 'Stephen would like to talk to you about a typescript. I've just finished typing it, and it's for a popular book about the universe.'"

It was the fall of 1983, and Hawking had been working on the typescript for months—partially cobbled together from previous lectures, partially novel material—and he was eager for Mitton to take a look. "Before I started reading it, I said to him, 'Stephen, what's your motivation

---

* Mark 13:30: "Verily I say unto you, that this generation shall not pass, till all these things be done" (KJV). Unlike Jesus, Hawking had to temper his optimism a bit as time passed. In 1998, he said, "What are the prospects that we will discover this complete theory in the next millennium? I would say they were very good but then I'm an optimist. In 1980 I said I thought there was a 50-50 chance that we would discover a complete unified theory in the next twenty years. . . . Nevertheless I am confident we will discover it by the end of the 21st century and probably much sooner. I would take a bet at 50-50 odds that it will be within twenty years starting now." "Remarks by Stephen Hawking," White House Millennium Council, 2000, https://clintonwhitehouse3.archives.gov/Initiatives/Millennium/shawking.html.

for writing this?'" Mitton says. "His response was moving, but also cli-chéd. He said that his motivation for writing the book was security for his family, and particularly the education of his three children." That wasn't a realistic goal for a book published with a university press; advances were modest and sales generally were, too. Even in the trade press, it's rare for a first-time author to make enough money to provide security for his or her family. When Mitton explained that this wouldn't be likely, at least with Cambridge University Press, "[Stephen] said, 'Well, please do me a favor and read it and give me some feedback, and together we can decide what to do next.'"[21]

Mitton was eager for a book by Hawking, and after digging into the manuscript, he thought the general idea of the book was good. Unfortunately, he found the draft far too technical for a popular audience. Not many years prior, Mitton had convinced another eminent cosmologist, Paul Davies, to write a cosmology book for the general audience. Though Mitton had been very excited about the book's prospects (and had come up with a striking title: *The Accidental Universe*), the press's marketing director had thrown cold water on the project. "He said, 'Simon, this book is unsellable. I hope you're not printing too many copies,'" Mitton recalls. "I said, 'What do you mean? Paul Davies is world famous! You can ship them by the bushel.' He said, 'No we can't. There are equations on dozens and dozens of pages and every one of those halves the market.'" The marketing director was right; when the book came out in 1982, sales were poor. The last thing Mitton wanted was a repeat performance.[22]

So his feedback for Hawking was rather disappointing. Hawking later told *Time* magazine, "Someone told me that each equation I included in the book would halve the sales." Mitton recalls that Hawking was resistant to making the book less technical, at least at first: "His initial response was, 'Well, you say that, Simon, but I've only used the kind of mathematics which seventeen- and eighteen-year-olds have to do in order to satisfy matriculation and university entrance.'"[23]

Despite his cavils, though, Mitton was convinced he could turn the book into a big seller, and he approached the powers that be at the university to get permission to offer Hawking a contract. "They agreed to the largest advance that I'd ever paid out to that time. . . . They agreed on £10,000, half on signature, half on delivery," Mitton says. "I knew that wouldn't be enough to satisfy him, but he was my friend, and I wanted to demonstrate to him that I was passionate about the book, and I did the best I could." Unfortunately for Mitton, as Hawking no doubt knew, he could get much more from a trade publisher. For a bit more than a year earlier, in late 1982, Hawking had found a literary agent. Or, more precisely, a literary agent had found him.

Daniel Z. Freedman was an esteemed physicist and one of the early pioneers of supergravity—at the time, what Hawking considered to be the leading candidate for a theory of everything—and he had learned that Hawking was thinking about writing a popular book. Freedman's sister was married to a high-powered literary agent, Al Zuckerman. "Danny asked if I would like to talk to Hawking, and I said sure, and I went to Cambridge," Zuckerman says. By the time Hawking approached Mitton, Zuckerman had already convinced Hawking that he would be able to get a sizable advance for the book.[24]

Zuckerman had very good reason to think so. Back in January 1983—in an odd coincidence—Zuckerman had had a lunch appointment with an editor at the Bantam publishing house, Peter Guzzardi. In Guzzardi's bag was a copy of *The New York Times Magazine*, the one with the image of Stephen Hawking on the cover. "It was just a wonderful article so beautifully written. And I was just struck by it to the point where I was carrying it around in my bag," Guzzardi recalls. "And it was a combination of the ambition [the author]* conveyed, the ambition of what Stephen was doing professionally, this quest he was on, and the contrast with his being trapped in his body, and that just struck a nerve with me. . . . Here's this guy exploring the outer limits of the cosmos and he can't tie his shoes." When Guzzardi mentioned the article to Zuckerman, "Al Zuckerman said to me, 'Hey, it's amazing that you should mention this because I'm actually on it.'"[25]

It took months of work, but by the end of 1983, the proposal, which included Hawking's one-hundred-page typescript, was ready. Zuckerman started approaching publishers, including Bantam. Bizarrely—and perhaps absentmindedly—Zuckerman didn't send the proposal to Guzzardi, despite their earlier conversation. "He must have forgotten about that because when the time came, he sent the proposal to another editor at Bantam, Jeanne Bernkopf, and she was a brilliant editor," Guzzardi says. Bernkopf remembered that Guzzardi was interested in Hawking and passed the proposal on to him rather than taking it to the editorial board herself. "It was an act of integrity and generosity and kindness that was

---

* Guzzardi described the *New York Times Magazine* article as having been written by Timothy Ferris, and even described the imagery which so struck him—that the physicist wore shoes whose soles were utterly pristine. However, that particular article appeared in *Vanity Fair* in mid-1984. Given the timing, however, it seems that Guzzardi's recollection of carrying the *New York Times Magazine* from January 1983 in his bag was correct; he just conflated its content with the later Timothy Ferris article. Such are the vagaries of memory.

rather remarkable in that kind of competitive environment," Guzzardi says. So Guzzardi would try to acquire the book for Bantam. However, Bantam wasn't the only publisher out there.

"In New York practically every publisher in town was interested in the book," Zuckerman told a trade journal shortly after *A Brief History of Time* was published. "I had six-figure offers from about six." Bantam came in at $225,000—a very sizable advance for its day, beating the Cambridge University Press offer by more than an order of magnitude. "Bantam was hungry for prestige. They had a prestigious success with [Chrysler CEO] Lee Iacocca's memoirs, and they liked the taste of that," Guzzardi says, adding that he hoped a Hawking success could show that Bantam could "play with the big boys." But, as it turned out, Bantam wasn't the highest bidder for Hawking's book.[26]

W. W. Norton was a publishing house best known for its anthologies, textbooks, and other academic-type books, but it had a few highbrow nonfiction successes as well. In its pipeline was a best-seller-to-be called *Surely You're Joking, Mr. Feynman*, a collection of anecdotes about Feynman's life and work. If Norton had grabbed the Hawking contract, it would have become the go-to house for scientists seeking to popularize their work. But even though Norton had tendered the highest bid, Hawking wasn't bound to accept its offer. He could choose whichever publishing house suited him best.

Guzzardi thinks that one sentence he put in a letter to Hawking clinched the decision. "He was always hungry, and was drawn to Bantam because of the promise I made in a letter that we could distribute the book in ways that other people couldn't at that point, which was true," Guzzardi says. "The sentence that we could put the book in airport bookshops was really the one that most captured his imagination." It was a technical advantage that Bantam had in those days; the publisher distributed its wares through paperback wholesalers that could put books into drugstores and supermarkets and airport bookshops. "And they could use that leverage to put their hardcover books there, too, when appropriate, which is not often but was certainly the case with Iacocca, for example." That clinched the deal. Bantam would be the publisher, at least in the United States and Canada, and Zuckerman eagerly set out to secure contracts for foreign and translated rights. Even if the book didn't sell well, Hawking stood to make a pretty decent profit from the advance alone.[27]

Mitton wasn't insulted when he found out that his friend and colleague had signed with a big publishing house; after all, good old Cambridge University Press couldn't come close to offering the sort of money or marketing or distribution that Bantam would be able to put behind Hawking's work. But Mitton gave Hawking a few words of wisdom:

> "Do be careful if you're dealing with those people, Stephen," he had said. "Do ensure that you are quite certain that, if the aim is to make money and sell lots and lots of books, you don't mind the marketing techniques."
>
> "What do you mean?" Hawking had asked.
>
> "Well, I wouldn't put it past them to market it as 'Aren't cripples marvelous?' You've got to go into it with your eyes open. If you don't mind that approach, okay."[28]

• • •

In addition to the book writing and media attention, Hawking was spending a disproportionate amount of time being showered with honors (including being named a Commander of the British Empire by the queen), traveling, and fighting for disability rights. Stephen and Jane (and Jonathan Hellyer Jones) helped the newly founded Motor Neurone Disease Association in its fundraising efforts and fought for access to services for the disabled; when one of the colleges at Cambridge planned to put in a library without provisions for wheelchair access, the Hawkings raised a public stink. With all this activity, it's a wonder that Hawking was able to get any physics work done at all.[29]

However, the distractions from work were welcome, at least to some extent. For all his enthusiasm about the no-boundary theorem, Hawking was worried that his great scientific achievements were behind him. "As far as theoretical physics is concerned, I'm already over the hill. Actually, quite far over the hill," Hawking told a reporter in 1982. "Well, you know most of the best work in theoretical physics is done by people at a very early age—usually by people in their twenties. So being forty is not a stage in life where one expects to make great discoveries in theoretical physics." Luckily for Hawking, he had his pick of the brightest twenty-something-year-olds that Cambridge had to offer—young physicists in training whom he could shape, and who could, in turn, shape his thinking and help him with his work.[30]

Ray Laflamme won Hawking as a mentor in 1984 thanks to his performance on the famously brutal Mathematical Tripos test. Not only was the Tripos long (administered over several days) and fiendishly difficult, but it was pretty much the only thing that a student was judged by. "I went to the exam and said, 'Shit, this is damn hard,'" Laflamme recalls. "Some questions I just have no clue how to answer them." And one tradition that Laflamme found particularly hard to swallow was that the results were announced publicly, at the Senate House, by a professor—to a gathered crowd, the prof would read out a list of names of people who had passed and who had failed. "To me, this is really barbaric," Laflamme says. "I cannot go there. I cannot go with my peers and sit and listen to this." So Laflamme skipped the readings. When he went to the department in the morning, he was pleasantly surprised to find out that there was a note for the people who had passed with distinction to go speak to the head of the group—and his name was among them.[31]

Laflamme duly knocked on the department head's door. "He said, 'Professor Hawking is waiting for you downstairs.' I was thinking, 'Is this real? Am I in a dream?' And I would pinch myself, just kind of try to wake up. Okay, this is really a fantastic dream, but it cannot be true." But Hawking was indeed there, along with a nurse, waiting. "'You'll be my student and tell me what you're up to.' So he asked me what I was going to do for the summer," Laflamme recalls. "He must have thought that I was a completely dumb idiot because I had a problem putting three words together and being coherent. Afterward, I learned that people in front of him would kind of freeze." Hawking gave Laflamme a tremendous amount of reading to do over the summer, and in the fall of 1984, Laflamme began working on the no-boundary proposal and the wavefunction of the universe.

"There wasn't much of a separation between his personal life and the students," says Raphael Bousso, who worked with Hawking a decade later. "He was very gregarious, and he kept us much closer than I think most advisers keep their students." And before 1985, when Hawking's tracheostomy forced him to hire nurses to watch him around the clock, some of his students would take on nursing duties—helping their adviser bathe and urinate and dress was as much a part of their lives as working on physics problems. "We would help him go to the bathroom and feed him and do things which makes a relationship with your [mentor] totally different," Laflamme says. "I depend on my students' brains, not physically. But Stephen was very strongly dependent on his students." Yet the students whom Hawking depended upon never resented their adviser or begrudged him his needs—quite the opposite. They seemed to treasure the intimacy that their unusual relationship with their adviser brought.[32]

Laflamme, for one, particularly enjoyed spending time with Hawking at airports. Travel for Hawking was quite a production—carting baggage, assembling and disassembling the wheelchair, carrying Stephen on board and dealing with his demands—which brought Jane each time to near her breaking point.

Laflamme also suffered from the drama of travel with the Hawkings, but from his point of view, not all of it could be laid at Stephen's feet. "I remember I'm traveling somewhere. And getting Stephen up in the morning, dressing him, giving him breakfast, and making ready to go somewhere, which took up a lot of time," he says. "And I remember sitting in a car waiting for Jane to show up. And saying, 'I applied to be a student in theoretical physics. I didn't apply to be part of a Hollywood drama.' But I could imagine it was damn hard for her." But to Laflamme, the rigors of travel were well worth it, for they provided an opportunity to have uninterrupted time with a great mind. "Just waiting for a flight turned out to be these magic moments for me. Because he was stuck there," Laflamme says. "So we're just waiting. And that's often where I would ask him questions. 'Why Stephen, do you think this? Like, why do you think summing over Euclidean geometry is the right thing to do for the wavefunction of the universe, like, where does that come from?'"[33]

And when a student shared Hawking's irreverent sense of humor—and fondness for pranks—road trips occasionally were punctuated by surreal moments of comedy. During one trip, Hawking told Laflamme that he had to go to the bathroom, so they wheeled over to the men's room—but it was closed for cleaning. Laflamme asked the physicist what he wanted to do. "Then he looked at the women's bathroom. So I said, 'You want to go there? Okay, fine.' So we go in the women's bathroom," Laflamme remembers. And so the pair go in, and the bathroom is empty. Laflamme pulls out a receptacle for urine and begins preparing Hawking for the procedure. "And so two women come straight in and he's in a wheelchair, his fly open, me with the bottle, and look at him. And he has this big smile on his face." Before the women could register their offense, Laflamme quickly turned toward them. "And I said, 'If you want to take my place, feel free!' And Steve had this big grin on his face. He was probably happy that the men's bathroom was busy; he thought it was a very good joke."[34]

Despite his health, Hawking traveled widely and often, with trips timed to take advantage of an academic schedule. Christmas breaks, spring and fall breaks, and summers were chock-full of travel, to the point where Jane—who was attempting to overcome a phobia of flying—complained that "travel had become an obsession with him and he regularly seemed to spend more time in the air than he did on the ground." Summer 1984: Chicago. Fall 1984: Moscow. Spring 1985: a tour of China, during which Stephen managed to drive his wheelchair on the Great Wall. Summer 1985: Geneva.[35]

Travel was central to how Hawking did physics; it was at scientific conferences that he and his colleagues forged collaborations, identified big problems to tackle, and even occasionally solved those problems. Travel tilled the intellectual soil, mixing ideas from the best minds from all over the world. Different scientific groups had different cultures, different strengths and weaknesses, and different philosophies. Over and over throughout Hawking's career, his thinking was shaped by the people whom he met at meetings and conferences—Kip Thorne and John Preskill and the rest of the Caltech crew, Andy Strominger and Sid Coleman in Massachusetts, Andrei Linde and Alexei Starobinsky and Yakov Zel'dovich in the Soviet Union. For physics, even theoretical physics, is a deeply social pursuit.

Of course, Hawking's time away from Cambridge wasn't entirely devoted to physics. In the summer of 1985, he took Laflamme to spend some time in Geneva, Switzerland—a short drive from CERN, the European center for high-energy physics. The pair were, in fact, working on a physics problem together, but it had to do with Hawking's wavefunction-of-the-universe work, which had almost nothing in common with the work that was going on at CERN. In truth, Laflamme suggests, Hawking was using Geneva as an excuse to set up shop on the European continent to satisfy his craving to listen to Wagnerian opera at the annual Bayreuth Festival, the mecca for Wagner fans. But the pair did, in fact, have physics to do first.

One of the main features of Hawking's no-boundary proposal was the underlying principle that any possible universe was compact, that it couldn't extend without limit in time or in space. At the time (until Turok's work a decade later, which gave a way out), Hawking thought this meant that the universe had to end in a Big Crunch: a reverse Big Bang. Just as a Big Bang is a birth of a universe from a point, a giant explosion that expands rapidly, a Big Crunch is the death of the cosmos, a rapid contraction of spacetime to a point and then to nothing.

To Hawking, this presented an interesting dilemma. The Big Bang takes our universe from a relatively simple state—something that looks roughly the same everywhere—to something highly disordered and lumpy and inhomogeneous, with stars and galaxies and clumps of gas scattered through vast tracts of vacuum. In physics terms, the entropy of the universe increases as our universe expands. (More on the concept of entropy in Chapter 13.) In fact, this increase in disorder is the main way that we can tell which way time is flowing: if we see a film where shards of glass suddenly assemble themselves into a vase, or cream suddenly separates itself from coffee due to the stirring of a spoon, we instantly know that the film is being played backward, because self-assembling vases and self-separating creamers require entropy to decrease rather than increase.

But if the universe stops expanding and eventually contracts into a Big Crunch, wouldn't that mean that the disorder of the universe would stop increasing? Right before the universe ends, it would have to wind up in a state as simple as it was right after the universe began—the entropy of the universe would have to decrease during the contraction, not increase. And if this is the case, wouldn't the arrow of time change direction? Would the universe run backward like a film run in reverse? Would dry bones in their graves suddenly start growing tendons and skin, and upon being lifted out of the dirt by backward-walking pallbearers, be infused once again with the breath of life? At the time, Hawking thought this was the only logical conclusion, despite the seeming paradoxes it might cause. As he wrote in his earliest draft of *Brief History*:

> People living in the contracting phase would measure time in the opposite direction to that we do and so they would also think that the Universe was expanding. One might ask what would happen to someone who was sufficiently long-lived that he survived from the expanding phase to the contracting one? Wouldn't he see jugs of water picking themselves off the floor and observe the universe to be contracting? The answer must be that he could not survive.[36]

That is, yes, jugs would leap off the floor and people would be coming back from the dead. But all intelligent life would be wiped out at the moment the universe begins to contract, ensuring that nobody would be around to perceive an apparent contradiction.

That was his intuition, anyway. Hawking hadn't yet done the detailed calculations to determine whether the arrow of time would, in fact, reverse like he thought it would. And that's where Laflamme came in. Sharing a small office in Geneva with Hawking, Laflamme was trying to work out the mathematics of what happens to entropy in a collapsing universe for Hawking. "And he really wanted to know the answer, and it was not going his way," Laflamme says. "And every half hour I would hear, 'Do you have an answer yet?' Not exactly the most relaxing. I was so tired. After two days, I was like a zombie walking around."[37]

In the days before his tracheostomy, Hawking didn't need round-the-clock nursing to keep him alive. But he still needed attendants to be around him twenty-four hours a day, because he was all but immobile in his wheelchair. And that duty fell to Hawking's nurse, assistant, student, or wife, depending on who was around at the time. Jane hadn't accompanied Stephen to Geneva; she was driving and camping through Belgium and Germany with Jonathan, her daughter Lucy, and her six-year-old son

Timothy, with plans to meet up with Stephen at Bayreuth later in the summer. For her, it was a welcome respite from caring for her husband, and a rare opportunity to let the "poor, sickly plant of [her and Jonathan's] relationship . . . come out into the open for an airing and to blossom," even though it was "watered with tears of tension and guilt."[38]

It was a heavy burden to care for Stephen, as Laflamme was quick to admit. After two exhausting days in Geneva, the weekend had finally arrived. "I can sleep late and just relax," Laflamme says, smiling. "But Stephen wants to go and look at the mountains. He wants to go downtown. Here goes the slave driver." So Laflamme and Hawking's nurse and his assistant piled into the car and drove to downtown Geneva, where Hawking decided he wanted to visit a record shop. After lifting Hawking up the stairs to get into the shop, they realized Hawking couldn't see the records from his wheelchair; all the displays were for people standing up. So Laflamme and the others had to pull down large bunches of records and show them to him one by one, waiting for him to say yes or no before flipping to the next one. It was an exhausting process, and then, after lifting him out of his wheelchair, getting him down the stairs, and positioning him in the wheelchair again, Laflamme was ready to collapse.[39]

"He's saying something. The nurse and the secretary have gone somewhere, so I'm on my own with him. And I cannot understand what he's telling me. I hear, 'Put me off,'" Laflamme recalls. "I said, 'Stephen, what are you saying?' He said, 'Put me off.'" After a few more repetitions, Hawking is getting extremely frustrated, and Laflamme is getting flustered, not knowing what to do. "I look at Stephen, and I said, 'I'm going to walk to the corner of the street, and I'm gonna come back. And by that time, I will have relaxed and hopefully I can understand you." Laflamme takes a deep breath and walks to the end of the block and back. "I can see smoke coming out of his ears," Laflamme says. "'Put me off!' I said, 'Yeah, you're telling me, 'put me off.' But then I see his eyes are turning, and he's looking at his hand. He wanted me to turn off the wheelchair—'Turn me off'—so that he could put his hand in a better way on the wheelchair. Because if he was trying to move his hand [while the wheelchair was on], the wheelchair would move." Everything that Hawking wanted to do, no matter how seemingly simple, meant that the people around him had to exhaust themselves overcoming barrier after barrier that stood in between the physicist and his desire. And Hawking was well aware of the burden that he placed upon those he knew and loved. By nature, Hawking was a fiercely independent man.

Hawking had been suffering from a cough, and that evening, the cough worsened. This, by itself, was not unusual; people with ALS often have problems keeping their airways clear as their chest and throat muscles deteriorate. Hawking had frequent coughing fits and occasional bouts of pneumonia. But this time was different. Hawking was turning blue, and Laflamme and the nurse tried to convince him to see a doctor. Hawking refused. "He said, 'I think I'll be okay.' I said, 'Stephen, I think you need to see a doctor,'" says Laflamme. "And he might have mentioned that he didn't want to waste anybody's time. 'It's just going to take an hour. And if you're okay, we're all good. If you're not okay, then there will be somebody.' And fortunately, he said yes." So at one in the morning, Laflamme—the only member of the party who spoke French—found a doctor willing to make an immediate house call. When the doctor saw Hawking, she was shocked. "And then she came back to the kitchen and said, if he doesn't go to the hospital, he will not make the night."

So they hustle Hawking into the car and drive wildly through the streets of Geneva, Laflamme behind the wheel and the nurse in the back seat trying to navigate with a map, and eventually make it to the hospital, where Hawking is immediately admitted. And then there they waited, not knowing what was going on. Sometime several hours later, Laflamme remembers, a doctor came in, a bit out of breath. "She said, 'I had to do it . . . to intubate him. He stopped breathing.'" In order to get air into Hawking's lungs, the doctors had to anesthetize him and put a tube down his trachea. He would remain unconscious so long as the tube was in place. It wasn't at all clear that he would ever regain consciousness.

According to her memoir, Jane was at a campsite in Germany and called Geneva to check in. She had no idea anything was wrong until she reached Hawking's assistant: "'Oh, Jane, thank goodness you called,' she almost shouted down the phone. 'You must come quickly, Stephen is in a coma in hospital in Geneva and we don't know how long he'll live,'" Jane writes. "The news was shattering. It plummeted me into a black pit of anxiety, misery and above all, guilt."[40]

Two days later, a doctor approached Jane with a terrible choice. "The question that he wanted to ask me was whether his staff should disconnect the ventilator while Stephen was in a drugged state or should they try to bring him round from the anaesthetic." To Jane, there was no question whatsoever; they must try to save her husband. But if Stephen survived, he would never be the same. To allow him to breathe, they would have to give him a tracheostomy, inserting a tube through an opening in the front of his throat, below the larynx. To keep him from choking to death, Stephen would require constant attention from nurses who could suction and maintain the tube, and keep him from acquiring a life-threatening infection. And since air would no longer be flowing past his larynx or into his mouth, he would no longer be able to talk.[41]

. . .

Nearly thirty years after his tracheostomy, Hawking told the BBC that the operation had driven him to attempt suicide. But without assistance, someone in his condition would be hard-pressed to figure out how. He tried holding his breath, he said. "However, the reflex to breathe was too strong."[42]

The physicist was almost entirely locked in. Unable to write, unable to talk, barely able to move, and entirely dependent on people whom he couldn't communicate with, in the early fall of 1985 Hawking was in a desperate state. Jane tried to take care of the immediate financial issues; she instructed his student Brian Whitt to work on revisions to the *Brief History* manuscript. On the advice of Kip Thorne, she applied to the Mac-Arthur Foundation to help get funding for the new, expensive nursing care Stephen needed. (There was little alternative because the NHS was so ill-suited to handle a patient like Stephen outside of an institutional setting.) But even these matters paled by comparison with Stephen's need to figure out a means of connecting to the world outside his head.[43]

At first, Hawking could communicate only by means of an alphabet card. An assistant would point to the letters and Hawking would raise his eyebrow to indicate which one was correct. However, this was such a slow means of communication that it would functionally mean the end of his career; there was no way that he could do physics or write books or give lectures at a pace of a word every two or three minutes, using the full capacities of two adults the whole while. Little better was a primitive buzzer system by which the physicist could indicate one from a small number of words or commands; this was suitable to indicate his immediate needs, such as whether he needed his limbs to be moved or something similar, but utterly useless beyond that. There had to be another solution.[44]

Half a world away, a Silicon Valley engineer, Walt Woltosz, had designed a communications system for his mother-in-law, Lucille, who had suffered from ALS and died in 1981. A colleague of Hawking's heard about the project and rang Woltosz up. "I got a call from a physicist, and he said 'I know you're working on computer systems for people with ALS, and I've got someone in England, he's a professor of physics who lost the ability to speak, and he needs a system,'" Woltosz later said in a video. The system, dubbed the Equalizer, allowed the user to apply very simple inputs—such as a button press or a click of a switch—to select elements displayed on a set of menus. It was laid out so that he could quickly indicate a number of frequently used words, or, if need be, to spell out words letter by letter. He could store sentences for later use or recall them when required, and the program would even attempt to reduce the number of clicks needed by anticipating his next word choice, similar in many ways to the predictive-text algorithms in modern cellphones. And then, on command, the program could send the sentence to a speech synthesizer. Once Woltosz discovered that the person in need was Stephen Hawking, he sent over the necessary equipment—an Apple-II computer linked to a hardware speech synthesizer made by the Speech Plus company—and Hawking took to it like a drowning man takes to a bit of flotsam. It was nothing short of liberation for the professor. His only complaint, which he frequently joked about, was that the speech synthesizer had an American accent.[45]

The computer system was big and bulky, far too large to move around with Hawking and his wheelchair. But, outside the Newnham Croft School, which Timothy Hawking had started attending around the time his sister graduated, Stephen had a lucky run-in. There was an engineer whose children also attended the school, and Stephen asked him whether he could help adapt the Equalizer system to fit on a motorized wheelchair. The engineer—David Mason—could, and did, and soon became the UK distributor for Woltosz' system. What's more, David's wife of ten years, Elaine, just happened to be a nurse.[46]

When she met Stephen, Elaine was freshly back in the work world (too soon for her taste) after taking a hiatus in her career as a nurse midwife to raise her two young children. After a bit more than a year working at Addenbrooke's Hospital in Cambridge, the opportunity to work with Stephen proved too tempting to pass up, and she soon became a mainstay of the nursing rotation. (Jane was having difficulty finding nurses and keeping them around.)[47]

Despite three difficult months of hospitalization, the loss of the ability to speak, and the financial strain that new nursing requirements put on the household, to all appearances the Hawking family was actually better off than it had been before the tracheostomy. Jane was no longer bearing the majority of the burden for Stephen's care; that role fell upon the nurses. And Stephen no longer needed a translator to be understood; paradoxically, having lost the facility for speech, he was able to communicate more efficiently and more quickly than he had in years, thanks to his new computer system. The difference was the most profound to Timothy. "The first four or five years of my life, my dad was able to speak with his natural voice, but it was very, very difficult to understand what he was saying. For me as a three-year-old, I had no understanding of what he was saying. So I didn't really have any communication with him for about the first five years of my life," Timothy told a BBC interviewer, Dara Ó Briain, in 2015. "It was only when he got the speech synthesizer that I was actually able to start having conversations with him. It was kind of ironic in a way that him losing his voice was actually the start of him and I being able to form a relationship, really."[48]

But Stephen still took the transition hard. He was now even more dependent on a support network of people around him. "For him it was a step downwards. He felt he was giving in to his condition," Jane told reporter Brian Appleyard. "There were so many instances when a practical step for everybody else meant, for him, giving in—an admission of defeat."

...

Stephen Hawking did not admit defeat easily, even when he would play with his children. "He used to play board games," Timothy told Ó Briain. "He wasn't the easiest opponent, particularly at chess."[49]

"Surely he let you win?" asked an incredulous Ó Briain.

"Well, no. There was no compassion there at all. He was hugely, hugely competitive."

Timothy wasn't the only child who experienced his father's fierce competitive streak. A number of years after Laflamme graduated, he visited Hawking with his two children, Patrick and Jocelyne, in tow—then about eight and six years old. At Elaine's suggestion, they decide to play a board game together, one called Avalanche. Each player, in turn, drops marbles into the top of a box full of little obstacles that swing back and forth; as the marble falls, it might get stuck on one of the obstacles, or, if dropped in the right place, it might jostle some of the obstacles just so and, trigger a cascade of falling marbles. As it's a physical game, Hawking couldn't do it alone; he directed Laflamme where to drop the marble. "So there is Stephen Hawking playing against my two kids. Patrick was not very good. He loses very quickly," says Laflamme. "But Jocelyne was six or seven. She has a good flair for maths and those kind of puzzles, and she's really competing against Stephen." Laflamme laughs. "I had to put a marble in a place, so I was going from one location to the other, and I'm waiting for Stephen to make a sign . . . and, I don't know, something caught him and he kind of winks, and I let it go down. And he loses. And Stephen looks really pissed. And then he says, 'I didn't! I didn't say yes!'"[50]

As Laflamme well knew, Hawking's stubborn refusal to admit defeat extended to his scientific work, too. During Hawking's convalescence, Laflamme kept beavering away at the arrow of time problem, and no matter what he did, the math was leading him in a different direction from what Hawking's intuition was saying. The collapse of the universe was not behaving like a time-reversed version of its expansion; the arrow of time would not simply switch direction as the universe begins to recollapse.

Upon Hawking's return, Laflamme showed the results to his mentor, who refused to believe him. "I was not getting that things were going to reverse. And Stephen would kind of challenge me back. And he would say, 'Well, you're doing *this*, but have you thought about *this* approximation?" Laflamme says. "And of course, I hadn't thought about it. So I would go back and calculate for a couple of weeks, and then come back and I say, 'Well, okay, I think this approximation is okay, because of this, this, this, this. And I still don't have the arrow of time reversing.' But he said, 'Oh, but what about *that*,' and then he would send me off for another two weeks." Hawking had an endless litany of objections, each of which could be defeated eventually, but bringing Laflamme no closer to convincing his adviser that his arrow-of-time assumption was wrong.[51]

Luckily, Don Page, a Hawking student from more than a decade back, came to town in early 1986, and sat in on one of the Hawking students' weekly Friday meetings. Page had been working on the same problem, but from a slightly different direction—and he was getting the same answer as Laflamme. "Don, being older and more mature than me, says, 'Stephen will never believe it unless it comes from himself," Laflamme says. "So our job is not to go kind of frontward, straight to his face, telling him he is wrong. We have to ignore the fact that he's wrong and build the case slowly, by moving things together until he will realize the idea is wrong." Page and Laflamme set out to do just that. They presented him with little results that they had proven, bit by bit, without any reference to the arrow-of-time problem, each of which Hawking accepted. Yet though these little results were seemingly unrelated, once you accepted them, together they excluded the possibility that the arrow of time would reverse. "And then, suddenly, one day, Stephen says, 'But the arrow of time thing—this idea will never work!' Then both Don and I said, 'Absolutely.'"

Hawking had to make a few tweaks to his draft of *Brief History*, deleting some of his earlier claims about jugs jumping off the floor and the like, and adding a paragraph in which he publicly admitted his error:

> At first, I believed that disorder would decrease when the universe recollapsed. This was because I thought that the universe had to return to a smooth and ordered state when it became small again. This would mean that the contracting phase would be like the time reverse of the expanding phase. People in the contracting phase would live their lives backward: they would die before they were born and get younger as the universe contracted. . . . However, a colleague

of mine, Don Page, of Penn State University, pointed out that the no boundary condition did not require the contracting phase necessarily to be the time reverse of the expanding phase. Further, one of my students, Raymond Laflamme, found that in a slightly more complicated model, the collapse of the universe was very different from the expansion. I realized that I had made a mistake: the no boundary condition implied that disorder would in fact continue to increase during the contraction.[52]

Laflamme's copy of *Brief History* bears a treasured inscription: "To Raymond, who showed me that the arrow of time is not a boomerang. Thank you for all your help. Stephen."

• • •

Considering how close he had been to death—and what his recovery had cost him—Hawking was functioning again in a remarkably short time. And thanks to his new computer system, and his round-the-clock nurses, subsidized, in part, by charitable organizations, Hawking in some ways was functioning better than ever. Though he was still totally dependent upon people around him, the nature of that dependency had shifted in subtle ways. By distributing his nursing needs over a larger, more professionalized team, he had reduced the burden on his students, his children, and especially, Jane. Indeed, he had mostly unyoked himself from his wife. And with his computer system, he could communicate more freely than ever before; he no longer had to rely upon a translator.

The computerization of Hawking's voice certainly came with its quirks and annoyances. It didn't slur words, which was important to him ("One's voice is very important," he would say. "If you have a slurred voice, people are likely to treat you as mentally deficient.") But he could not inflect his intonation to express emotion or emphasis. And because the screen was visible not just to Hawking but to anyone looking over his shoulder, there was no longer any privacy in the act of composing a sentence; it is as if people could tap into his premotor cortex and divine what he was going to say before he actually said it. A number of his friends and colleagues describe chatting with Hawking as something very different from speaking to anyone else. "It was participatory," says director Errol Morris. "You could sit next to him, and as he was writing, you could read what he was writing on the screen. And it became a kind of guessing game. How is he going to finish that sentence? What's going to come next? What is he saying?" ("He almost always wanted to complete his own sentences," Thorne recalls. "I soon learned not to complete his sentences for him. There was one exception: if we were in a huge hurry and I had a taxi [en route] to take me to the airport, then I was permitted to complete a sentence so we could move rapidly.")[53]

But the voicebox and its voice quickly became so deeply associated with Hawking that he himself couldn't imagine replacing it even as better synthesizers came around. Neither could his family; that robotic voice was as much a part of Stephen as his corporeal body. As Jane put it, "it was now so closely identified with Stephen's personality that I found it quite upsetting if one of the children played around with it."[54]

The computerized voice—the most visible aspect of the disaggregation of his bodily functions across a network of external people and devices—also helped cement the physical-infirmity-begets-genius leitmotif. It raised Hawking once and for all to the realm of transhuman, which, unfortunately, sometimes made it harder for him to engage in the ordinary social interactions of humanity. As he told BBC interviewer Ó Briain in 2015: "I sometimes get very lonely because people are afraid to talk to me or don't wait for me to write a response."[55]

But those who knew Hawking were relieved that the physicist had so quickly recovered his strength from his near-death experience. By early 1986, Hawking had turned back to book-writing with renewed vigor. There was a lot of work yet to do, and Guzzardi and Hawking and his students set about revising the manuscript, kneading it and kneading it until it was ready to go to production.* Nor did Hawking's scientific production slacken. He and his students were publishing at the same sort of pace in 1987 and 1988 as they had been before Geneva. On the surface, at least, everything seemed back to some semblance of normal.

And normalcy suited Hawking just fine. Despite his disability, his deepest desire was not to be perceived as anything abnormal, just as a particularly gifted scientist. He didn't want people to concentrate on his personal struggles, or to pity him. Most of all, he didn't want to be turned into the star of some movie-of-the-week sob story. "Neither I nor my family would have any self-respect left if we let ourselves be portrayed by actors," he wrote.[56]

At this point in his life, for Hawking, it was all about the physics.

---

* At the very last moment, just a few weeks before publication, Guzzardi left Bantam for another publisher. Editor Ann Harris took over and shepherded the book to the very end of the process.

CHAPTER 11

INFLATION (1977–1981)

Christmas season, 1977. Cambridge is cold and dark. Very dark. At this time of year, the sun begins its retreat before tea-time, and has surrendered the field by 4:30. Jane Hawking had spent the past few hours caroling—she had recently joined the choir of St. Mark's Church in Newnham—and, with Lucy in tow, headed back toward her home where Stephen was waiting for her. Alongside walked Jonathan Hellyer Jones, the choirmaster. In her memoir, Jane later wrote, "I talked as I had not in years. I had the uncanny sensation that I had met a familiar friend of long acquaintance." Talk of music, of travel, of faith. Of loss. Jonathan had lost his wife, Janet, to cancer just a year and a half prior. As the three walked through the darkness, illuminated periodically by the headlamps of a passing car, an intimate friendship bloomed.[1]

Jonathan, an organist as well as choirmaster, began giving weekly piano lessons to the seven-year-old Lucy. "At first he came strictly for the length of the lesson, then he stayed a little longer to accompany me in the Schubert songs I was learning," Jane writes. "After a few weeks of this routine, Jonathan began to come early for lunch or stay for supper afterwards and to help with Stephen's needs, relieving Robert of all the chores which had oppressed him for so long."[2]

Robert, the eldest Hawking child, was then only ten years old, and had been put in an unusual and difficult position for a kid his age: caretaker to his father. And Jane was rightly worried about the effect this was having on her son, who was growing moody and withdrawn. "Much though he loved and respected his father, it was obvious that Robert needed a male role model, someone who would romp and tussle with him, someone who would ease him out of a childhood already lost into adolescence, someone who would not expect anything of him in return, least of all assistance with their own physical requirements." (Many years later, when asked how motor neurone disease had diminished his life, Stephen said, "When my children were young, I missed not being able to play with them physically.") Jane soon learned that Jonathan was willing to fulfill that role. "When we had got to know him a little better, Robert would lie in wait for him by the front door and pounce on him on his arrival, throwing him to the floor and wrestling with him. Jonathan took this unconventional form of greeting in good part and responded in kind to a growing boy's need for a good rough-and-tumble to release his excess energies."[3]

In May 1978, when the pair were sitting in one of the small chapels that ring the perimeter of Westminster Abbey, Jonathan made clear that he was committed to Jane and to her family, "come what may." Though the two had not yet consummated their relationship, Jonathan had become a pillar for her, and he was to become a part of the Hawking household.[4]

Quite naturally, Stephen resented the relationship and was hostile at first. Stephen eventually acquiesced, telling her that "if there was someone who was prepared to help me, he would not object as long as I continued to love him."[5]

When Stephen and Jane were married in 1965, they were barely able to hope it would last thirteen months, much less thirteen years. They were just as unaware in 1978 that this unconventional arrangement would mark the midpoint of their marriage: the next twelve years would be the story of Jane and Stephen and Jonathan, not just Jane and Stephen.

• • •

In his late thirties, Stephen Hawking was no longer a wunderkind. The blush of youth had faded from his work on black holes—as revolutionary as his insights had been, the Golden Age of Black Holes was over.

The blush was off of his marriage as well. The idealism of the first decade and the changes in Stephen and Jane's life together had given way to practicalities—as much as Stephen wanted to hold fast against the progress of his condition, day by day he had to make concessions to the people who had to care for him, and surrender little bits and pieces of his independence and his dignity.

Even so, Hawking adapted. He realized there was more to physics than coming up with new theorems—and more to life.

...

"The acknowledged leader in black hole theory is Stephen Hawking," the announcer declares, as the camera focuses on the physicist in his wheelchair. It was Hawking's first appearance on the BBC—and on international television.[6]

Hawking made only a brief appearance in 1977's *The Key to the Universe*, a two-hour documentary that surveyed the latest developments in particle physics. But Hawking's story was given pride of place. "Although the gentle gravity of the planet Earth confines him to a wheelchair, in his mind, he masters the overwhelming gravity of a black hole."

As Nigel Calder, who produced the show along with the BBC's Alec Nisbett, later wrote, "For Nisbett and me, Hawking was not only an up-and-coming physicist but an image of the frailty of *Homo sapiens* confronted by a confusing and often violent cosmos. After describing his then-recent suggestion that small black holes could explode, producing new particles, we incited Hawking to use stirring words to climax our show." Which Hawking did with aplomb.[7]

It was before the days of his speech synthesizer, so Hawking's words were unintelligible. The producers had to use a voiceover—but that didn't diminish the power of his statement one bit: "On one view, we're just weak, feeble creatures at the mercy of the forces of nature. When we discover the laws that govern these forces, we rise above them and become masters of the universe."[8]

The metaphor of Hawking was almost too perfect. The physics he so loved was always at risk of being obscured by his personal story. Yet he couldn't ignore the media because he desperately wanted to bring his science to a popular audience, and he was beginning to figure out how.

Back then—as now—*Scientific American* magazine was one of the favorite choices for scientists who want to popularize their ideas. Each month, they run a handful of long articles written by researchers, and their staff of editors labors mightily to ensure that the final versions are at a level that a sophisticated but nontechnical audience can appreciate. The list of *Scientific American* contributors was like a who's who of important scientists—including Albert Einstein, J. Robert Oppenheimer, Paul Dirac, and Erwin Schrödinger. Even Frank Hawking, Stephen's father—a prominent parasitologist—had published not just one article, but two. And the time was ripe; in 1972, Roger Penrose had written an article about black holes, and in 1974, Kip Thorne published one about black holes as well. By

1977, *SciAm*'s biannual black-hole article was therefore overdue. Hawking decided to give it a go.

As far as science communication goes, "The Quantum Mechanics of Black Holes" isn't a particularly compelling or memorable piece of writing. Hawking does a solid job of summarizing the present state of knowledge about black holes, but the prose is generally rather weak, to the point of obscuring Hawking's trademark wit. For example, the article ends with a somewhat turgid joke:

> It therefore seems that Einstein was doubly wrong when he said, "God does not play dice." Consideration of particle emission from black holes would seem to suggest that God not only plays dice but also sometimes throws them where they cannot be seen.[9]

Hawking did better a few months earlier in an important technical paper in the journal *Physical Review D*, which was read only by hard-core physicists:

> Einstein was very unhappy with the unpredictability of quantum mechanics because he felt that "God does not play dice." However, the results indicated here indicate that "God not only plays dice, He sometimes throws the dice where they cannot be seen."[10]

(Some two decades later, by the time he was selling out the Albert Hall, the joke had finally reached a state of high polish: "Thus it seems Einstein was doubly wrong when he said, God does not play dice. Not only does God definitely play dice, but He sometimes confuses us by throwing them where they can't be seen.")[11]

Nevertheless, Hawking had reached a wide audience on his own terms. There was not a single word in his *Scientific American* article about motor neurone disease, or wheelchairs, or disability. It was all about the physics.

• • •

Since the beginning of Hawking's career, his work had centered on singularities—the infinities that mar the otherwise smooth manifold of spacetime. The singularities at the heart of a black hole had occupied the past few years, and in the late 1970s, once again, he was beginning to concentrate his mind's eye on the singularity at the beginning of the universe. Hawking's scientific publications at this time show him thinking about path integrals and Euclidean quantum gravity—presaging his no-boundary proposal—as well as some work on what spacetime looks like on the very smallest scales. So he was primed for a new theory that first emerged in 1979, perhaps the most exciting development in cosmology since the discovery of the cosmic microwave background radiation a quarter century prior.

The discovery of the CMB, the faint microwave afterglow of the Big Bang, marked a turning point in cosmology; scientists were forced to accept that the universe was born in an explosion that took place billions of years ago, and that it has been expanding ever since. But the idea of a universe that suddenly grew to enormous size from a tiny cosmic seed failed to solve some serious puzzles that were vexing cosmologists. Two of the biggest were the so-called flatness problem and horizon problem.

Spacetime as a whole could have a number of different shapes. Indeed, if you imagine a "random" universe—something arbitrarily created from a Big Bang, most likely it would wind up very curved in one way or another. It could be curved like a very small sphere and collapse again very quickly; or it could be extremely warped and saddle shaped, and fly apart so quickly that stars and galaxies might never get a chance to form. Yet our universe looks very, very flat—balanced on the knife-edge between immediate collapse and too-fast expansion. It seems an unlikely coincidence that of all the possible universe shapes out there, we wind up with this flat or nearly flat geometry that perfectly balances expansion and collapse. Scientists don't like such coincidences. That's the flatness problem.

The horizon problem is a little more subtle. As an analogy, imagine that in the distant future, our species discovers space travel, and finds a couple of dozen intelligent—but primitive—civilizations scattered across the universe. None of them have discovered space travel or radio communications, or have ever had contact with alien races before. Yet for every single one of these civilizations, the language is classical Latin. How could all of these civilizations have such a profound similarity without being in contact with each other? The natural conclusion is that something happened in the distant past that somehow connected these seemingly unconnected civilizations. The horizon problem is the cosmological equivalent. It's not that our universe is totally homogeneous—there are patches of stars and galaxies and there are voids—but overall, one patch of sky looks pretty much like every other patch of sky. They are profoundly similar. Yet those patches are separated by such vast distances that light hasn't had a chance to travel from one patch to the other—they were so far apart that they were never able to interact or influence each other. So how can they possibly look so similar? The natural conclusion is that somehow, in the early universe, they were connected in some way.

Two vexing cosmological problems,* and in late 1979, Alan Guth, a cosmologist at the Massachusetts Institute of Technology, began to figure out a possible way to solve them both: an idea now known as *inflation*. Hawking would soon play a pivotal and controversial role in the development of inflation theory.

The most natural way to think of a Big Bang is like a huge explosion: a sudden, one-time eruption of force that starts spacetime expanding and sends the parts of the newborn universe flying in all directions. After that single eruption, the cosmos essentially coasts, expanding entirely from its initial momentum, but slowing down all the while like a speeding car gradually rolling to a stop.† Guth proposed a more complicated picture: that a tiny fraction of a second after the Big Bang, the universe's coasting expansion was momentarily interrupted: Nature suddenly pressed down on the accelerator pedal. The fabric of spacetime expanded more and more rapidly, doubling and doubling and doubling again in size, taking the universe very quickly from subatomic size to roughly the size of a basketball, until, another fraction of a second later, nature takes its foot off the accelerator and the expansion of the universe coasts once more.‡

As complicated as this scenario is, it solves the horizon and flatness problems in one go. Thanks to an almost unfathomable growth spurt, an inflationary universe can start off much, much smaller than a Big Bang universe could otherwise—and because the seed starts off so tiny, all of its different parts could interact with each other for a short time before being blown apart by inflation. It's no longer such a mystery how distant regions of the universe can look so similar; they weren't really isolated from each other until inflation forced them apart. And the flatness problem goes away, too: the initial rapid expansion stretches the fabric of the universe, making it appear almost flat, no matter what the initial shape of the cosmos

---

* There was actually a third, known as the *monopole* or *relic* problem, which was Guth's initial motivation, but it's not necessary to go into that particular detail to understand the reasons for inflation's success.

† With the discovery of dark energy, scientists came to realize that the universe wasn't really just coasting, but slowly accelerating in its expansion. But that came two decades after Guth proposed inflationary theory.

‡ The fabric of the early universe was trapped in an uncomfortably energetic state—something akin to a supercooled fluid, which is trapped in the liquid phase well above its freezing temperature. When the universe was trapped in this unstable state, the fabric of spacetime expanded exponentially quickly. But after a certain point (and the precise details of how this transition occurs are a major part of what distinguishes different variants of inflationary theory), there was a phase transition akin to the supercooled fluid suddenly freezing, where the rapid expansion stops.

might have been. It was a very attractive proposal, and gave scientists hope that they could begin to understand the conditions of the universe in the tiniest fraction of a second after the Big Bang. As Guth's idea began to circulate in early 1980 and through 1981, the excitement in the cosmological community began to build. But there was one problem. It didn't *quite* work.

As is usually the case, the devil was in the details. Guth had proposed a mechanism for describing how and when the inflationary phase of the universe stops, but however he did the calculations, the transition happened too quickly, too suddenly. The kinds of universes that Guth's model created collapsed soon after birth and were too small to create stars and galaxies, or too asymmetric to match real observations. The model was broken—it couldn't be quite right—and Guth knew it. But an increasing number of cosmologists were convinced that there had to be some nugget of truth there, that, even if the details weren't quite working out, Guth was onto something.

Guth knew almost from the start that his model of inflation was wrong. He wasn't the only one. On the other side of the iron curtain, a number of physicists, including Alexei Starobinsky, Andrei Linde, and Gennady Chibisov at the Lebedev Physical Institute in Moscow, had come up with similar ideas. They were encountering the same sorts of difficulties that Guth was, so they decided not to publish. "We just stopped," says Linde. Then, some time later, Linde got a phone call from Lev Kofman, also at Lebedev. "I think it was somewhere in the winter or the early spring of '81, and he told me, 'Did you see the preprint by Alan Guth? He suggested a way to solve the flatness problem. . . . ' And I told him, 'No, I didn't see the preprint, but let me explain to you what is inside. And let me tell you why it doesn't work.'"[12]

But the thirty-three-year-old Linde kept thinking about the problem. He plugged equations into an "old, grotesque" computer that the theorists had on hand, and in the early summer of 1981, realized that the mechanism for stopping inflation might be subtly different from what Guth had initially thought. "I understood that the process may go differently from what Alan suggested," Linde says. He realized that the different mechanism might fix the model; a slight modification might start generating realistic universes rather than tiny little squib universes that promptly recollapse, or asymmetric messes. "I got extremely excited," he says. "I wake up my wife and say, 'I know how the universe was born.' Because it was really simple." Linde decided to write up his idea (later dubbed *new inflation* or *slow-roll inflation*) and sent it to the peer-reviewed journals to try to get it published. It was not an easy process.

The iron curtain was a major barrier to information flowing from Soviet scientists to their counterparts in the West and vice versa. It was an ordeal for a Russian physicist to publish something for international consumption, says Linde:

> You get the permission, first from our internal authorities of the theoretical group, then permission from the [Lebedev Physical] Institute, then it was sent somewhere else, and we finally get the permission to submit it somewhere. If it is submission for a Soviet journal, this would not be a problem; it was very quick. But if it is submission for anything going abroad, then it would go first to the Academy of Sciences. Then from Academy of Sciences, it would go to a special place where censors were checking that we did not say anything unallowable about the beginning of the universe. All of these would take maybe two months, three months, depending. Then it will return to us with permission, hopefully. Usually, yes. After we get permission, if it is something to be sent abroad, then we need to translate it into English and somebody will translate. . . . If we first make a preprint, then we actually got two separate permissions, one for preprint and another for sending it, for example, to *Physics Letters*. And it was not possible to make a Xerox copy; we must type it twice, we must insert all equations twice. And then we send it to *Physics Letters* or somewhere abroad, then usually it will travel from us to the journal a month, two month[s] depending, and when they reply, the replies would come to us with the same time delay.[13]

Much better were conferences, where scientists could meet face to face, share ideas, and make friendships. However, the cold war was still very much underway in 1981—Ronald Reagan, Margaret Thatcher, and Leonid Brezhnev gave little hope for a détente anytime soon—and relatively few Western physicists ever went to conferences in the Soviet Union, much less forged relationships with Russian physicists.

One rare exception: Stephen Hawking.

Hawking's Russian connection came through his longtime friend Kip Thorne. In the mid-1960s, Thorne had attended a London lecture by a prominent Russian physicist, Igor Novikov. "After Novikov's lecture, I joined the enthusiastic crowd around him and discovered, much to my pleasure, that my Russian was slightly better than his English and that I was needed to help with translating the discussion," Thorne later wrote. "As the crowd thinned, Novikov and I went off together to continue our discussion privately." The two struck up a friendship, and Thorne had an entrée into the world of Russian physics. Thorne visited Moscow for several weeks in 1969 and again in 1971 as the guest of Novikov's collaborator, Yakov Zel'dovich.* In 1973, he brought Hawking along with him. Since then, every few years, Hawking made it over to Moscow. Along with Thorne, he had become one of the main conduits for Russian cosmological ideas to flow to Europe and the Americas.[14]

In the fall of 1981, Linde's work—he had written up three papers to explain his idea—was still bouncing around somewhere inside the Soviet bureaucratic machine. But it so happened that there was a conference on quantum gravity in Moscow, and Hawking was in attendance. Linde presented his idea at the conference, and Hawking, in the audience, listened attentively, but didn't make any comment.

The next day, Hawking was invited to give a lecture at Moscow State University, and Linde, who had come to hear the talk, found himself unexpectedly drafted to act as translator. So Linde got up on stage—in front of all the finest minds in Moscow—and began the laborious act of translating. Hawking would say a word or two. Then, since these were the days before Hawking's computer, Hawking's student would have to interpret what Hawking was saying, and then Linde would translate the student's words into Russian. After half an hour of this, Hawking started discussing Linde's idea of slow-roll inflation: "Steve said, 'There is this interesting suggestion of Andrei Linde,' and I happily translated," Linde recalls. "And then he said, 'But the suggestion is completely wrong.' And for the next half hour, in the presence of all the best people in Moscow, I was explaining what's wrong about my paper."[15]

After the excruciating talk, Linde told Hawking that he had been a faithful translator, but he disagreed on a number of crucial points—and asked whether the esteemed professor would like to discuss the matter further in private. The two repaired to a smaller room with a blackboard and started arguing about the details of Linde's proposal. "Apparently, all the institute was in a panic, because a famous British scientist had just disappeared in the middle of Moscow," recalls Linde. Eventually, Hawking asked Linde whether he would like to continue the discussion at his hotel. "Oh, my God! I did not have any permission," Linde remembers thinking. "To hell with the permission!" Linde followed Hawking to the hotel. The

---

* Zel'dovich, like Thorne's PhD adviser, John Wheeler, had worked on his country's thermonuclear weapons project. The physics of compressing matter at extremely high temperatures is interesting in fields other than black-hole theory.

discussion turned from physics to family—they showed each other photographs—and they were soon fast friends. Before they parted, Hawking told Linde he was planning a conference in Cambridge for the summer of 1982, and invited the Russian to attend. It would be a conference to remember.[16]

• • •

A conference in Corsica was memorable for an entirely different reason. Every summer, a scientific institute in Cargese, a little town on the west coast of the Mediterranean island, had been running a two-week-long conference-slash-summer-school-slash-retreat devoted to cutting-edge issues in physics. In July 1978, the subject was gravitation, and Stephen was invited. Jane and the two children tagged along; the relationship with Jonathan was brand new, so he did not accompany the family. During the day, Stephen and his student Don Page attended physics talks and discussions while Jane and the kids ran on the beaches and splashed around in the sea.

"For some time I had ceased to bother about contraception in the already fraught marital relationship as it hardly seemed relevant and simply added to the complications," Jane writes in her memoir. But the Corsican vacation made it all too apparent that it was still relevant. "The combination of the powerfully sensual effects of the Mediterranean climate and my genuine desire to love Stephen, reinforced by the glowing knowledge that I was no longer fighting all the battles singlehandedly, had brought us together in fleeting moments of carefree abandon." By Jane's account, this led to a predictable, if unexpected, outcome. Shortly after the pregnancy was confirmed, she writes, she was suffering from morning sickness so acute that she decided not to travel to another physics conference in Moscow. Stephen's mother, Isobel, accompanied him instead.[17]

Jane would later write that the pregnancy was unwelcome, and not just because she found the prospect of "another being totally dependent on me" to be "terrifying," but also because it would destroy the fragile balance that she was trying to build with her family life. "To expect [Jonathan] to take on another small Hawking, especially when he had no children of his own and no prospect of ever having any as long as he associated with us, was inconceivable," Jane writes. "I was resigned to losing him and with his loss, to losing all hope for the future. I would be alone again."[18]

Jane couldn't find solace in her family. She hadn't been getting along with the Hawking clan for some time; according to her memoir, they had always been cool to her. The unexpected pregnancy, coupled with her association with Jonathan, brought the already fragile relationship with her in-laws to a breaking point. Shortly after her second son, Timothy, was born in April 1979, Jane writes, she found herself alone with Isobel. "She looked me straight in the eye. 'Jane,' she said, adopting a stentorian tone, 'I have a right to know whose child Timothy is. Is he Stephen's or is he Jonathan's?'" Jane was shocked and dismayed; all her hard work to maintain a discreet relationship "was being trampled under the elephantine tread of Hawking insensitivity." Jane insisted that there was no possibility that anyone other than Stephen was the father, but Isobel didn't accept those reassurances. According to Jane's account, in the heat of anger, Isobel said, "We have never really liked you, Jane, you do not fit into our family."[19]

Jane also hadn't been finding solace in her research. She had been laboring, on and off, to write a PhD thesis since the late 1960s, since before eleven-year-old Robert was born. At the time, she recognized that she would not be able to cope at Cambridge merely as the spouse of a professor; she had to come up with some other purpose for herself to maintain her happiness. "From my observations of the dynamics of life in Cambridge, I could see that the role of a wife—and possibly a mother—was a one-way ticket to outer darkness," she wrote. "Stephen's work was beginning to bring him such acclaim that unless I too had some academic identity, I was in danger of becoming a mere drudge." An ordinary job—Jane had once thought about working in the Foreign Office—was scuppered by the rigors of caring for Stephen; once Robert was born, the concept went from merely implausible to ridiculous. "A doctorate seemed to be the ideal solution. I could easily adapt my hours of study in the University library and my work at home to Stephen's schedule," Jane wrote.[20]

Jane's idea for a thesis had to do with small sections of poetry written in Mozarabic—the popular, vulgar language in the Muslim-controlled sections of medieval Spain. These poems, called *kharjas*, were tales of longing and betrayal, of wives and husbands and lovers, such as this one from the eleventh century:

Mew sidi Ibrahim	My lord Ibrahim,
Yanuemnedolze	Oh sweet name,
Fen-temib	Come to me
De notje.	To-night.
In non, si non keris	If not, if you don't wish,
Yire-me tib	I'll come to you
-Gare-me a ob	Tell me where
afer-te.	To find you.[21]

By studying the themes and imagery in these kharjas, and in various other medieval poetic traditions, Jane thought she could help untangle their evolution throughout different regions of Spain and their spread to other parts of Europe.

But Jane's hopes came crashing into reality soon after Robert was born, and she gave up "whatever illusions I might have held about combining motherhood with some sort of intellectual occupation."[22]

Though she could occasionally find snatches of time to work on her thesis, Jane didn't find a great deal of support from her husband, whose "contempt for medieval studies was unrelenting" as she labored on and off for more than a decade to complete it. Now, with a baby on the way, Jane pushed to finish her dissertation—which she did with just days to spare. Thirteen years of effort, but it paid off; she defended her thesis a year later and then was officially a doctor of philosophy. It scarcely mattered that the degree didn't lead to a university teaching position, or to any further work in medieval literature. She started teaching French to primary-school students after-hours, and eventually wound up as a part-time tutor for high-school students. "I won the respect of my pupils and gradually discovered a professional identity for myself," Jane writes. "I was awakening from an intellectual coma."[23]

She also found an outlet in monthly meetings of a new local antinuclear group, Newnham Against the Bomb. At the same time that Stephen visited Moscow in October 1981, the Hawkings mailed a letter to their friends and colleagues around the world expressing "alarm at the acceleration of the arms race." The pair had been antinuclear activists early in their relationship; it was one of the activities that had drawn them together, and Jane believed that it was doing so again. "The tendency for us to slip into the roles of master and slave was arrested. We were companions and equals again, as we had been in our campaigning in the 1960s and early 1970s," she writes in her memoir. However, even as she found fulfillment in her activism, in teaching, and eventually, by helping Jonathan plan concerts, the nagging doubt remained that Stephen didn't take her pursuits seriously. As Jane put it, she had "become used to regarding [her]self as inferior and [someone] whose endeavours were usually either met with disdain or ignored."[24]

...

At almost the exact same moment in October 1981 that Jane sent the antinuclear letter to everybody in the Hawkings' Rolodex, Stephen and fellow Cambridge physicist Gary Gibbons were sending out invitations to a select group of physicists around the world.

The Nuffield Foundation—a charitable organization set up in the 1940s by automobile magnate William Morris, Lord Nuffield—had agreed to sponsor a series of scientific conferences. Hawking and Gibbons had decided that, given all the excitement surrounding the theory of inflation, the next one should be devoted to the "very early universe": the first second after creation. The workshop they planned would stretch over three weeks at the end of June in Cambridge—and all the leading lights in the field would be invited.

Since the news of inflation had begun to break in 1980, physicists around the world had been playing around with the idea in hopes that they could make the idea work; even though Guth's model was broken, it so neatly solved a number of key problems nagging cosmologists that it sparked an international effort to try to shore it up.

Andrei Linde's paper describing his "new inflation" idea emerged from the Moscow bureaucracy in late 1981 and circulated informally as a preprint at the same time that it was submitted to *Physics Letters B*, which then sent it out for peer review. One of the reviewers was none other than Stephen Hawking. "As a friend of Linde's, I was rather embarrassed, however, when I was later sent his paper by a scientific journal and asked whether it was suitable for publication. I replied that there was this flaw," Hawking writes in *A Brief History of Time*. The problem was that like Guth's inflation, Linde's new inflation would produce universes unlike our own. But Hawking and his student Ian Moss suggested that by fixing some of the oversimplifications of Linde's model, they could generate a realistic universe—not perfectly realistic (it would have too much matter), but no longer suffering from gross flaws. As a reviewer, Hawking made an unusual suggestion to *Physics Letters B*: "I recommended that the paper be published as it was because it would take Linde several months to correct it, since anything he sent to the West would have to be passed by Soviet censorship, which was neither very skillful nor very quick with scientific papers." Linde's paper came out in February 1982, and a Hawking/Moss paper fixing the model up a bit was published a month later.[25]

In late 1981, Paul Steinhardt and his student Andreas Albrecht had independently come up with a similar idea to Linde's new inflation scenario; they submitted their work to *Physical Review Letters* in January 1982, and it was published in April. Hawking despised Steinhardt as a result—Hawking had come to believe that Steinhardt and Albrecht had plagiarized Linde, having gotten the idea from a lecture Hawking gave in Philadelphia shortly after the Moscow conference. This is even though Steinhardt and Albrecht referenced Linde's work—which had since circulated in preprint—in their *Physical Review Letters* paper. At the time, though, they were blissfully unaware of Hawking's (and Linde's) anger, and happily toiling away at trying to figure out what sort of universe would emerge from a new-inflation-type model.

One important feature of any inflationary-type scenario is that the way our present universe looks is closely linked to the roughness, the foaminess, of spacetime on the smallest scales. The laws of quantum mechanics ensure that very tiny objects (like the tiny proto-universe prior to inflation) aren't completely uniform. When inflation hits, those non-uniformities, those fluctuations, get blown up incredibly rapidly. Places where the proto-universe was slightly more dense would become vast swaths of space where galaxies form in abundance; less-dense places would become voids, empty spots where few stars or galaxies will ever be born. So, one big challenge for cosmologists was to try to use the model to predict precisely the effect of the inflationary blow-up on those density fluctuations—to explain exactly how the quantum-mechanical fluctuations in the proto-universe would leave their imprint on the universe after inflation finally turned off. And scientists around the world were getting vastly different answers with their calculations. Nobody could tell whether new inflation was working or not.

This was a solvable problem, and the Nuffield workshop was the ideal place to solve it. Most of the key players were going to be attending. Alan Guth at MIT. Mike Turner and Jim Hartle from Chicago. Paul Steinhardt at the University of Pennsylvania. Jim Bardeen at the University of Washington. Frank Wilczek at Santa Barbara. John Preskill from Harvard, to name a few. And, thanks to Hawking's Russian connections, a large group of Soviet physicists (now including Andrei Linde) were there as well.

"There was somebody who never said anything, posing as a scientist and wearing a black suit—that was the KGB guy—but they had a very distinguished Russian delegation with [Alexei] Starobinsky and Linde," Preskill recalls. "And I remember two things about them. One, the World Cup was going on, and they'd be glued to the TV watching the soccer game. And the other was they would xerox papers like mad, because it was very hard to get a xerox machine back then in the Soviet Union if you wanted copies of scientific papers. . . . They were always at the xerox machine, in the xerox room making copies."[26]

At the start of the conference, nobody could agree on the proper value for the density fluctuations. "People come and start talking with Steve and telling him that the amplitude of the fluctuations is large," Linde says. "But they did not know, because there was a group of people who were saying that, yes, they are very small. And then a group of people were saying, we don't really know what they look like, but they are large. And Starobinsky did not discuss much with anybody because he has problems with his speech—stuttering."

For three weeks, the physicists attended lectures and then broke up into groups to calculate. Hawking, Guth, and Starobinsky worked solo while Steinhardt and Turner teamed up with Bardeen. (Steinhardt apparently didn't know why Hawking was giving him the hairy eyeball.) And after a time, the different groups—and their different approaches—revised their calculations. (Except for Starobinsky. "He just made the statement which he made before. He did not care," says Linde.) The answers began to converge—Guth and Starobinsky and Steinhardt's team all were getting the same answer, more or less. Hawking was a late holdout; he had published a preprint which made an initial calculation that disagreed with the consensus by several orders of magnitude. Toward the end of the workshop, Stephen gave a lecture, and Guth, Steinhardt, and other physicists "were poised to challenge him when it came to the crucial, controversial, step" in his calculations, Guth writes. "But it never happened! Hawking's lecture followed the preprint until the end, but at the last step he substituted a new calculation. . . . True to Hawking's style, he did not mention that he was correcting his preprint, or that his new answer was basically in agreement with Starobinsky or me."[27]

By the end of the conference, everyone was in agreement. They had an answer. And the answer was: new inflation didn't work; it would create a universe populated with black holes rather than one filled with stars and galaxies.

It was a bittersweet ending. New inflation was dead, having lived for barely six months. But in its failure, it pointed the way forward. With their detailed calculations, the scientists had figured out what was going wrong—and realized what conditions would be necessary for a new-new-inflation-type scenario to work. By the end of the workshop, in fact, Turner and Steinhardt had proposed just such a model; new inflation might be dead, but a newer inflation would soon take its place. Better yet, the Nuffield cosmologists had managed to generate some predictions about what kind of universe one might expect after inflation—predictions that could ultimately be tested by careful observations of the cosmic microwave background and other astrophysical phenomena. Despite the early failures, inflation would become a mainstay of modern cosmological thought.[28]

Stephen Hawking was at the very center of the inflation revolution. Even though he himself didn't come up with the idea, he was a key catalyst that rallied the community around the idea, developed it, and triggered its widespread adoption. To be sure, Hawking had done some important work on inflation at Nuffield, and with his student Ian Moss, he had not just tweaked Linde's model, but incorporated inflationary ideas into his nascent no-boundary scenario.

Hawking's contribution to inflation was not that of a junior scientist performing novel calculations, but that of a senior scientist spreading ideas, helping to develop them, and connecting communities of scientists so they could work together. At forty, Hawking was no longer a wunderkind. But Nuffield proved that Hawking was still at the center of the cosmological universe.

The early 1980s was an exciting time for Hawking. Even though he suspected that he wouldn't come up with many more big original ideas like the no-boundary scenario, much less Hawking radiation, it scarcely mattered. He was beginning to establish a new persona as an established, mid-career physicist. Stephen Hawking had made a name for himself in the scientific community, and he had amply proven that he deserved his reputation.

• • •

One downside of being a mid-career scientist is that you have to care more about politics and relationships—you have more weight, and you're expected to throw it around to support your allies. And in the wake of Nuffield, Hawking did just that to help his friend Andrei Linde. Hawking had been muttering about Steinhardt for a few months before Nuffield, and the problem came to a head shortly thereafter.

Michael Turner and John Barrow (who, like Hawking, was a former student of Dennis Sciama) were writing up a summary of the workshop for *Nature*. When they passed a draft along to Hawking for comment, Hawking apparently took exception to a passage that gave Steinhardt and Albrecht co-credit with Linde for coming up with the idea of new inflation. Hawking reportedly suggested that the pair drop the credit for Steinhardt and Albrecht, or, failing that, dilute their claim by giving Hawking and Moss credit as well. They took the latter course, but told Steinhardt about the behind-the-scenes maneuvering. After a heated back and forth with Steinhardt, Hawking backed down, seemingly accepting that Steinhardt and Albrecht had come up with their result independently. The matter was closed. For the moment, at least.[29]

Hawking also began to enjoy the platform that he was beginning to acquire as a world-renowned expert on relativity. He was getting used to delivering public lectures; with all the awards he was receiving, he had to speak more and more frequently. And with more and more gravitas. Some of his lectures, such as an address he presented in 1980 to inaugurate his tenure as Lucasian Professor, were considered important events in their own right—the Lucasian talk, entitled "Is the End in Sight for Theoretical Physics?," was reprinted in *Physics Bulletin*.

Hawking, ever the optimist, seemed to be growing more convinced that the "end" of physics—the formulation of a grand unified theory—was likely to happen before the turn of the century. Making such predictions is always dangerous. Indeed, Hawking recalled a few cases where scientific soothsayers had been proven wrong by history; for example, he quoted quantum scientist Max Born as declaring that "physics, as we know it, will be over in six months" in the late 1920s, upon learning about Dirac's equation.* But Hawking made predictions with all the confidence of a mid-career scientist flushed with success. Though some of the details would change over the years, and some of the more out-there suggestions (such as that physicists might soon lose their jobs to computerized scientists) would disappear entirely, Hawking's thoughts were gaining an audience beyond the small circle of physicists who truly understood his work.[30]

Like Jane, Stephen was trying to find fulfillment as his circumstances shifted. His physical condition was slowly but surely getting worse. He had lost the ability to write in the mid-1970s; a few years later, he was barely able to scrawl a signature. When Hawking was elected to the Lucasian chair in 1979, he was asked to sign a big book that all the teaching officers at Cambridge University sign. "They brought the book to my office, and I signed with some difficulty," Hawking explained a number of years later. "That was the last time I signed my name."[31†]

---

* Interestingly, I have not yet been able to find any credible source for this quotation. Other literature that uses this quotation ultimately seems to trace it back to the Lucasian lecture or *A Brief History of Time*. A few months before the Lucasian lecture in 1980, Hawking wrote to Born's son, Gustav, stating that he had "heard" the quotation in question: "I wonder if you have any information on the accuracy of this story or any further details that would be relevant." Gustav's reply: "I am afraid that the assertion attributed to my father is news to me." Gustav then suggested that Hawking contact a professor at Gottingen University. Letter from Stephen W. Hawking to Gustav Born, February 14, 1980, Churchill College Archives, BORN 1/4/2/25; letter from Born to Hawking, February 19, 1980, Churchill College Archives, BORN 1/4/2/25. Unfortunately, I have found no further correspondence on the matter or any indication of where or from whom Hawking might have heard the quotation. "I'm not familiar with the quote," Born biographer Nancy Greenspan told me in a series of e-mails." She added, "I don't believe Born would have said it."

† As fitting as this would have been, it appears not to be quite true. A 1981 book (the proceedings from a different Nuffield conference held in 1980) bearing Hawking's signature—with solid provenance—was sold at a Christie's auction in 2019. (It went for £8,750.) See "*Superspace and Supergravity* Signed by Hawking," Stephen Hawking, 1981, Lot 51, Christie's, https://onlineonly.christies.com/s/shoulders-giants-making-modern-world/superspace -supergravity-signed-hawking-51/70082.

In addition, despite his outward confidence, the nagging doubt remained about whether his best days were behind him mentally as well as physically—whether, as he told a reporter in 1982, he was over the hill with respect to generating brilliant and novel ideas in physics. Indeed, the mathematical precision that marked Hawking's early work was already beginning to give way to a vagueness that was only partially masked by the boldness of his ideas. "I've given up being rigorous," he told an interviewer in 1983. "All I'm concerned about is being right."[32]

In retrospect, there was a seed of truth in Hawking's fears; the early 1980s marked the last time that Hawking was producing truly novel ideas in physics. The Nuffield conference and the no-boundary proposal were the last really significant scientific contributions of his career. While he would continue to do good work, nothing that Hawking would produce would come close to what he had done before. His reputation from here on would rest on something other than his physics.

CHAPTER 12

BLACK SWAN (1974–1979)

I n California, everything was outsized. Gargantuan cars on mammoth freeways, towering skyscrapers in the city and enormous homes in the suburbs, furnished with huge sofas and massive television sets. Jane estimated that their new home in Pasadena—where they would be staying for the next year—could fit four of their little Cambridge bungalows inside.[1]

The personalities at the Caltech Physics Department were just as oversized, particularly the two Nobel laureates, Richard Feynman and Murray Gell-Mann, who were engaged in a never-ending pissing match. Both men were New Yorkers, children of Jewish immigrants, and both had wound up at posh Ivy League institutions despite their background. When admitted to the Ivory Tower, Gell-Mann fought back against being perceived as *nekulturny* by trying to prove himself smarter than everyone else; Feynman did the exact opposite, getting into bar brawls, banging on the bongos, and speaking in an unashamed Brooklyn-cabbie accent.

The two loved tweaking each other—publicly, if possible. Jane tells of one such occasion in her memoir:

> Stephen was present when Feynman turned up at the first of a course of lectures by Gell-Mann. Noticing him in the audience, Gell-Mann announced that he would be using his lecture series to conduct a survey of current research in particle physics and proceeded to read from his notes in a monotone. After ten minutes, Feynman got up and left. To Stephen's great amusement, Gell-Mann then heaved a sigh

and declared, "Ah, good, now we can get on with the real stuff!" and proceeded to talk about his own recent research on the cutting edge of particle physics.[2]

The pair's rivalry was good-natured, and they occasionally teamed up when it came to physics or pranks (at one point, the two snuck a peacock into a friend's bedroom). But there was a bit of sharpness at times—it was a rollicking, complex, love-hate relationship that was totally unlike anything the Hawkings would have seen at ultraconservative Cambridge.[3]

Even in the world of physics, where professors were expected to be a wee bit quirky, Feynman, in particular, stood out. He not only flouted the norms, but he made sure everybody knew he was flouting them. For example, Feynman spent many of his afternoons at a Pasadena strip club— five to six days a week, by his own account—and he was proud to serve as an expert witness when the proprietor of the club was charged under obscenity laws. "When my calculations didn't work out, I'd watch the girls," Feynman testified. The *Los Angeles Times* headline blared, "Bottomless [Club] Helps Nobel Physicist with Figures."[4]

Feynman's seemingly effortless ability to attract attention apparently grated on Gell-Mann. One Feynman friend wrote:

> Almost without fail, whenever Feynman's name came up in private conversation, Murray Gell-Mann would inevitably remark, He's always concerned with generating anecdotes about himself. In fact, there was some truth to Gell-Mann's remarks. On one occasion Feynman and I attended a physics lecture by a visiting professor. We got there early and took the front row seats. Feynman noticed that the lecturer had left his notes on the seat beside him. Feynman proceeded to look through the notes, and I could see that he was registering what he was reading. He put the notes back down and the professor came back in. During the course of the lecture, the professor stated, I have spent a considerable time working out the derivation of this particular formula. . . . Feynman stated, Ahh, the solution is obvious! It's . . . The professor, and the rest of the audience for that matter, was dumbfounded as Feynman, who appeared to be giving an answer off the cuff, gave the solution. As we left the lecture, I turned to Feynman and gave him a knowing look. He smiled back.[5]

Gell-Mann was every bit as good a physicist as Feynman, but Feynman got almost all of the attention. As journalist George Johnson put it a number of years later, "Gell-Mann knew how to package ideas, and he had a knack for giving whimsical, and unforgettable, names to the most abstract concepts in science. Feynman had a more vital gift: he knew how to package himself."[6]

. . .

Cambridge was reserved, tradition-bound, stuffy; its professors were expected to be polite and understated even when asserting their superiority over others. Pasadena was brash, iconoclastic, even a bit crass; its professors were not just able, but expected, to put their oversize egos on display. No wonder that a year's fellowship at Caltech transformed Hawking.

Under the Cambridge veneer, Hawking began to develop a Pasadena center; at the very least, he fell in love with the area and began to spend as much time as he possibly could beneath the palm trees of California, particularly when the soot-gray skies of England and the constraints of his everyday life became too much to bear.

Pasadena brought Hawking a change in scientific perspective, too. Even though Caltech had only a single general relativist, it was exceptionally strong in particle physics—and it had Feynman, whose intuition about physics was almost as visual, and arguably even stronger, than Hawking's. It was inevitable that Hawking would return to England with a brain full of new ideas.

. . .

When Hawking arrived at Caltech at the end of August 1974, he was there to work not with Feynman or Gell-Mann—who had both made their names in particle theory—but with his general-relativist friend Kip Thorne. Thorne, like Feynman a generation before him, had been a student of John Archibald Wheeler before landing at Caltech. But despite the recent advances in the field, Thorne was the only one there at the time doing general relativity. ("I do not think that I have the independence, as Kip does, to forge relativity out of the wilderness," another of Wheeler's students wrote his adviser.) So, in addition to his general relativity work, it was inevitable that Hawking got a great deal of exposure to particle physics and particle theory—including Feynman's path integral method.[7]

Hawking's ability to use his hands and arms was continuing to deteriorate; by the time he reached Caltech, he was no longer writing equations or drawing pictures, and by the time he left a year later, he was unable to feed himself. Feynman's path integral method gave Hawking a tool well suited to his very visual way of doing physics problems—and, thanks to his prodigious memory, Hawking not just absorbed the method, but mastered it, to the point where Feynman himself seemed to be a little touchy about it. "I remarked to Feynman that I was impressed by Stephen Hawking's ability to do path integration in his head," a Feynman friend wrote. "'Ahh, that's not so great,' Feynman replied. 'It's much more interesting to come up with the technique like I did, rather than to be able to do the mechanics in your head.'"[8] A few years later, Gell-Mann couldn't resist throwing a bit of shade as well:

> [Hawking] dazzled students at a Caltech (California Institute of Technology) seminar when he dictated a forty-term version of an important equation from memory. As his assistant finished writing the last term, his colleague, Nobel laureate Murray Gell-Mann, who happened to be sitting in on the talk, stood up and declared that Hawking had omitted a term. Gell-Mann was also working from memory.[9]

Caltech was a nakedly competitive, dynamic, liberal environment, in its way as unlike Cambridge University as Moscow State was. In some ways, the cynical, funny, practical-joking physicist fit in better with the merry pranksters at Caltech than he did with his more staid colleagues back home. So long as Hawking was able, both physically and financially, to travel, Caltech was a second home of sorts; he went back over and over again, as frequently as he could—typically in the winters when the weather in Cambridge was crummy.

Just as Hawking became an important conduit for the flow of interesting physics from the Soviet Union to the United Kingdom, he was tapping into the dynamism of the US physics scene. "He went there and got his rush of testosterone in Caltech," says Neil Turok. "And I think it helped him when he came back to Cambridge. He didn't take the Cambridge politics too seriously. He tended to sort of look down upon the faculty, you know, disputes about this, that, and the other, because it was all very Cambridge-focused."[10]

Jane, too, found the atmosphere at Caltech to be a refreshing break from her uncomfortable position at the very bottom of the Cambridge pecking order, from getting snubbed constantly at events and dinners. "Over the years we had become used to sitting trapped at the ends of tables or in corners on our own, never really expecting anyone to speak to us," Jane wrote. She was surprised by the openness and friendliness of the Americans toward her in California, to the point where she began to suspect that it was only because of her husband's success that she was being treated so generously.[11]

California also gave her some respite from the relentless grind of caring for Stephen. Two of Hawking's students, Bernard Carr and Peter D'Eath, had accompanied the family to Caltech, and Carr would be the first student whose academic duties would be formally combined with some degree of caretaking for Stephen. In return for lodging—and for extraordinary access to their thesis adviser—Carr and his successors would help feed, bathe, and dress Hawking; they would get him settled in bed at night and help perform other basic functions for him. Feynman visited Hawking's office on a number of occasions—he very much wanted to understand Hawking radiation, and the two had several discussions about that and other matters—and Carr remembers having to act as a go-between, translating Hawking's increasingly difficult-to-interpret speech so that Feynman could understand.[12]

Even though the Hawkings didn't have easy access to family members who could help in a pinch, thanks to the students the burden was somewhat lighter than it was back home. What's more, despite the higher curbs in the United States, Caltech seemed more interested in ensuring wheelchair access for Stephen than Cambridge ever was. Stephen also reveled in a spiffy new electric wheelchair—much faster than the model he had in Cambridge—which gave him an extra degree of independence. Even as Hawking's condition slowly worsened, Caltech came to represent a kind of freedom that couldn't be found so easily at Cambridge.[13]

The Hawkings even got to watch television. At Cambridge, they didn't even own a TV set. In California, they spent the evening watching *Upstairs Downstairs* and science historian Jacob Bronowski's *The Ascent of Man*—a smash-hit thirteen-part BBC series about the evolution of human culture and society.[14]

If there was a downside, it was that Caltech—so fiercely devoted to the sciences—didn't have even a hint of medieval lyric poetry in the libraries; there was no way that Jane could continue working on her thesis. Despite a flurry of socialization and bouts of voracious reading, by the end of 1974 she was laboring "in the certain conviction that my brain is turning to putty." And she suspected that she wasn't the only one: as she wrote at the time, she couldn't "get away from the suspicion that there are quite a lot of disaffected wives who are wretchedly demoralized by their husbands' obsession with science and their (possibly subconscious) disdain for any other occupation, walk of life, or interest." Even so, Jane found her element—by February 1975, she had discovered her love for chorale singing, and was finding comfort in people who really seemed to understand her.[15]

In April, Stephen heard that he was to receive a prestigious medal from the pope. One of Jane's new friends presented her with a pearl brooch, "because, she said, I should be given something too."[16]

· · ·

There's a photograph of Stephen Hawking receiving the Pius XI Medal from Pope Paul VI—a medal given by The Pontifical Academy of Sciences every other year to an outstanding scientist under the age of forty-five. The pope is dressed in all-white regalia, his white cassock spilling outward across the marble floor. The heir to St. Peter kneels before Hawking, who sits in his wheelchair, arms crossed rigidly in his lap, head bowed. His stick-thin legs are swallowed up in the fabric of his suit pants. Hawking was just thirty-three years old.

Hawking's reputation had grown rapidly from his work on singularities and black holes, but at the time, in 1975, nobody had definitively spotted a black hole, even indirectly. Unlike typical stars, black holes don't shine brightly with energy released from fusion reactions. Instead of emitting light, they swallow it. Hawking knew better than anyone that black holes weren't *totally* black—that they emit photons of light and other particles, but that radiation would (almost always) be far too faint to detect from Earth.* "How could we hope to detect a black hole, as by its very definition it does not emit any light?" Hawking wondered in *A Brief History of Time*. "It might seem a bit like looking for a black cat in a coal cellar."[17]

One might be able to detect black holes almost directly—by watching massive stars wheel around an empty spot in space. But this only works well with relatively long observations with high precision for nearby or massive black holes, like the one at the very center of our galaxy. Nothing like that had ever been done before; indeed, these sorts of observations were still fifteen to twenty years away at the time.

Another possibility would be to detect the gravitational waves coming off of a massive object circling and being swallowed by a black hole. But in 1975, the first direct measurement of gravitational waves was still a pipe dream. There had been a few false alarms, as a few researchers had convinced themselves that they had detected the rippling of spacetime, but all such detections turned out to be fantasies—at least until LIGO finally did it in 2014. Four decades away.

A third option would be to spot light coming off of a black hole's lunch. Black holes happen to be messy eaters; even as they swallow matter falling across the event horizon, their monstrous, spinning magnetic fields eject hot gas in enormous jets. The matter in these jets ought to be so hot that they radiate X-rays. Even without the jets, the disk of matter swirling around the black hole, the so-called *accretion disk*, was also extremely hot

---

* The nature and amount of radiation emitted by the black hole depends on how small it is, so, in theory, a very small black hole—much smaller than anything known to exist—might be visible as it emits light or, eventually, explodes.

and radiating X-rays as well. So an X-ray telescope ought to see something. And, starting in the mid-1960s, detectors flown above the X-ray-absorbing atmosphere began to look at the heavens with X-ray eyes.

One of the first handful of X-ray sources astronomers found was a brilliant little spot in the constellation Cygnus, the swan. Unlike many of the other sources that shone brightly with X-rays, this one had no known star—nothing that was visible in the optical or radio spectrum—that could possibly be a source of the X-rays. Could this new object, Cygnus X-1, be a black hole?

Scientists were hopeful, but proving it was a difficult matter. There were lots of possible objects that could be responsible for X-rays, even ones without an obvious source; before concluding that Cygnus X-1 was a black hole, astrophysicists had to whittle down the other possibilities until nothing else remained. So they did what they did best: observe. By 1974, they had nailed down its position and showed fairly convincingly that the nearby visible star couldn't be responsible for the X-ray emissions. They had also proved that the source was compact rather than diffuse. And it seemed that something massive—several times the size of our sun—was pulling on a nearby star. It was getting harder and harder to argue that Cygnus X-1 was anything but a black hole.

Thorne had been doing a lot of thinking about Cygnus X-1. Along with a former student, Richard Price, he had made some detailed predictions about what future observations, with better telescopes, should be able to see, if in fact the X-ray source were a black hole. They submitted their paper in November 1974, hoping they weren't building a castle on sand.

Like Thorne, Hawking had spent much of his career pondering the properties of an object that—at the time—hadn't ever been seen. "I have done a lot of work on black holes, and it would all be wasted if it turned out that black holes do not exist," Hawking noted wryly in *A Brief History of Time.*

Over the years, Thorne had picked up a habit of making scientific bets with colleagues. At first, they were unwritten—like one in the late 1960s with Yakov Zel'dovich over the properties of spinning black holes (the stakes: a bottle of liquor). But recently, Thorne had taken to writing them down and framing them. At the time, there was only one such wager hanging on the wall next to the door of his office: a bet with Bill Burke, one of Thorne's graduate students, about certain properties of gravitational radiation.[18]

So, right after Thorne submitted his Cygnus X-1 paper, the two scientists decided to make a fun little wager—Hawking would bet *against* Cygnus X-1's being a black hole. "Whereas Stephen Hawking has such a large investment in General Relativity and Black Holes and desires an insurance policy, and whereas Kip Thorne likes to live dangerously without an insurance policy," the document read, "therefore be it resolved that Stephen Hawking bets 1 year's subscription to 'Penthouse' as against Kip Thorne's wager of a 4-year subscription to 'Private Eye,' that Cygnus X-1 does not contain a black hole of mass above the Chandrasekhar limit." Thorne had the document framed and hung it in his office. (*Penthouse* was a very Caltech choice—according to Thorne, a "tongue-in-cheek sort of thing," a sort of counterculture reaction to his very conservative Mormon upbringing.)[19]

For Hawking, the bet was a win-win scenario: either astronomers would have confirmed the existence of black holes, or he would get the consolation prize of a four-year subscription to *Private Eye*, a beloved British satire magazine. The 4-to-1 ratio of *Private Eye* to *Penthouse* reflected the two scientists' estimate that the evidence was already 4-to-1 in favor of Cygnus X-1 being a black hole.

It would be a decade and a half before Hawking would concede the bet. "Late one night in June 1990, while I was in Moscow working on research with Soviet colleagues," Thorne later wrote, "Stephen and an entourage of family, nurses, and friends broke into my office at Caltech, found the framed bet, and wrote a concessionary note on it with validation by Stephen's thumbprint."[20]

As nice as it was to have experimental evidence that black holes really do exist, in truth, it wasn't terribly important to the sort of work that Hawking was doing. His analysis of black holes was less about figuring out the details of an astrophysical object than about divining the laws that govern how space and time behave. And perhaps something even more fundamental than that.

• • •

Upon his return to Cambridge in 1975, Hawking was promoted and at last given a permanent position—a "readership," a junior form of professorship—and some stability. Previously, Hawking had bounced around Cambridge with a series of fellowships and research assistantships—not by itself unusual for a person his age, but more and more untenable given his condition and his increasing renown. After all, Hawking had become one of the world's leading experts on black holes; any scientist interested in those odd collapsed stars would encounter Hawking's work. As a result, Hawking's name was likely to come up in any scientific discussion about black holes, no matter how farfetched.[21]

For example, in 1973, two physicists at the University of Texas at Austin suggested that the 1908 Tunguska event—a mysterious blast that flattened hundreds of square kilometers of Siberian forest—had been caused by a miniature black hole smashing through the Earth. Miniature black holes had been dreamt up by none other than Stephen Hawking, so any story that mentioned the Tunguska theory had to mention him. (In fact, Hawking's second appearance in the pages of the *New York Times* was in a 1974 article about Tunguska.)[22]

There were maybe a dozen or two theorists—at Cambridge, at Caltech, at Moscow, at Princeton—who knew that the recent work on black holes heralded more than just a better understanding of a new and bizarre type of object in the heavens. It was a key to reconciling the clash between quantum theory and relativity. And one discovery by Hawking, in particular, promised to blow the field wide open. As a front-page story in the *New York Times* put it a few years later:

> Many scientists believe a new master theory, one uniting existing physical field theories, is needed to look back to the instant of creation and beyond.
>
> The noted British cosmologist, Dr. Dennis W. Sciama, believes such a theory may already be taking form in the hands of one of his former research students, Dr. Stephen Hawking.
>
> Dr. Hawking discovered in 1974 that black holes, celestial objects of such enormous density that neither light nor matter may escape them, actually radiate energy and finally explode.[23]

When Hawking discovered that black holes could radiate and evaporate in a burst of energy (more on this discovery in the next chapter), it had all sorts of unexpected consequences—consequences that Hawking and his colleagues were just beginning to grasp.

The past decade of work, what Kip Thorne dubbed the Golden Age of Black Holes, had proven beyond a doubt that black holes had incredibly troublesome properties. In 1964, Roger Penrose showed that at the center of a black hole, there had to be a singularity, a point where the fundamental assumption that spacetime is a smooth manifold no longer holds; the laws of relativity must break down when a black hole is born. However, that singularity is shielded from view, entirely blocked off from the rest of the universe. The singularity is shrouded by the event horizon. Beyond that imaginary boundary, there is no means of escape from the black hole—not even a means of communicating anything to an outside observer.

Shortly thereafter, Werner Israel, Brandon Carter, Stephen Hawking, and other physicists formulated the no-hair theorems: the idea that a black hole has almost no distinguishing features whatsoever; once you know how much a black hole weighs, how fast it is spinning, and how much electrical charge it carries, there's nothing else to say. There is nothing else you *can* say; mass, charge, and angular momentum are the only information that someone outside of the event horizon can extract from a black hole. Everything else about the black hole is lost, inaccessible to anyone unwilling to cross the event horizon after it. That's what general relativity says, anyhow.

Quantum mechanics also has something to say—its laws dictate absolute limits on extracting information from a system. It's impossible to know both a particle's position and its momentum with perfect accuracy; after a certain point, getting information about one means losing information about the other. Just as important, although less well explored by scientists at the time, was the implication that information cannot be totally annihilated—it can be moved from place to place, stored, transformed, gathered, scattered—but it can't be totally erased. This principle is hardwired in the mathematical structures that describe quantum objects, thanks to a property known as *unitarity*. Without this property, the quantum mechanics doesn't really make sense; the logic of quantum interactions is no longer perfectly consistent.

So quantum mechanics says that information can't be lost under any circumstances, while general relativity says that when information falls into a black hole, it can't escape. These two statements aren't necessarily in contradiction; perhaps the black hole is storing the information, keeping it out of sight forever behind the event horizon. That was perfectly acceptable to both the relativists and the quantum theorists. Everything was copacetic. Until Hawking messed it all up.

Hawking's 1974 discovery that black holes radiated energy was counterintuitive and troubling. And he soon discovered that it had a very disturbing implication. When a black hole radiated energy, the energy had to come from somewhere; in fact, it came from the mass of the black hole itself. So as a black hole radiates, it gets less massive—smaller—over time. As the black hole gets smaller, Hawking's calculations implied that its radiation becomes hotter and brighter, making the black hole shrink even faster. Smaller. Brighter. Smaller. Brighter. The process can take billions of years, but eventually, the black hole, in theory, shrinks to a point and explodes in a flash. Black holes aren't permanent; they explode.

It took many of his colleagues a few years to realize how important this revelation was. Hawking—perhaps because of his exposure to particle physics at Caltech—quickly understood that the impermanence of black holes drove a major wedge between the rules of general relativity and quantum theory.

Quantum mechanics says that information can't be lost under any circumstances. Relativity says that when information falls into a black hole, it can't escape. And now, Hawking argued, black holes must eventually themselves be destroyed: they can't store information indefinitely behind an event horizon. Someday, far in the future, the event horizon must finally be ripped away, leaving nothing behind. So where did the information go? It couldn't have been stored; it couldn't have escaped past the event horizon; it couldn't have been encoded in the Hawking radiation (for the nature of Hawking radiation implies that it couldn't carry information).

That's really the essence of the black-hole information paradox. Quantum theory says that information cannot be destroyed. Yet Hawking's description of a black hole implies not just that information is lost to an outside observer during a black hole's lifetime, but that a black hole's lifetime is finite—and no information survives its demise. It's a dilemma that sits squarely at the intersection of quantum theory and relativity. Two theories giving contradictory answers—both can't be correct. Perhaps, just perhaps, by figuring out what's going wrong in one or the other analysis (or both), Hawking could take a large step forward in reconciling these two mathematical frameworks. By delving into the heart of a black hole, Stephen Hawking might have pointed the way to a new master theory: one that would surpass quantum theory and relativity just as those theories had surpassed Newtonian mechanics.

The first step, though, would be to convince others that there was a contradiction. Toward the end of his stay at Caltech, Hawking wrote up a paper outlining the information theory paradox and submitted it to *Physical Review D*. It was published more than a year later—likely the sign of a battle among peer reviewers who took issues with Hawking's argument. And though the work was noticed (and cited) over the next few years by a handful of relativity theorists, particularly those in Dennis Sciama's and John Wheeler's circles, the black-hole information paradox took a relatively long time to diffuse into the wider scientific community. Even Leonard Susskind, who would spend much of his career wrestling with the implications of the paradox, didn't hear about it until 1981—fully half a decade later.

As Susskind tells it, he first heard about the black-hole information paradox at a quirky little gathering in the attic of a very controversial millionaire: Werner Erhard. Erhard was the leader of a self-improvement movement that grew rapidly, gained high-profile celebrity followers like John Denver and Yoko Ono, and was soon likened by some to a cult.

Erhard pumped some of the money from those self-improvement trainings, known as "est," into a foundation—and with that foundation, he managed to attract some of the world's leading physicists around him. Starting in the mid-1970s, Harvard's Sid Coleman, Caltech's Dick Feynman, and a number of other leading physicists* had been periodically wined and dined and entertained at a physics conference paid for by the largesse of the est Foundation. Coleman, who helped recruit attendees, had no illusions about how participants might feel about the "self-improvement courses" that bankrolled Erhard's fortune:

> The proposed conference will no more be devoted to promoting Erhard Seminars than the activities of the Ford Foundation are to pushing Pintos. I have received explicit agreements to this effect from the responsible parties, and I promise you that at the slightest sign these agreements are not being kept, I will throw a tantrum and cancel the conference.[24]

The agreements were kept—kept well enough, at least, to convince some of the physicists—particularly Feynman, Coleman, and Susskind—to come back multiple times. "[Erhard] would have the three of us up regularly to come to San Francisco and dine," Susskind recalls. "He would bring in famous chefs and he had this whole stable of servants . . . not servants. They were his cult followers, and they would bring us food and wine. It was very strange." Like Coleman, Susskind had no illusions: "You know, I knew he was a faker. I knew you call him a cultist or whatever. I was completely aware of that. And yeah, I was a little bit worried that this is going to come back at us but it somehow never did." But, from Susskind's point of view, Erhard never abused the relationship: "We were always concerned that he was going to use us in some nefarious way. And he never did. He never did," he says. "He just wanted us to sit and talk with physicists. He's very smart."[25]

Stephen Hawking was invited to an Erhard conclave in 1981, and it was there, in an attic on Franklin Street in San Francisco, that Susskind

---

* The list of people attending one 1977 gathering reads like a who's who of cutting-edge particle physics. It includes several future Nobel winners: Curtis Callan, Geoffrey Chew, Sid Coleman, Ludwig Fadeev, Dick Feynman, Jeffrey Goldstone, Yoichiro Nambu, David Gross, Roman Jackiw, T. D. Lee, Larry Susskind, and Gerard 't Hooft, to name a few.

first understood just how important Hawking's paradox was. It was, in fact, the first time that Susskind met Hawking. "Of course, your impressions are dominated by the wheelchair and the simple fact that this man can function as a theoretical physicist without being able to move a muscle practically. And that was astonishing. There was no question of his extraordinary talent," Susskind says. At that point, Susskind was not an expert on black holes; he was mostly interested in straight-up particle physics, looking at the behavior of the strong force and quarks, rather than anything astrophysical. But when Hawking presented his argument about information loss in black holes, Susskind was thrown for a loop. "When Stephen presented his conclusions about information being lost, that just felt wrong. [Future Nobel winner Gerard] 't Hooft was also there, and 't Hooft was very perturbed by it. I would almost say angered by it," says Susskind. "The three of us stood in a room in Werner's attic, only the three of us. And we stood there at the blackboard with a Penrose diagram of a black hole. And Gerard was standing there with the veins sticking out of his neck." Hawking sat in his wheelchair with his infuriating, impish smile. "He was amused that we were so puzzled and confused. And I was shocked that he had such a convincing argument that quantum mechanics was wrong. . . . I don't think either 't Hooft or I could see a way out of it."[26]

"I could not let go after that," Susskind says. It would be a slow burn, but the fuse was lit. To Susskind, 1981 was the beginning of a "war" with Hawking over the fate of information in a black hole, a war that would eventually draw in physicists from around the world: nothing less than the integrity of quantum mechanics was at stake. And though Hawking himself would concede in Dublin in 2004 that he had been wrong all along, physicists are still struggling with the implications of Hawking's insight. And Susskind both resented and admired the man who had posed such a baffling paradox.

"I didn't like him at first. More than that, I disliked him," says Susskind, who found Hawking to be a stubborn, frustrating intellectual adversary. "But there was a positiveness to the man. Real positiveness to not feel sorry for himself. I never thought in any way that he felt sorry for himself. So that was kind of a revelation that somebody could be that strong."

. . .

Hawking seemed to enjoy getting under people's skin; a conversation with him could quickly take a turn into an attempt to irk someone, even establish dominance over him or her. He would even try to needle his closest colleagues. Roger Penrose once told of a dinner party in the 1970s: "Sitting at the head of my table, seeming intent on getting a rise out of those present, Hawking made three remarkable assertions," Penrose wrote.[27]

Hawking's first assertion had to do with subatomic particles; he suggested, contrary to what particle physicists tended to believe, that protons decayed over long time scales. While the suggestion was "unorthodox," it didn't particularly trouble Penrose. "Hawking's proposal entailed an unobservably slow decay rate, not disturbing to my world-view."

His second assertion was related to the black-hole information paradox. Hawking asserted that the laws of quantum mechanics were fundamentally broken, and in a way that people who spent their lives studying quantum mechanics couldn't accept. But this, too, failed to get a rise out of Penrose. "Far from displeasing me, this was appealing, as I already had other reasons for believing quantum mechanics must someday require profound modification. I was rather glad to have Hawking's new line of reasoning. . . . Again I nodded my acceptance and exclaimed approval," recalled Penrose.

"I fear that this was far from the reaction that Hawking wanted, so he launched into 'Black holes and white holes are the same thing!'" A "white hole" is a hypothetical time-reversed black hole—an object that vomits up matter rather than swallowing it. "That did it, for me."

Despite his mental strength—and combativeness—Hawking's physical condition continued to deteriorate. Though friends and assistants tried to get him out of his wheelchair for a couple of hours a week for exercise—supported on each side as he would attempt to move his legs—his limbs were badly atrophied, and no amount of therapy could prevent the inevitable loss of function. Then there were the coughing fits. His trachea would become irritated, and he would choke and sputter and wheeze, unable to catch his breath for hours on end. Jane varied his diet in an attempt to minimize the irritation to his throat, and eventually settled on one without sugar, gluten, and dairy products. That was when he was well; if he happened to catch a bug, Stephen could quickly wind up in the hospital with a respiratory infection—often life threatening, given his fragile state of health.[28]

In addition to raising Lucy and Robert, Jane was primarily responsible for taking care of the physical needs of her husband—feeding him, washing him, helping him void himself. The National Health Service wasn't providing nearly enough support; nurses would come round two mornings a week, far less than his condition required. Stephen's needs were relentless, and getting bigger all the time, not just as his illness progressed, but as his renown increased. Stephen had to do more travel and attend more conferences, more parties, more ceremonies, and more receptions, all of which interfered with whatever routine Jane could establish, even with the assistance of a live-in graduate student. By 1976, Jane wrote, the nine-year-old Robert had been called upon to "fetch and carry, lift and heave, feed and wash, and even take his father to the bathroom when I was overwhelmed with the weight of other chores or just too exhausted to respond."[29]

Stephen's strength of will was in some ways a curse as well as a blessing. In stubbornly clinging to whatever independence he still had, he refused to seek outside assistance. "I was at breaking-point but still Stephen was determined to reject any proposals which might have suggested that he was making concessions to the illness," Jane wrote in her memoir. On a radio program a number of years later, she elaborated: "I think that was the great difficulty in our marriage, because he just clammed up at any mention of motor neurone disease. . . . He really clammed up and gave the impression that it was too hurtful for him to talk about it, so we couldn't talk about it and we didn't talk about it," she said. "He was in denial about his illness, about the disease, and therefore he refused . . . every possible means of help."[30]

When things got bad, Stephen wouldn't let Jane call a doctor; he was "inherently mistrustful of the medical profession," she wrote, because of the doctors' "shabby treatment of him in 1963 at the time of his diagnosis." He would instead rely on his father's occasional advice, and vitamins, and try to tough it out. He wouldn't even take cough syrup, likely for fear that it would suppress his breathing. Jane did her best to keep Stephen out of the hospital, but it often wasn't enough.[31]

In 1979, his condition was so grave that Jane had to put Stephen in a nursing home for a brief spell. Sensing that the family needed assistance, one of his colleagues, astrophysicist Martin Rees, stepped in to help find a bit of money from charitable organizations to pay for some home care. With that help, Jane managed to find a nurse. At first, Jane wrote, Stephen resented the nurse's ministrations: "When he came for his first shift, Stephen adamantly refused to acknowledge his presence, to look at him or to communicate with him in any way, except by running over his toes with the wheelchair." But over time, as with the other concessions that Stephen had to make as a result of his disease, he swallowed his pride and got used to the changes forced upon him.[32]

· · ·

Hawking didn't let anyone outside his intimate circle see even a hint of his struggle; he would wheel through the streets of Cambridge day after day to work, and was proud of how seldom he failed to show up. He labored for hours together with his students or by himself, completely absorbed by the physics problems swirling inside his head.[33]

There were plenty of problems to work on. His time at Caltech had left him with greater interest in particle physics and the techniques peculiar to that field, especially the Feynman path integral method. Starting in 1976, along with Jim Hartle and Gary Gibbons, he spent several years exploring and refining the ideas that would eventually lead to his no-boundary proposal in 1981. Then there was the black-hole information paradox; he had convinced himself that information had to be lost when it fell into a black hole, but he hadn't yet come up with an explanation for where it might go. Hawking would think about this on and off for years, eventually coming to the conclusion that the particles falling into a black hole "go off into a little baby universe of their own"—that the information wasn't truly lost, but was carried off into a region that bubbled off our own sheet of spacetime, forming its own tiny cosmos. The black hole, the ultimate engine of destruction, was also a source of creation. (Of course, Hawking was forced to abandon this notion when he conceded that information wasn't lost in a black hole.) So Hawking was not showing any signs of running out of ideas.[34]

It was those ideas—that brain—that made people take notice of Hawking, and allowed him to climb the academic hierarchy so quickly. Alas, the perks of such a high academic position weren't financial; even at its peak, Hawking's Cambridge salary was never enough to sustain him in a comfortable manner, especially given his medical expenses. The chief benefit was respect. And Hawking was getting plenty of that. Sometimes that respect even translated into a cash dividend.

The Einstein Award, for example, was a sign of respect that came with $15,000 and a gold medal. It was a somewhat controversial honor that had been given out, off and on, since the 1950s. Controversial because it had been put together by former US Atomic Energy Commission chairman Lewis S. Strauss, who was reviled by much of the scientific community for engineering the downfall of J. Robert Oppenheimer, the scientist who, during World War II, had led America's project to design the atom bomb. With few exceptions, the award had gone to bomb designers, mostly hawkish ones. (When presented with the award in 1954, Richard Feynman didn't want to accept it until physicist Isidor Rabi—an outspoken dove—convinced him to. "You should never turn a man's generosity as a sword against him," Rabi reportedly said.)[35]

In August 1977, Strauss' son sent out a letter to former winners of the prize, including Feynman and John Archibald Wheeler, asking for advice.* "Do you think that Hawking's work is of sufficient importance to justify the Einstein Award, quite apart from the humanitarian aspects of the matter?"[36]

Wheeler's response left little doubt. He explained that he had co-written a textbook on gravitation in the early 1970s that had included biographical sketches of the important scientists in the field of gravitation, from Galileo onward. Only three of those sketches were of living people: Bob Dicke of Princeton University, Roger Penrose, and Stephen Hawking.[†] "I have no question he belongs among the immortals," Wheeler wrote. (Feynman's response was less effusive. His whole response to Strauss' letter reads: "In reply to your letter of August 2, I would most certainly agree that Dr. Stephen W. Hawking's work is deserving of the Einstein award.")[37]

The following January, Hawking was awarded the prize. "The work being done by Hawking and his group may lead to a Unified Field Theory which has been the Holy Grail of 20th century physicists," the award announcement declared. "It was sought, unsuccessfully, by Einstein for the last 40 years of his life."[38]

Like Einstein, Hawking would never set eyes on the grail. But it's not necessary to complete a quest to achieve immortality. And as Wheeler had predicted, Hawking was already being counted among the immortals of science.

In late 1979, officials at Cambridge decided that Hawking would become the next Lucasian Professor of Mathematics. The Lucasian Professorship was created in the 1600s by order of Charles II to fund, in perpetuity, a "man of good character and reputable life . . . soundly learned and especially skilled in mathematics." The second person to hold the position was none other than Isaac Newton—and from then on, the Lucasian Professorship was the most prestigious academic position in the world. Its holders were instantly considered (rightly or wrongly) to be the intellectual heirs to Newton himself. Many of the subsequent Lucasian Professors— Charles Babbage, Gabriel Stokes, Paul Dirac—were legendary figures in mathematics and physics, adding to the already nearly mystical prestige

---

* Wheeler, despite being a hawk on the Lewis Strauss and Edward Teller side of the Oppenheimer schism, never suffered as a result. He was almost universally beloved by the physics community.

† When Hawking sent Wheeler some biographical information in 1972, it may have been the first professional use of his famous "born-300-years-after-Galileo died" line. Letter from Stephen Hawking to John Archibald Wheeler, June 12, 1972, Wheeler Papers, Box 11.

of the office. And now, Hawking was counted among them. Nobody could any longer question Hawking's position as a first-rate scientist, or doubt that his growing fame was based upon his important contributions to theoretical physics.[39]

Nobody, that is, except Hawking himself.

"I think I was appointed as a stopgap to fill the chair as someone whose work would not disgrace the standards expected of the Lucasian chair, but I think they thought I wouldn't live very long, and then they could choose again, by which time they could find a more suitable candidate," Hawking told an interviewer a number of years later. "Well, I'm sorry to disappoint the electors."[40]

PART III

INSPIRAL

But at my back I always hear
Time's wingèd chariot hurrying near;
And yonder all before us lie
Deserts of vast eternity.

—ANDREW MARVELL,
"To His Coy Mistress"

The salmon-falls, the mackerel-crowded seas,
Fish, flesh, or fowl, commend all summer long
Whatever is begotten, born, and dies.
Caught in that sensual music all neglect
Monuments of unageing intellect.

—WILLIAM BUTLER YEATS,
"Sailing to Byzantium"

CHAPTER 13

BLACK BODY (1970–1974)

———————————

S tephen Hawking's greatest scientific discovery was born from spite.
August 1972. Leading physicists were gathered at a summer school—
this one at Les Houches in the French alps, not far from the borders
of Italy and Switzerland. The subject was black holes,* and in attendance
were two of the people who had recently been making such a splash in
the subject in the past few years: Stephen Hawking and Brandon Carter,
academic siblings who had come out of Dennis Sciama's shop at Cambridge. Kip Thorne was also there, along with his Russian colleague Igor
Novikov, as was Feynman protégé (and son of a double Nobel Prize winner) Jim Bardeen.

Seeking to learn from these experts, a young student of John Archibald
Wheeler's, fresh from completing his PhD, was also there. But this student, Jacob Bekenstein, had put his foot in it—he had suggested something
about black holes that Hawking found outrageous. Irritating. Offensive,
even. Bekenstein, he felt, had taken one of Hawking's key insights and
twisted it so that it no longer made any sense. An angry Hawking gathered
up Carter and Bardeen and the three experts faced off against Bekenstein.
"These three were senior people. I was just out of my Ph.D. You worry
whether you are just stupid and these guys know the truth," Bekenstein
later said. But Bekenstein was not to be bullied out of his idea; despite the

———————————

* At the time, the French term for black holes was *astres occlus* (occluded stars); the French
were still resisting the obscene-sounding name *trous noirs* (black holes) that Wheeler had
coined—made even worse by Bekenstein's observation that "black holes have no hair."

hammering he took from Hawking and his friends, Bekenstein was convinced he was on to something, and he would keep pressing forward. So Hawking, Bardeen, and Carter did the only thing they could do—write their own counter-paper. "I must admit that in writing this paper I was motivated partly by irritation with Bekenstein, who, I felt, had misused my discovery," Hawking wrote in *A Brief History of Time*. It was the beginning of a two-year struggle to understand a seeming paradox that would unveil a totally unexpected property of black holes, a property that would ensure Hawking's name among the immortals of physics.[1]

Hawking's magnum opus was an attempt to prove Bekenstein wrong, and, against all expectations, he proved his adversary was right after all.

· · ·

The years from 1970 through 1974 were the peak of Stephen Hawking's scientific life. Two major results, two big insights about black holes, were enough to establish him as a physicist of the first rank.

Hawking was laboring in an increasingly crowded field. There were several other physicists, from the United States, from Russia, from England, and from Israel, who were all working together—and against each other—in divining bizarre and unexpected properties of black holes. But it was Hawking who finally came to the key realization, one so strange that even he didn't believe it at first. It was Hawking whose name would be forever linked with the understanding that black holes weren't really black. And this discovery became the crowning achievement—and the coda—to the Golden Age of Black Holes.

· · ·

The soul of a black hole is its singularity. The idea of a point of infinite curvature where the laws of physics no longer make sense—that's the puzzle that first drew physicists into studying black holes. They wanted to understand the pathological heart of these collapsed stars. And in the late 1960s and early 1970s, nobody understood singularities like Roger Penrose and Stephen Hawking. Hawking had made his name on singularities. But by 1970, Hawking was beginning to turn his attention not to the singularity, but to another part of black-hole anatomy: the event horizon.

An event horizon isn't a physical thing any more than the equator of the Earth is a real line; it's an abstract mathematical concept. Luckily, physicists were more or less agreed on the definition of that concept. An event horizon was a boundary around the black hole that marked the point where even light wasn't moving fast enough to avoid crashing into the singularity. That is, the event horizon was an imaginary surface where the escape velocity—the speed an object needed to be moving to get out of the black hole's clutches—was light speed. This was the natural, obvious definition. But it had subtle problems. For example, as a star collapsed into a black hole, the full-sized event horizon would suddenly pop into existence once the surface of the star got small enough. Instantaneous, discontinuous changes lead to problematic math.

In 1970, Hawking realized that a subtle change in the definition of the event horizon could eliminate those problems. Instead of thinking of the event horizon as a place where the escape velocity was equal to the speed of light, he thought of it as a boundary between two separate realms: the realm where objects were doomed to hit the singularity versus the one where objects could still belong to the outside universe. Hawking's definition, which he dubbed the *absolute horizon*, wasn't *quite* the same as the old escape-velocity definition, now termed the *apparent horizon*. It takes a bit of playing around with Penrose diagrams to see the differences, but they're there. For example, when a star collapses into a black hole, an absolute horizon doesn't suddenly pop into existence fully grown like an apparent horizon does; it starts off as a point in the middle of the star and grows and grows as the star collapses, eventually bursting outward beyond the surface of the shrinking star.*

It's a somewhat confusing and very subtle difference that doesn't mean a heck of a lot to a nonspecialist. But to Hawking, who was painting Penrose diagrams of collapsing stars in his head, it was a terribly important distinction that forced him to look at black holes in a new way; it changed his mental picture of the geometry of black holes. And one night in late 1970, not long after his daughter, Lucy, was born, he was exploring his new mental map while getting ready for bed. All of a sudden, he was struck by a geometric insight.

Hawking was picturing how light rays behaved when on the absolute event horizon, and his new definition made it clear to him that those rays could never approach each other. They could move parallel to each other, or get farther apart, but they couldn't get closer together, because if they did, at least one of them would be captured by the black hole. This is a really tough concept that Hawking struggled to try to explain to nonscientists. In *A Brief History of Time*, he tried to give his readers a sense of

---

* Another way to think of it: the old definition of event horizon was, roughly speaking, the boundary between light that can escape and light that can't escape at a given moment in time. Hawking's new absolute horizon is the boundary between light that could escape and light that couldn't escape not just at one point in time, but throughout all of time. As a star collapses, light in the center of the star might be unable to escape the black hole that will form in the future, but *doesn't yet exist*, and thus doesn't have an old-style event horizon!

this geometric insight by likening it to robbers trying to escape from cops: "It would be like meeting someone else running away from the police in the opposite direction—you would both be caught!"[2]*

That geometric insight led to another. The behavior of light rays on the event horizon determines the shape that the event horizon can take, not just through space, but through time. Saying that light rays on the event horizon can't approach each other is equivalent to saying that the horizon can never get smaller over time; it must always get bigger, no matter what. And when two small black holes merge to make one big black hole, the event horizon of the big black hole must be at least as big as the combined areas of the smaller black holes. That is, the geometric rules of general relativity ensured that black holes could never shrink; they had to grow, or at the very least stay the same size. It was a new law of nature, a novel insight not just into the behavior of black holes, but also into the nature of space-time itself. This was what became known as Hawking's *area theorem*.

"I was so excited with my discovery that I did not get much sleep that night," Hawking later wrote. "The next day I rang up Roger Penrose. He agreed with me." By Hawking's telling, Penrose had, in fact, been thinking along the same lines, but hadn't yet grasped its significance.[3]

"The thing about the area theorem, this needs to be straightened out, because I don't know," says Penrose. "You see, I'd been in Cambridge to give a talk, and after the talk, I went to Stephen's office, and we talked about various things. And I talked about the area of a black hole in general, and the fact that the area had to increase over time." The two discussed the idea for a while, and puzzled a bit about the nature of event horizons, because, in general, their shape could be quite complicated and might defy analysis. Penrose then went back to his hotel room and went to bed. "Early the next morning, Stephen phoned in, and he said he got a new idea. And he said, 'It's really your idea.'"[4]

Nearly half a century on, Penrose finds the story of Hawking's evening revelation a bit confusing. The idea that black-hole areas had to increase was already well in hand before Hawking put his daughter to bed. "Exactly what his epiphany was, I don't know. I thought it was that the areas of black holes increase, but that wouldn't be completely consistent because I have a

---

* A slightly different way of looking at it is to remember that a particle of light on the event horizon is just on the edge of getting caught—it has to be traveling as fast as it can (at light speed, naturally!) on the shortest path away from the singularity just to avoid getting captured. If two photons, A and B, are frantically trying to escape and cross each other, then it means there are two "shortest" paths away from the intersection: the path that A takes after meeting B, and the path that B takes after meeting A. But there can only be one shortest path—which means that the situation is impossible.

clear memory that he said it was your—he's talking to me—idea," Penrose says. "I suspect it may have been a misunderstanding. . . . He might have thought I was talking about apparent horizons. . . . I certainly wasn't talking about that." Penrose credits Hawking with first thinking about the combined area of two black holes colliding and combining into one, rather than merely looking at the horizon of an individual black hole increasing over time—but that distinction isn't really the key insight of the area theorem, nor does it fit in with the story of Hawking's bedtime epiphany. "I don't know what he thought. Maybe he thought I had the idea but didn't quite have it. It's not clear. I don't know what the story was, really," Penrose says. "I never wanted to bring it up. Because it was a big thing for him."

It was a big thing. The area theorem was a breakthrough; it allowed theorists to constrain the behavior of black holes in ways they never could before. For example, by applying the area theorem to the merger of two black holes—precisely the sort of event that the LIGO gravity-wave detector started seeing forty-five years later—Hawking was able to calculate the maximal amount of gravitational radiation that gets released in the merger. Too much, and it would cause the black hole to lose mass and the horizon to shrink, violating the area theorem.

It was a profound discovery. But even Hawking didn't yet realize just how important it would turn out to be.

• • •

A cross the Atlantic, the inspiration came from a cup of tea. Two, actually. At Princeton, John Archibald Wheeler was pondering what happens when you bring two cups of tea into contact—one hot and one cold. You don't need an advanced degree in physics to know that heat spontaneously flows from the hot one to the cold one until they both reach the same temperature. But to a physicist, the question of *why* nature behaves that way, why heat always flows from the hot cup to the cold one and never in reverse, is the gateway to some fundamental insights into the rules that govern our universe: rules about how energy behaves and how time flows. It all has to do with a concept known as *entropy*.

Imagine that you've got one hundred marbles—fifty red and fifty blue. Pick one at random and toss it, plunk, into a shoebox a few feet away. Then do the same to another marble. And another. Plunk, plunk. Continue until you've tossed all the marbles into the shoebox. Go over and take a look; what do you see?

You'll (obviously!) see a jumble of red and blue marbles all mixed together. When you dump those marbles randomly into the shoebox, they wind up randomly arranged in the box—there's no particular order to how they fell, and no particular order to how they lie. Indeed, if you looked in the box and all the red marbles were sitting on the left side and the blue marbles on the right, you'd be astonished. The world doesn't work that way: if you toss marbles randomly into a box, they won't spontaneously self-segregate by color. Of course, it's not impossible to have a box with red and blue marbles neatly separated into each corner, but even in such a case, it probably won't last long. As soon as someone picks up the box, the marbles will jostle and mix with each other once again. The nice neat order imposed from without will quickly dissipate into chaos once more.

That's entropy in action. Entropy, in some sense, is the amount of "disorder" in a system. A shoebox with all the red marbles on the left and all the blue ones on the right has low entropy; when someone moves the shoebox around and the marbles mix, the entropy increases and increases until the marbles are randomly distributed once again. It's possible to restore order to the system if you want; you can reduce the entropy in the shoebox by sorting the reds from the blues. However, doing that requires you to spend energy—and when you stop spending that energy to sort the marbles, the system's entropy will increase once again.

Entropy isn't just about colored marbles in shoeboxes; the same principle applies to all matter and energy everywhere in the universe. Red marbles might represent "hot" molecules in a cup of tea, and the blue marbles represent "cold" ones. Or the red marbles could stand for nitrogen in a room full of air, and blue ones for oxygen. It doesn't really matter. In each case, the system will tend toward maximum entropy—total, random mixing of red and blue—unless you expend energy to reduce the entropy. And if you don't have a way of pouring energy into the system to sort things out, the entropy must inexorably increase until it reaches its maximum.

This principle is one of the most important scientific discoveries of the nineteenth century, and it forms one of the pillars of the field of physics known as thermodynamics: the study of energy, of heat, of temperature, of work. The tomb of Ludwig Boltzmann, who first expressed entropy in terms of the different configurations of marbles/atoms/molecules/objects in a system, is inscribed with his formula for entropy:

$$S = k \log W$$

where S represents entropy, k is a number now known as the Boltzmann constant, and W represents (roughly speaking) the number of possible configurations of the objects in the system. And the second law of thermodynamics—the rule that entropy must increase in a system unless energy is poured in from an outside source—is considered by physicists to be immutable bedrock. It's the foundation of the scientific understanding of not just energy and heat and work, but much more than that besides.

Entropy is closely related to the arrow of time. Entropy always increases in our universe, so "future" can be defined as our universe when the entropy is greater than it is now, and "past" is our universe when there was less entropy than now. Entropy also determines the fate of the cosmos. Since our universe was created in a one-time event, the Big Bang, there's no more energy pouring into the system. This means that the entropy of the universe must unceasingly increase over time until everything in the cosmos is at the same, cold, lifeless temperature.* So, in some sense, anything we do that increases the entropy of the universe is bringing it one step closer to its eventual demise.

"You put a hot teacup next to a cold teacup and they come to a common temperature. That makes me unhappy," John Wheeler told an interviewer in 1996. "I have contributed to the entropy of the universe, the degree of the disorganization of the universe, by putting the two together." Simply by putting these two cups of tea near each other, Wheeler reasoned, he was helping destroy the universe in a small but very palpable way. The second law of thermodynamics says it must be so—and the second law of thermodynamics brooks no exceptions.[5]

Except . . . Wheeler thought he had come up with a possible way to cheat, using a black hole. Since a black hole has no hair, it seemed to be an eraser that wiped out all properties of infalling matter beyond its mass, its charge, and its spin around the black hole's axis. But if that was the case, then the entropy of the infalling matter had to be wiped out as well. If mixing the hot and cold tea increased the entropy of the universe, then, Wheeler realized, dumping the mixed cups of tea into the black hole afterward would essentially negate the act of mixing them.

The idea tortured him, and since it tortured him, he made sure it made his graduate students unhappy too. "I said when Jacob Bekenstein came in the room, then a graduate student, 'Jacob, if a black hole comes by, I could drop both teacups in the black hole and conceal the evidence of my crime.' Well, he looked uncomfortable, but he didn't have any positive objection then."[6]

Bekenstein was stumped. The second law of thermodynamics was inviolable, yet a black hole seemed to provide a loophole. Something had to be wrong.

· · ·

---

* Entropy is also closely tied to the concept of information. A fuller explanation of this relationship between entropy and information can be found in my book *Decoding the Universe*.

Around the time of Lucy's birth in 1970, Stephen resigned himself to using a wheelchair; he had been able to walk short distances, albeit unsteadily, with crutches, but that was becoming too difficult and too dangerous to maintain. He no doubt knew that sitting in a chair constantly would hasten the atrophy of his muscles and render him increasingly reliant on others to get about. But he tried to get exercise in other ways. He would insist on getting himself ready for bed—apart from untying his shoelaces and undoing his buttons, he was able to undress himself and get into pajamas. And he insisted on climbing the stairs to the bedroom, too, even after his legs were no longer able to support him.[7]

"The way he got up the stairs was, he grabbed hold of the pillars that support the bannister and pulled himself up the stairs with the strength of his own arms, dragging himself up from the ground floor on up to the second story in a long effort," Kip Thorne told director Errol Morris a number of years later. "Jane explained that this was an important part of his physical therapy to maintain his coordination and strength as long as possible. At first it was sort of heartrending to watch what appeared to be the agony of pulling himself up the stairs until I understood it was part of life. Pulling himself up the stairs like that." ("There I give credit to Jane," says Roger Penrose, who also saw the struggle with the staircase on more than one occasion. "I can imagine that many people would have insisted on helping him, but she felt that this was absolutely necessary.")[8]

The Hawking household was scraping by on sheer determination, but barely. Stephen's salary wasn't great, and it wasn't perfectly secure, as he wasn't tenured or on the tenure track. It was a stable fellowship with a renewable six-year term, yet it was clear that Stephen was destined for something bigger. The twenty-eight-year-old physicist was in the middle of a burst of white-hot creativity that filled him with energy.

Just a week after Lucy's birth, Stephen submitted a paper to the *Monthly Notices of the Royal Astronomical Society* suggesting that the universe might be filled with microscopic black holes.

Ordinary black holes—the ones that astrophysicists had been thinking about for decades—are born when large stars, several times the mass of our sun, collapse. Smaller, sun-sized stars simply don't have enough gravitational oomph to collapse down to a singularity; these become white dwarfs or neutron stars rather than black holes. So black holes formed in this way have a minimum mass, a bit more than twice our sun's mass. But Hawking realized that in the hot, dense early universe, it might be possible to create black holes that have much smaller masses, perhaps merely as heavy as a mountain. That sounds large, but a black hole of that size would be roughly as small as a proton, and might even work its way into the guts of an ordinary star and sit in its core, slowly eating it from the inside.

It was a really interesting idea, and theorists liked it for a few reasons—applied judiciously, mini black holes could account for miscellaneous unexplained phenomena that had been troubling astrophysicists.* Unfortunately, Hawking based his theory on a number of faulty assumptions. And part of his motivation was to explain some gravitational waves that an experimentalist, Joe Weber, was detecting with his instruments. Those detections turned out to be phantoms; his experiments were discredited a few years later. Even so, the idea was so compelling that Hawking's early appearances in the popular press were, as often as not, related to his mini black holes.

Hawking was also thinking about the no-hair theorems—he was helping to solidify the mathematics underneath the hood. At the same time, he was working with George Ellis, a friend and fellow former student of Dennis Sciama's, to finish up a treatise on spacetime, relativity, and black holes—a book that would become a classic in the field. He even won the Gravity Research Foundation's essay prize in 1971, which gave him a bit of cash in pocket.

That prize was something of a running joke in the relativity theory community: it had been set up by eccentric millionaire Robert Babson (best known for founding Babson College), who had a lifelong grudge against gravity, as he held it responsible for having drowned his sister. Despite the prize's cranky origins, impoverished physicists entered the contest year after year in hopes of winning the fairly generous purse. (Hawking entered half a dozen times over the years, but only won it once. When congratulated by post on winning such a prestigious prize, Hawking reportedly wrote back, "I don't know about the prestige, but the money's very welcome.")[9]

It was the most productive time in Hawking's life. Even without the area theorem, it would have been notable. However, the magnum opus was just over the horizon.

· · ·

Jacob Bekenstein had been hard at work for several months trying to understand what was going wrong with the scenario posed by his adviser. Wheeler was right: if black holes truly erased all evidence of what they

---

* For example, Hawking suggested that it could be used to explain a lack of a certain type of particle, known as a neutrino, created during fusion reactions in the sun. The deficit was later proved to be due to neutrinos having mass, rather than mini black holes at the core of the sun swallowing them up.

swallowed, it was irrelevant whether the infalling matter had high entropy or low entropy. The ultimate outcome was the same. And if that were the case, a black hole could serve as a mechanism for getting rid of entropy, for avoiding the thermodynamic consequences of your actions. You could mix hot tea and cold tea together without increasing the entropy of the universe, in violation of the second law of thermodynamics.

The second law of thermodynamics says that in a closed system—one with no energy flowing in—entropy must, without exception, increase, or at least stay the same. And in 1971, when Hawking published his area theorem, Bekenstein was struck by how it paralleled the second law: the size of a black-hole event horizon must, without exception, increase, or at least stay the same. Could there be some connection? Could the area of a black hole's event horizon be related to entropy? And could that harbor the answer to Wheeler's teacup mystery?

There were other reasons to think that the similarities between black holes and thermodynamical systems were important. For one, another student of Wheeler's, Demetrios Christodoulou, had worked out some rules regarding spinning black holes that were parallel to well-known principles in thermodynamics.* There were just too many coincidences on too many fronts for it to be an accident; the rules which dictate how black holes behave simply had to be intimately related to the rules of thermodynamics. So Bekenstein did some calculations and confirmed that, indeed, the area of a black hole's event horizon behaved very much like a thermodynamical entropy. Indeed, by May 1972, Bekenstein realized that it made sense to define a black hole's entropy as:

$$S_{bh} = \eta \ (kc^3/G\hbar) \ A$$

where A is the area of the black hole's event horizon (and k is Boltzmann's constant, c is the speed of light, G is the universal gravitation constant, $\hbar$ is Planck's constant [a value that occurs and reoccurs in quantum mechanics], and η was an as-yet-undetermined number). This equation formalized the tight relationship between the event horizon's area and the entropy of a black hole—and since a black hole had entropy, it would no longer serve as a bottomless pit where one's thermodynamic "crimes" could be dumped, as Wheeler had suggested.[10]

---

* Specifically, Christodoulou was looking at pulling energy out of a spinning black hole by means of gravitational radiation; he discovered that the most efficient techniques for doing so were precisely the ones that were reversible. This is analogous to the idea of a Carnot cycle in traditional thermodynamics.

Take a hot cup of tea and a cold cup of tea and pour them into a black hole. The area of the event horizon will increase by a certain amount. Do it again, but mix the hot and cold tea first, increasing the entropy of the system. This time, the area of the black hole will increase *more* than it would if you dumped the hot and cold tea separately. That is, the increased entropy of the teacup system isn't lost when you throw it into a black hole; it's preserved as an extra increase in the event horizon area. As Bekenstein told Wheeler, "You have not avoided the entropy increase. You've just put the entropy increase in a new place. The black hole itself has entropy."[11]

Bekenstein's formula worked beautifully; it closed the apparent loophole in the second law of thermodynamics that was so troubling Wheeler. But there were disturbing consequences to assuming that a black hole could have entropy. From a thermodynamic point of view, an object with entropy had to have a temperature. But what could it possibly mean for a black hole to have a temperature? If you put it next to something "cooler," would heat flow out of the black hole? Could a black hole cool down somehow? Worse yet, anything that has a temperature must radiate energy; after all, quantum mechanics was born when Max Planck came up with the correct formula for what kinds of light had to shine from an object at a given temperature. But a black hole can't emit light! It swallows it—that's the very definition of a black hole.

Even so, even knowing the seeming contradiction that his formula created, Bekenstein decided to publish it in the physics journal *Nuovo Cimento*. The article came out in August 1972, just as he was heading to Les Houches in the French Alps for a summer school on black holes. He was eager to discuss his ideas, as many of the world's leading experts on black holes would be there. Surely they would be as excited about black-hole thermodynamics as he was.

...

At Les Houches, Bekenstein didn't get the sort of reception he thought he would. Hawking not only rejected the idea of black-hole entropy, he was openly scornful of it. In Hawking's eyes, Bekenstein had twisted Hawking's beautiful area formula into something ridiculous, something ugly, something obviously aphysical. "Black holes couldn't have entropy or temperature; that simply didn't make any sense. One would have to be crazy, or at least a little dim-witted, to claim that the area of a hole's horizon in some sense is the hole's entropy, Hawking thought," wrote Kip Thorne, who was also at Les Houches, a few decades later. And without entropy or temperature, there couldn't be any sort of "thermodynamics" of black holes. So Hawking recruited Jim Bardeen and Brandon Carter—both of whom, like Hawking, were lecturers at the summer school—to gang up on Bekenstein, who had handed in his PhD thesis just a few months earlier. Thorne, like almost the entire community, sided against Bekenstein.[12]

There was one notable exception. "In those days in 1973 when I was often told that I was headed the wrong way," Bekenstein recalled in *Physics Today* a few years later, "I drew some comfort from Wheeler's opinion that 'black-hole thermodynamics is crazy, perhaps crazy enough to work.'" Heartened by the support of his mentor, Bekenstein returned from Les Houches more determined than ever to prove himself right. By October, he had figured out that the unknown constant in his equation, η, was about 0.028, or something "very close to this, probably within a factor of two."[13]*

By the early fall of 1972, Bekenstein was ready to submit to *Physical Review D*, but as an impoverished postdoc, he was unable to scrounge up the money to defray the cost of publication. Wheeler paid the bill out of his own budget and encouraged his former student to keep plugging away. "It has been a deep pleasure for me to be associated with you. It raises my spirits to know someone who has such a deep love for the truth, and for discovering new truth, and such ability and energy in doing so, as you have shown," Wheeler wrote to Bekenstein in October 1972. "Your work on black hole entropy continues to fascinate and disturb me. It is urgent to find a resolution of the questions still left that will be decisive and convince everybody."[14]

Hawking, for his part, was as unconvinced as he could possibly be. Along with Bardeen and Carter, he decided to nip Bekenstein's idea in the bud before it could do any more damage. So as Bekenstein's *Physical Review D* paper was going through the months-long peer-review process, Hawking, Bardeen, and Carter penned a counterattack.

When they wrote their paper, "The Four Laws of Black-Hole Mechanics," Bardeen, Carter, and Hawking pointedly didn't use the term "thermodynamics" in the title—even though much of the paper was devoted to demonstrating the parallels between the four laws of thermodynamics and the behavior of black holes. By using "mechanics" rather than "thermodynamics," the trio were emphasizing that these parallels were mere illusions rather than anything deeper.

True, the area of a black hole's event horizon was analogous in some ways to entropy. True, a quantity known as the *surface gravity* of a black

* Bekenstein's guess for η, as the equation is written in this chapter, was $(1/2 \ln 2)/4\pi$. As Hawking later showed, the true value is $1/4$.

hole behaved like a "temperature" of sorts.* The parallels were striking, but they didn't mean that black holes actually had temperature or entropy in any meaningful way. "It can be seen that [surface gravity] is analogous to temperature in the same way that [area] is analogous to entropy. It should however be emphasized that [surface gravity] and [area] are distinct from the temperature and entropy of the black hole," the three argued. Indeed, as they pointed out, a black hole simply couldn't have a temperature in any meaningful way. Objects that have temperature have to radiate, and black holes, they said, couldn't do that. "In fact the effective temperature of a black hole is absolute zero . . . because no radiation could be emitted from the hole."[15]

It was a solid-looking argument. Black holes don't radiate, so they don't have a temperature. Any attempts to describe the "thermodynamics" of black holes is therefore wrong. *Quod erat demonstrandum*. Anyone who had been paying attention to the scientific literature over the previous few months would recognize that the paper was meant to put the upstart Bekenstein in his place—as would anyone who noted the lack of Bekenstein's name in the paper's acknowledgments. The trio had thanked "Larry Smarr, Bryce De Witt, and other participants of the [Les Houches] school for their valuable discussions."

• • •

Bekenstein was still convinced that he was correct. But he couldn't get more traction for his ideas; he couldn't figure out a way to resolve the objections regarding black-hole temperature, or even figure out an exact value for that troublesome constant η, which was still hanging out in his entropy formula. But nobody else was working on the idea of black-hole thermodynamics, even its opponents.

Having launched his salvo against Bekenstein, Hawking moved on to other things. He did more work on mini black holes, and briefly explored how the cosmic background radiation would be affected if the universe were rotating while it was expanding. He was still plugging away at his textbook with George Ellis—a long project that was finally coming to its end; it would be published by Cambridge University Press in 1973. Contained in that work—and in a paper that came out a wee bit earlier—Hawking presented a mathematical proof that (given a few assumptions) a black hole had to be symmetric about the axis about which it spins. Hawking's so-called *rigidity theorem* put a bow around the famous no-hair theorems

---

* It's easiest to think of surface gravity in terms of mass (the larger the mass, the smaller the surface gravity, broadly speaking).

for black holes—it ensured that there were no exceptions as a result of bizarre distributions of matter in a collapsing star. As a consequence, after Werner Israel and Brandon Carter, who were the main people behind the no-hair theorems, Hawking and mathematical physicist David Robinson are also considered co-creators.

That September, Stephen accompanied Kip Thorne to Warsaw and then to Moscow for the first time since his student days. Jane overcame her fear of flying and joined her husband overseas. It was a glorious visit; his suite in the Hotel Rossiya overlooked the parti-colored onion spires of St. Basil's Cathedral, and Thorne, who spoke fluent Russian, showed them the town, including Tchaikovsky ballets and Borodin operas at the Bolshoi. The only substantial complaint was the suspicion that the hotel was infested with bugs—of the electronic variety.[16]

On that trip, Hawking met Alexei Starobinsky for the first time—taciturn because of his stutter, but happy to share what he was working on. As it happened, Starobinsky and his mentor, Yakov Zel'dovich, had a piece of research that Hawking found fascinating.

Hawking well knew that it was possible to extract energy from a spinning black hole; indeed, the no-hair theorems were based, in part, upon how spinning black holes have to emit gravity waves under certain circumstances, carrying energy away from the black hole. But Starobinsky and Zel'dovich realized that a very rapidly spinning black hole could lose energy not just by radiating gravity waves, but also by radiating particles of all varieties—photons, electrons, neutrinos, and the like.

It was an intricate argument, but the idea was inherently quantum mechanical. As described in Chapter 5, the uncertainty about energy and time built into quantum mechanics ensures that on the tiniest scales and for the shortest times, particles are constantly winking in and out of existence. These *vacuum fluctuations* are the reason why the quantum theoretic picture of space is frothy and constantly churning, rather than smooth and continuous as relativity would have it. These quantum fluctuations aren't figments of mathematical imagination; under certain circumstances, such as if you pour enough energy into a small enough space, these evanescent, now-they-exist-now-they-don't particles are given enough oomph to fly away and be detected by a particle counter.* Starobinsky and Zel'dovich argued that in the vicinity of a fast-spinning black hole, the violent twisting

---

* Pouring enough energy into a small enough space is precisely what "atom smashers" are for. When scientists at the massive LHC machine at CERN see a spray of particles in their collider, they're really seeing particles created out of the vacuum. You don't even need to have the atoms smash into each other to create those particles; if they pass by close enough to each other, there's still a spray of particles even though the original atoms are intact.

of spacetime in the region around the event horizon would do just this, and send particles shooting outward in all directions.

Hawking was intrigued by the Muscovites' argument; he tended to believe it, unlike Kip Thorne, who had already made a friendly wager with Zel'dovich that the calculations were wrong.* But he was unconvinced that the pair had done the math quite right. They were venturing into almost uncharted territory. The important work on black holes thus far was pretty much classical physics. Gravitational radiation and the like could be handled purely with the equations of general relativity without trying to tangle with the clashing principles of quantum theory. But once you started mucking about with particles, you needed to take quantum behavior into account. Classical physics wouldn't cut it, but because relativity and quantum theory clash, it wasn't easy to make the two work together. So, while Hawking believed the result, he didn't think that Zel'dovich and Starobinsky had done the calculations in a sufficiently rigorous way. "I didn't like the way they derived their result," Hawking said in 1984, "so I set out to do it properly."[17]

Hawking spent weeks absorbed in thought, blasting Wagner operas at high volume on the record player, and ignoring all distraction. As Jane later wrote, he was "transported to another dimension, lost to me and to the children playing around him." He often wouldn't respond to Jane's attempts to elicit a response out of him, so absorbed was he in his mathematics. Jane soon grew to hate Wagner.[18]

Stephen's physics took a "semiclassical" approach: he chose some elements of classical general relativity and added quantum theory when necessary to describe the behavior of matter. Almost by definition, semiclassical gravity is wrong on some level; it's a mishmash of sometimes internally contradictory theory, but with luck, it's good enough to yield some interesting insights. And so Hawking spent the late fall trying to figure out whether spinning black holes would emit particles. By the end of November, he had done enough work to get an answer: yes. There was radiation. But he hadn't yet figured out how much.

Lots, it turned out. Too much. Way too much. "I was expecting to discover just the radiation that Zel'dovich and Starobinsky had predicted from rotating black holes," Hawking wrote in *A Brief History of Time*. "However, when I did the calculation, I found, to my surprise and annoyance, that

---

* This was, in fact, Thorne's first big scientific wager. He had bet Zel'dovich a bottle of White Horse scotch against one of Georgian cognac that a spinning black hole wouldn't radiate particles. He lost.

even non-rotating black holes should apparently create and emit particles at a steady rate." Something was wrong.[19]

Hawking's calculations showed that even when a black hole wasn't rotating, the mere existence of an event horizon would affect the vacuum fluctuations nearby; the black hole would swallow some of those frothy, evanescent particles, while others would be liberated, freed from the confines of the black hole's gravity, and zoom off into the cosmos. A distant observer would see those liberated particles as radiation—it was as if the event horizon were emitting particles in all directions. Even though the black hole itself would never allow a particle to escape, it could shine brightly nonetheless. As Hawking wrote in *Brief History*, "particles do not come from within the black hole, but from the 'empty' space just outside the black hole's event horizon!"[20]

Hawking checked and rechecked his calculations in late December and early January, convinced that he had made an error somewhere, that perhaps the imperfect half-quantum, half-classical mathematical approach was letting him down. He simply didn't believe that a motionless black hole could radiate particles, even though his calculations were coming to a remarkable conclusion: the particles streaming away from the black hole had precisely the characteristics predicted by Max Planck in 1900 at the dawn of the quantum revolution.

As described in Chapter 5, Planck was trying to figure out the properties of the light that was emitted by an object at a given temperature. To get the right answer, he had to make an uncomfortable assumption—that light came in discrete packets known as quanta. But after Planck did so, his formula did a beautiful job; it predicted exactly the sort of radiation that a nonreflective object, known as a black body, would emit at any given temperature. Planck's derivation of the so-called black-body spectrum was the first triumph of quantum theory. Whenever it appeared in nature, it was a sign of the quantum-mechanical underpinnings of the universe—it had profound connections to the ultimate workings of nature.

Hawking's calculations were showing that the radiation coming off of a black hole had a black-body spectrum. More, the black-body spectrum was precisely that predicted if the black hole had a "temperature" defined by the black hole's surface gravity—just as Bekenstein had predicted!

Hawking, Bardeen, and Carter had objected to Bekenstein's "thermodynamics" arguments because black holes didn't have temperature—they didn't have temperature because they couldn't radiate. Now Hawking's calculations seemed to show not only that black holes *do* radiate, but that they radiate just like any other object with a fixed temperature does. Bekenstein had been right all along. At first, Hawking didn't want to tell anybody about his results, for fear that Bekenstein would find out—that Hawking's error would give his young rival ammunition "to support his ideas about the entropy of black holes, which I still did not like." Consequently, in early January 1974, he told only his closest friends about the puzzling result he was getting.[21]

Hawking was having a hard time believing what his calculations were telling him, but Martin Rees immediately saw their significance. Dennis Sciama was heading toward his office in early January when he ran into a pale and trembling Rees. "He was shaking with excitement," Sciama later recalled. "And he said, 'Have you heard? Have you heard what Stephen has discovered? Everything is different. Everything is changed!'"[22]

...

On January 8, 1974—Hawking's thirty-second birthday—Dennis Sciama wrote a letter to John Wheeler, who was scheduled to visit England the following month. "I should perhaps warn you that we expect to hear expounded another bombshell from Steve Hawking," Sciama cautioned. And then Sciama explained what was so shocking about Hawking's discovery: black holes are unstable. That is, a black hole can't last forever; because it radiates energy, it has to get smaller over time. What's more, the shrinking gets more and more rapid as the black hole melts away. As a black hole gets smaller, its surface gravity—its "temperature"—goes up, which means that it shines more brightly and shrinks faster. It is a feedback loop that gets faster and faster as time goes on. "The lifetime for one solar mass black hole would be much longer than the age of the universe, but for a mini-black hole the effect could be astrophysically interesting. Finally the black hole itself would disappear," Sciama continued.[23]

Disappear is a tame word for what would happen as a black hole dies. The temperature of the object would increase faster and faster and the black hole would blaze brighter and brighter with radiation, until . . . until nobody really knows what, precisely, would happen, but almost certainly it would be an enormously violent burst of energy shooting forth in all directions. "Black holes are not eternal," Hawking explained. "They evaporate away at an increasing rate until they vanish in a gigantic explosion." (Once Hawking's work became public, scientists quickly seized upon the idea of mini-black-hole explosions as being the source of the mysterious gamma-ray bursts that telescopes had been seeing.)[24]

Nobody realized it yet, but the discovery of Hawking radiation also set up the black-hole information paradox. So long as black holes were stable, they could gobble and store information without bound. But once they had a finite lifetime, they could no longer serve as information storage devices indefinitely; they had to disgorge it eventually. As it turns out, black-body radiation—the very type of emission that Hawking radiation seemed to be—cannot carry information. All it ever can reveal is the temperature of the object that emitted it. So where did the information go as the black hole evaporated? Hawking spent much of his career grappling with that question hidden implicitly in the theory of Hawking radiation.

Most problematic of all, at least for Hawking, was that he was going to have to eat a hearty helping of crow. His intuition had failed; it was not he, but a younger chap with a freshly minted PhD, who had made the great intellectual leap connecting the properties of black holes with the laws of thermodynamics. Worse, he had not just rejected the connection, but had actively fought Bekenstein's ideas about the relationship between entropy and temperature with event-horizon area and surface gravity. "It was just the relationship he had been sneering at," Sciama commented a number of years later.[25]

It was a lot for Hawking to swallow. At first, he simply didn't accept the implications—as Sciama made clear in his letter to Wheeler. "Steve does not believe that Nature really behaves this way," he wrote; instead he suspected that the fault lay with the kludgy approximations inherent to his semiclassical approach. But as Hawking pondered the matter further, he accepted the inevitable. "One always has a certain reluctance to give up a point of view that one has invested in," Hawking told journalist Timothy Ferris in 1983, "but this was an occasion when one had to do it. It all fitted together so perfectly that it *had* to be right. Nature wouldn't have set up something that elegant if it were wrong."[26]

By the evening of January 8, Hawking had finally overcome his reluctance and embraced the idea of black-hole evaporation. Penrose, who had heard about the idea from Dennis Sciama, called Hawking that evening. "It didn't take him long to convince me that he was right," Penrose says. Hawking recalled the call very well—for other reasons. "Roger Penrose phoned up on my birthday. He was very excited and went on so long that my birthday dinner got quite cold," Hawking said a decade and a half later. "It was a great pity because it was goose, which I'm very fond of."[27]

Emboldened by the reaction from his colleagues, Hawking submitted an article summarizing his calculations to *Nature* on January 17; perhaps betraying some lingering misgivings, the title of his submission ended with a question mark: "Black Hole Explosions?"

That February, there was a quantum gravity meeting at the Rutherford-Appleton labs near Oxford. It had attracted an international contingent: Wheeler was there—having arrived in England as promised—and many of the local physics luminaries were in attendance as well. But the star of the show was Hawking. Assisted by his student Bernard Carr, Hawking dropped his bombshell on the audience: black holes have entropy, they have temperature, they radiate particles, and they must eventually explode. After a few questions from the gathered scientists, physicist John G. Taylor—who had recently written a lay-audience book about black holes—stood up and brought the session to a close. As he did, he declared, "Sorry, Stephen, but this is absolute rubbish."[28]

Hawking radiation was such a radical idea, and had such dramatic consequences, that Taylor was far from the only one who couldn't accept it at first. Even Zel'dovich, who came up with the idea that spinning black holes emitted particles, had a hard time. Apparently, it took him two years. In 1976, Kip Thorne saw Zel'dovich throw up his hands in surrender: "I give up! I give up! I didn't believe it, but now I do." After a time, there were no holdouts. Hawking had triumphed.[29]

Most of the scientific world learned of Hawking's work with the publication of his *Nature* paper in March 1974. The follow-up paper, which described the process of particle creation by black holes, came more than a year later, in August 1975 (the journal he submitted it to apparently lost the manuscript, and only published after Hawking resubmitted). And Hawking was suddenly the toast of the scientific community.[30]

Hawking had been a standout before that—his work on singularities with Penrose had been incredibly important, and the black-hole area theorem and his work with the no-hair theorems put him as one of a small handful of people who were at the cutting edge of developments in general relativity. But the discovery of black-hole radiation and evaporation was something different, something that seemed to set Hawking apart from his black-hole peers like Brandon Carter and Jim Bardeen. He had done an incredibly complex mathematical calculation that married principles from both quantum theory and relativity—two mutually incompatible mathematical frameworks—and emerged with a brand-new physical principle that had the aroma of Truth with a capital T. It felt like Hawking had made a great step toward the dream of finally reconciling quantum theory and relativity, that he had taken on Einstein's mantle of coming up with a single, overarching theory that described everything in the universe. Hawking had, all of a sudden, performed a feat that catapulted him to the first rank of scientists.

"Hawking's discovery surprised everybody; of all, I was probably the most pleased for it provided the missing pieces of black-hole thermodynamics," wrote Bekenstein. Nor did he begrudge his rival his discovery.

"I had no idea how a black hole could radiate," Bekenstein later said. "Hawking brought that out very clearly. So that should be called Hawking radiation." Wheeler didn't hide his enthusiasm, either, even though Hawking, rather than his former student, carried off the prize. "The interface between general relativity and quantum theory is the most important frontier of modern theoretical physics," he wrote in a letter to another physicist in late 1975. "You and I also know that there has been no more important development in this area in the last five years than the 'quantum radiance of the black hole.'"[31]

Of course, Wheeler had more to add:

> Recent developments in this field have been associated with the name of Hawking, although there have been many other significant contributors. However, Hawking has acknowledged in his papers, and everyone close to the subject knows, that it was Bekenstein who initiated the field. He emphasized on physical grounds that the surface area of a black hole is not only analogous to entropy; it is entropy; and the surface gravity of a black hole is not only analogous to temperature; it is temperature.[32]

Nor would Bekenstein's contribution be forgotten. The final form of the entropy equation is:

$$S_{bh} = (kc^3/4G\hbar)\ A$$

Though "S_{bh}" initially meant "black-hole entropy," thanks to a happy coincidence with last initials, it now stands for "Bekenstein-Hawking entropy." The formula for black-hole temperature, however—

$$T = \hbar c^3/8\pi GMk$$

—typically bears no subscript. It is Hawking's alone. It is the formula inscribed on his tomb at Westminster Abbey. It represents his greatest scientific achievement, and it was his ticket to immortality.

...

In March 1974, just as Hawking's magnum opus was being printed in the pages of *Nature*, he learned that he had been elected to be a Fellow of the Royal Society. His signature would forever be enshrined in the society's Charter Book alongside those of Newton, Darwin, Faraday, Einstein, and other luminaries of science. At just thirty-two, Hawking was the youngest member to be inducted into the society in recent memory.*

The evening when the election became public, the college threw a party in his honor. Dennis Sciama gave a long speech about his former student's accomplishments. Stephen, in turn, spoke at length in his slow, soft, halting way about his gratitude—he thanked Sciama, he thanked the gathered audience. But he didn't thank Jane. "It may have been a mere oversight in the heat and excitement of the moment that he did not mention us at all," Jane writes. "He finished speaking to general applause while I blinked back prickly tears of dismay." Unaware of his wife's discomfort, Stephen basked in the glow of recognition.[33]

It was a high point in Hawking's career, even though there might have been some doubt rattling about in the back of his mind. He must have known that his election to the Royal Society couldn't have been triggered by his discovery of Hawking radiation. The timing simply didn't work out. Given that the business of nomination and election involves the process of peer review and takes many months, his discovery of Hawking radiation was simply too new. He must have been nominated on the basis of his prior work, principally his black-hole area theorem and his work with Roger Penrose, particularly his singularity theorem (which is described in detail in the next chapter.) There was no question that it was a solid body of work—but Hawking must have wondered: Was it truly worthy of such an honor at such a young age?

Hawking also would have suspected that his friend Roger Penrose must have had a hand in the process; two years earlier, Penrose had been elected to the Royal Society, and thus was able to nominate candidates. As it happened, Hawking was the very first person Penrose had nominated, and in mid-November, Penrose had helped secure letters from outside experts to support the election, including one from John Archibald Wheeler. (Wheeler's recommendation certainly helped make the case: "His deep geometrical insight and physical understanding, the incisiveness of his proofs and his instinct for the central point establish him as an unsurpassed leader in the relativistic astrophysics of our time.") When Penrose's request for recommendation letters went out, even he had no inkling of the work on Hawking radiation. The election process was well underway

---

* A number of sources say that Hawking was the youngest member to ever have been inducted. However, there are quite a number of counterexamples. Isaac Newton, for example, was named a Fellow in 1672 at the age of twenty-nine, Robert Hooke in 1663 at the age of twenty-eight, and Arthur Cayley in 1852 at thirty-one, to name a few.

by the time Hawking's great triumph became known. It's likely that the buzz about Hawking radiation helped seal the deal—and helped him beat out the other two astrophysics candidates of that year, Donald Lynden-Bell (who was elected in 1978) and Leon Mestel (who was never elected). However, Hawking radiation was not the *primum mobile* for Hawking's honor.[34]

Despite his youth, and even without what would become his signature discovery, Penrose had no doubt that Hawking had done work that earned him elevation to the Royal Society. "It was sure worthy of an FRS," says Penrose. But, of course, there was another consideration. "You see, we all thought he didn't have long to live," Penrose says. "It seemed quite likely that it wouldn't, you know . . . usually, it can go on for ages; you could be nominated and it could be five years or longer before the person finally gets in. In his case, he got in immediately."[35]

Hawking had finally arrived. The media took notice. The local newspaper, the *Cambridge Evening News*, ran a large spread on Hawking, and splashed his photo on the front page. "A 32-year-old Cambridge scientist, Dr. Stephen Hawking, today became the youngest ever Fellow of the Royal Society," the teaser read. "Dr. Hawking has overcome physical handicaps in his rapid rise to prominence in astronomy and the study of Black Holes. Full report—page 17."[36]

CHAPTER 14

BLACK HOLE (1965–1969)

———————————

L ong before humans set foot on North America, glaciers crept south from what is now Canada. These mountains of ice gouged out great chunks of soil and rock as they advanced. New York was almost entirely covered with a thick sheet of glacial ice until, a mere ten thousand years ago, the climate began to warm and the ice made its final retreat. The scars gashed out by the glaciers filled with water and became deep and narrow lakes hemmed in by high walls. It was here, surrounded by the post-glacial beauty of upstate New York, that Jane and Stephen Hawking spent the second half of their honeymoon.

It was the summer of 1965. Stephen's PhD thesis was almost finished; he had done most of the intellectual heavy lifting, and all that remained was to put it into final form and submit it. Better yet, he had also secured a research fellowship at Cambridge that would begin in the fall. It didn't pay well, but it came with some degree of stability, providing a decent foundation for a newly married couple. Their honeymoon had started with a week in Suffolk—and then, their marriage just a few days old, Jane and Stephen had bundled into an airplane and crossed the Atlantic. A car was waiting at the airport and drove the couple to their lodgings amid the gorges and waterfalls of Ithaca, New York. They had arrived at Cornell University, where Stephen was to participate in a summer school devoted to general relativity.[1]

"That was a mistake," he later wrote, rather laconically. "We stayed in a dormitory that was full of couples with noisy and small children, and it put a strain on our marriage." The screaming babies were a problem, but so was the fact that they had little money for food, and wound up dining on whatever inexpensive ingredients they could figure out how to prepare in a cheap saucepan over a hotplate. After helping her husband—who could walk, albeit slowly, with a cane—navigate the mile from the dorm to the lecture hall, Jane would find a typewriter so she could type out a draft of Stephen's PhD thesis. As Stephen reveled in connecting with new colleagues, Jane looked with trepidation at what the scientists' wives had become: "They were already, to all intents and purposes, widows—physics widows." The Hawkings' honeymoon was over.[2]

• • •

Stephen Hawking's PhD thesis rested on one really important idea—one exceptional insight about the early universe. It was more than enough to get him his doctorate and establish him as a young physicist to be taken very seriously indeed. It was obvious to those who understood his work that Hawking had a tremendous amount of potential.

However, potential doesn't put food on the table. Just as, soon after his graduation in 1965, Hawking had to adjust to being a husband and (shortly thereafter) a father, Hawking also had to adjust to being a professional physicist. It wasn't a seamless transition. He had to prove that he was worthy of a permanent position at a university—a professorship, or at least some sort of stable appointment. That meant coming up with new ideas to follow on his success with his thesis. But they weren't quick in coming.

As creative as Hawking was, the years right after his graduation in 1965 didn't yield the new ideas that he needed to establish himself. He extended the work from his thesis as much as he could, but that train of research had run its course within a few years. The excitement in Hawking's field was elsewhere; the hot gravitational physics was in black holes. For the Golden Age of Black Holes had already begun.

• • •

To Hawking, the summer school was definitely a net positive, even if it did put a strain on his marriage. "In other respects, however, the summer school was very useful for me because I met many of the leading people in the field," he recalled. And there were a few friends from back home as well; there was Brandon Carter, another of Dennis Sciama's students a couple of years behind Stephen, and there was Roger Penrose, with his wife and sons in tow.

For Hawking, Penrose's arrival at the summer school was most welcome. About a decade older than Stephen, Penrose had already established
himself as a faculty member at Birkbeck College (part of the University
of London). And he was taking an interest in the twenty-three-year-old
Hawking, much as Hawking's adviser Dennis Sciama—who knew Penrose's
brother—had taken an interest in Penrose a number of years earlier. In fact,
it was Sciama who got Penrose into physics in the first place. "Although I
was doing pure mathematics, he was trying to convert me to do cosmology," recalls Penrose.[3]

As a pure mathematician, Penrose had a different set of tools than
physicists did, tools that turned out to be very powerful when used on
problems in general relativity. Einstein's equations of general relativity are
used to understand the geometry of spacetime; they describe the fabric
of space and time and how they curve under the influence of matter and
energy. But they say nothing about the overall shape, the topology, of the
universe—any more than saying that an object made out of flannel will tell
you whether it's a shirt or a dress or a bathrobe or a tablecloth or a ball, for
that matter. The mathematical description of the fabric itself doesn't give
you a picture of the overall shape that that fabric takes. Penrose started
using topological tools that hadn't been used by the relativity community
before, and they happened to be extremely effective in places where spacetime gets unruly. Like near singularities.

In late 1964, Penrose used topological arguments to prove that a collapsing star of sufficient size would always wind up creating a singularity. It
didn't matter how lumpy or asymmetric or irregular the star was as it collapsed; the end result was always the same. A singularity was unavoidable.

Penrose's discovery caused a huge splash in the general relativity community; he had proven something that had eluded physicists for years. For
there had long been doubts whether the singularity at the heart of a black
hole was an accident of the oversimplified physics of a perfectly symmetrical implosion, an artifact of trying to describe messy real-world systems
with the idealized language of mathematics. Penrose had answered that
question once and for all: the monstrosity at the heart of a black hole was
no figment of the imagination. It had to be real.*

At that time, Hawking was struggling with his PhD thesis. He had
come up with a bunch of minor results—poking a hole in an alternative

---

* This insight would win Penrose the Nobel Prize in 2020, more than half a century later;
the committee described his proof as perhaps "the most important contribution to the general theory of relativity since Einstein."

theory of gravitation, exploring the formation of galaxies in the steady-state model of the universe—but nothing really particularly notable or important. Finding a good idea for a PhD thesis is the hardest part of the entire endeavor; it has to be a problem that's solvable by a relatively inexperienced researcher, yet interesting enough to be worthy of solution. And Hawking hadn't found that sweet spot—until Penrose announced his black-hole singularity theorem in early 1965.

Hawking wasn't at the lecture where it was announced, but Brandon Carter was, and Carter told Sciama and Hawking and George Ellis and all the rest of the Cambridge crew about the result. Sciama asked Penrose to come to Cambridge and present the results there. "That's when I first met [Hawking]," Penrose remembers. "We went into a side room where I described more details to Stephen and George Ellis." And then Hawking suddenly made a connection that nobody else had. As Sciama recalled in 1990: "I remember Stephen Hawking, who was then approaching his third year as a research student, saying, 'what very interesting results. I wonder if they could be adapted to understanding the origin of the universe?'"[4]

Penrose's topological techniques had revealed the presence of a singularity during the collapse of a star. Hawking realized that the same techniques could be applied to the Big Bang model of the universe once he made one tweak: he needed to reverse the flow of time. A Big Bang cosmos begins from something very, very tiny and expands rapidly into something enormous. Played in reverse, the physics looks almost the same as an enormous star shrinking into something very, very tiny. And just as Penrose proved that a black-hole collapse must end in a singularity, Hawking saw that the same mathematics implies that the expansion must begin in one—with all the difficult implications that conclusion entails.

"I realized that if I reversed the direction of time so that the collapse became an expansion, I could prove that the universe had a beginning," Hawking later told director Errol Morris. "But, my proof, based on Einstein's general theory of relativity, also showed that we could not understand how the universe began because it showed that all scientific theories, including general relativity itself, break down at the beginning of the universe."[5]

It was an important and disturbing insight—and would make a superlative PhD thesis. "It was a nice idea, and I was struck by his picking up the idea very quickly, and being able to use it in a very general way, in a general context," Penrose says. However, from Penrose's point of view, Hawking's real insight was not merely in reversing the direction of time—something Penrose describes as "obvious"—but in a very technical mathematical point: Hawking took a key element of Penrose's proof and figured out how to apply it not to the region near a black hole, where Penrose did, but way, way out at cosmological distances.[6*]

Back then, Hawking was indeed as much mathematician as physicist. ("I was trying to be as rigorous as a pure mathematician," Hawking wrote toward the end of his life. "Nowadays, I'm concerned about being right rather than righteous.") Hawking, who had a natural talent for mathematics, picked up the techniques from Penrose quite quickly. Penrose, for his part, was thrilled to have Hawking use his approach. He was no longer the lone topologist in a field of geometers.[7]

Over the next few years, Penrose and Hawking would collaborate on a number of projects. The most important was refining and extending the singularity theorems. Sometimes working separately and sometimes discussing matters by telephone, they chipped away at the underlying assumptions of the theorems and drew more powerful conclusions. Despite Hawking's prodigious mathematical talent, he made some significant mistakes. On several occasions, Hawking lofted a new mathematical proof in a preprint or even in a published paper that wound up being wrong, only to be corrected by other physicists. However, the errors were never severe enough to invalidate the overarching—and correct—insights he'd had. "I call these mistakes of the first kind: that some mistakes which are correctable and they don't affect the general argument," Penrose says.[8]

Penrose and Hawking also wrote separate but complementary papers that, in 1966, won each of them a version of the prestigious Adams Prize—awarded by Cambridge for distinguished research in mathematics. (Technically, Penrose had won the prize, but Hawking was—unusually— given a supplementary award.) The pair's collaboration culminated in 1970, when they combined their ideas about singularities into a more general result now known as the Penrose-Hawking singularity theorem.

But back in July 1965, during that summer school session at Cornell, Hawking's relationship with Penrose was just beginning—and Penrose's presence was no doubt appreciated as Hawking put the finishing touches on his thesis. When Hawking turned it in that October, it was his singularity theorem that carried the day. The first three chapters were respectable, if not terribly innovative, but the fourth—the proof that an expanding

---

* The element is what's known as the *trapped surface* condition. For the proof to work properly, Penrose had to invent a precise way of describing a curve in spacetime that enclosed the singularity of a black hole into the future. Hawking figured out how to adapt that condition and apply it to the universe as a whole, which was not obvious. "It's just the observation that you have an anti-trapped surface if you go far enough away in some standard open cosmology," Penrose says.

universe must begin in a singularity—that was exceptional work. It was that theorem that got Hawking his PhD.[9]

It was also what began Hawking's life-long honeymoon with physics.

• • •

When the Hawkings got back to England, the newly married couple immediately had to tangle with the unpleasant practicalities of setting up a household with very little money in their pockets. Stephen had luckily landed a research fellowship with one of the residential colleges at Cambridge—Gonville and Caius—which would give the two just enough money to live on for the time being. But it didn't provide them with any housing, and the pair struggled to find a place to live. Not that they could really settle in even if they had found a domicile; Jane still had to finish up her undergraduate degree in London, so she would take the train from Cambridge on Monday morning, spend the week there, and then hustle back to Stephen on Friday afternoon.[10]

Stephen could still get along on his own reasonably well; he could walk to and from work with his cane, and though his speech was deteriorating, he could make himself understood. But there was always the knowledge that the disease would be tightening its grip. It was already happening.

During the summer school at Cornell, Stephen was seized by an uncontrollable paroxysm of choking, a coughing fit that left him rumpled and exhausted—the first such fit Jane had witnessed. Even though she knew her husband was battling a terminal illness, her helplessness in the face of the onslaught was a shock. Hawking's father, Frank, was a physician—actually, one of the world's most renowned experts in tropical medicine—and though it was a bit out of the area of his expertise, he put his son on a regimen of oral and injectable B vitamins. Sadly, there was little else he could do for his son.[11]

Despite the shadow cast by the disease, Jane and Stephen began to carve out a life for themselves, with a little bit of help from their parents here and there. And Stephen was beginning to make a reputation for himself as a physicist. His thesis launched his career in style. It resulted in three peer-reviewed papers in quick succession, the first of which was published in *Physics Letters* the day after his wedding. His pride in the accomplishment was probably only slightly diminished by the fact that the primary author was one "S. Hawkins." The Hawking singularity theorem was winging its way around the world, and people began to take notice of the young physicist.[12]

Hawking was definitely noticeable, even to people who hadn't met him in person. He continued to extend his work on singularities, and published quite a few papers on that subject, as well as some of his other research on the early universe. By the end of 1966—at the tender age of twenty-four—he had nine publications under his belt, including one in *Nature*, the most prestigious scientific journal in the world. Even his father, Frank, didn't get his first *Nature* article until his late thirties, shortly after Stephen was born. Though Frank had wanted the young Stephen to be a doctor (and had tried to steer the boy away from mathematics), he must have been proud of his son's work, even if it had nothing to do with medicine. And like Frank, who traveled all over the world for his research, Stephen spent a great deal of time away from home, networking with the world's experts in relativity at every meeting he could convince his college to send him to. At first, it was mostly the United States—Florida and Texas were his first two trips abroad as a postdoc—but he would soon be visiting more exotic locales.[13]

Now, for the first time, people were making pilgrimages to visit him. Princeton's John Archibald Wheeler, like Cambridge's Dennis Sciama, encouraged his students to visit other research groups to forge friendships and collaborations that would help shape their careers; Wheeler scrounged funding to allow his young charges to imbibe the "vitamins" that come with exposure to a new group of scientists. In January 1967, for example, Wheeler used his own money—honoraria for lectures he had given—to send a promising young grad student, Bob Geroch, to Cambridge, so that he could "review his work and discuss recent developments for a few days with Stephen Hawking, who, along with Roger Penrose, is tops in the field of singularities."[14]

Writing from England in February, Geroch expressed gratitude to Wheeler for the opportunity. "Dr. Hawking has been very nice—both in sacrificing his time to talk with me, and in finding lodgings, food, entertainment, etc. I can now appreciate your remark that no volume of letters can substitute for one personal visit in producing a good relationship." As Geroch knew firsthand, it's not just because face-to-face meetings create stronger bonds than letters ever could, but also because letters can be misunderstood.[15]

Half a year earlier, in June 1966, a furious Hawking had sent a telegram to Sam Goudsmit, the editor of *Physical Review Letters*, in an attempt to block Geroch from publishing one of his papers. Hawking had misread one of Geroch's letters—he initially thought Geroch was attempting to publish some of Hawking's work as well as his own—but the letter merely suggested that the two publish in tandem. Since Geroch's work depended on Hawking's, Geroch didn't want to go to press before Hawking did. After sending out a "telegram of protest" or two, Hawking reread the letter and realized that he had misunderstood Geroch's intent. Hawking didn't quite apologize for the misunderstanding, but he did smooth things over with Goudsmit, writing up a little summary of his work so that the two could publish in the same issue of the journal. But, clearly, by the time Geroch visited Cambridge in 1967, there was no ill will. Geroch imbibed plenty of "vitamins" from Hawking and from Brandon Carter, who brought him up to speed on the latest developments in the field.[16]

Travel was the lifeblood of Stephen's early career, but travel was expensive, and his salary was already stretched to the limit. Especially since, in the early fall of 1966, Jane found out that she was expecting a baby.

· · ·

Stephen was losing control over the muscles in his fingers and was having difficulty writing. Yet the young couple couldn't afford physiotherapy to try to stave off the inevitable loss of function. Dennis Sciama intervened with the powers that be at Cambridge to get the university to pay for twice-weekly visits from a physiotherapist, which helped a bit. But it was clear that most of Stephen's writing—and all of his typing—would have to be done by others. As he was too junior to have an assistant, that burden would fall upon the increasingly encumbered Jane.[17]

Luckily for Jane, late 1966 and early 1967 marked a lull in Stephen's research; he had parceled off all the publishable work from his thesis into papers already, and was looking for another line of research to tackle. As he mentioned to Bob Geroch in the winter of 1967, research into singularities was "asymptotically approaching a fixed number of results"—in other words, the big questions that could be tackled by the Penrose-Hawking approach were pretty much in hand; it was really a matter of tying up loose ends with the singularity theorems rather than making big new breakthroughs. It was nigh time to pull up stakes and move on to more fertile soil. It would be three years before he made the crucial shift in his investigations from black-hole singularities to their event horizons, leading to his next great insight. Before that, though, Hawking would have to wander the wilderness in search of the next big idea. For a young researcher on a contract—even one without a baby on the way and engaged in a battle to slow down the effects of a degenerative disease—it's a perilous and nerve-racking time in one's career.[18]

Even in the absence of a next big idea, though, Hawking wasn't spinning his wheels so long as he could travel and make connections. He applied in February to go to a small physics meeting, a "Rencontres," in Seattle. Many of the leading lights in physics would be there: Roger Penrose, Richard Feynman, Charlie Misner, Tullio Regge, Bryce and Cecile DeWitt. And it was run by John Archibald Wheeler. So Jane typed up the application (in duplicate) and Stephen gathered the necessary recommendation from a more senior faculty member; then he sent it overseas, fully expecting to spend the summer in Seattle—presumably with an infant in tow.

In March, Hawking's plans were dashed. His younger colleague Brandon Carter was invited to the Seattle Rencontres—but Hawking was not. The number of spots available were limited, and Wheeler had chosen Carter over Hawking.

Carter had asked Penrose for a recommendation, but Hawking had not, leaving Penrose in a somewhat awkward situation. Penrose didn't want to undermine Hawking's chances of acceptance while helping Carter's. So in the recommendation, he sang Hawking's praises as well:

> [Hawking's] important work on the question of singularities in cosmology is perhaps known to you and I would certainly also give him a very high recommendation, if a recommendation from me is required! If I were asked to give a comparison between Hawking and Carter a year or two ago I should have certainly rated Hawking more highly, but with Carter as a very promising and somewhat unknown quantity. Today I would find the comparison more difficult.[19]

Penrose went on to say that Hawking's condition—presumably his slurred speech—would make communication with the other participants more difficult for Hawking than for Carter, but made clear that he thought both should be invited: "My own feeling is that it would be a shame if Hawking could not come, and that Carter, who is a good friend of Hawking's and who is familiar with most of Hawking's work, could serve as a bridge in more ways than one!" Carter, in fact, was doing impressive work at the time, not just with black holes, but also helping Penrose and Hawking develop the singularity theorems—and fixing some of the errors in Hawking's proofs. "Stephen would often make mistakes, that Brandon would correct," recalls Penrose. So when it came to a choice between Hawking and Carter, Penrose clearly thought that both were worthy.[20]

Upon hearing that Hawking had been rejected, Penrose immediately wrote to Wheeler to plead Hawking's case. After making clear that he didn't mean to give the impression that Carter was a better choice than Hawking, for whom he had "exceptionally high regard," Penrose urged Wheeler to put Hawking at the very top of the waitlist: "I realize also that people who are turned down this time will in general have other opportunities to apply for future Rencontres," he wrote. "I am unsure whether this is likely to be the case with Stephen."[21]

"Wheeler clearly didn't want Stephen," Penrose says. "I don't know what the reason was." But it turned out all right in the end. Penrose prevailed upon Wheeler, who somehow found a way for Hawking to attend, and Hawking was invited to spend the summer in Seattle after all. However, Wheeler's choice of Carter over Hawking was not a one-time fluke. At the time—and at least until 1971—Wheeler apparently judged Carter to be the stronger physicist. For example, in describing Sciama's students to an Oxford professor, he sang the praises of "such men as Ellis and Hawking and particularly Brandon Carter [who] also have striking achievements to their names and promise ahead for their futures."[22]

At the end of 1970, he wrote a frank appraisal of the state of the field in gravitational physics and relativity:

> In every assessment of any scope I continue to name as the six most promising people I know Zel'dovich of Moscow, Misner of Maryland, Penrose of London, Carter of Cambridge, Thorne of Cal Tech, and Geroch of Texas. If Zel'dovich is the most versatile and Misner the most imaginative and Penrose the most geometrically minded, Carter is the deepest.[23]

The Golden Age of Black Holes was well underway, and as Wheeler was aware, Brandon Carter was playing a major role. Hawking wasn't—at least not yet. He was still awaiting a second great insight, something that would push his career forward beyond what he had done with his PhD. As important as it was, the singularity theorem alone wouldn't sustain his reputation for long. And back in 1967, his bedtime flash of inspiration about the area theorem was still three years in the future.

•••

With a baby months away, Hawking had other things on his mind than black holes. Luckily, Jane got her bachelor's degree and no longer had to commute to London, but Stephen was increasingly dependent upon her—from the daily walk to and from the department to the necessaries of maintaining a household. On top of that Jane had, somewhat optimistically, secured an adviser for her PhD at the University of London, Alan Deyermond. The thought was that she would write her thesis while raising a baby. And in early 1967, she was rushing to put the finishing touches on her first research article.

Jane's paper had to do with a fifteenth-century Spanish tragicomedy, *La Celestina*, in which an elderly procuress' scheming leads to a shocking number of characters dying after jumping out of windows or falling from great heights. The central argument of Jane's work was to lay the cause of the tragedy to a frustrated servant's unfulfilled longing for a mother figure. "The idea was a fascinating one which won Alan Deyermond's amazed approval: he was even more amazed when I confessed that the idea was Stephen's," Jane wrote. Stephen had idly thumbed through *La Celestina* while waiting for his wife to finish her final examinations, and mentioned the idea to her on the drive back to Cambridge. All Jane had to do was "to flesh out the argument and justify the Freudian concept when applied to a text dating from 1499."[24]*

Jane thought Stephen's contribution to the *Celestina* paper was a sign of the "harmony" in their relationship, the mutual respect for and contribution to each other's intellectual pursuits. It didn't seem to occur to her that the respect might have been one-sided. Stephen didn't hide his scorn for medieval studies, likening it to "collecting pebbles on the beach," and the ease with which he had come up with a publication-worthy idea probably reinforced that scorn rather than lessening it.[25]

Regardless, the question was moot, at least for the time being. The baby's arrival in May put Jane's intellectual ambitions on hold. Barely two months after Robert's birth, the three packed up for the Seattle workshop. "That, again, was a mistake," Hawking later admitted. "I was not able to help much with the baby because of my increasing disability, and Jane had to cope largely on her own and got very tired." It was a long four months, and though Stephen had earned some money in speaking fees, he hadn't yet figured out his next line of research. Jane, for her part, had begun to develop a fear of flying.[26]

Over the summer, though, there was an important development in black-hole research: the object that everybody had been studying finally had a name. John Wheeler was at a meeting in New York where he was discussing black holes with the audience—but there was no official name for these objects. Unlike white dwarfs or red giants or neutron stars, there were only generic-sounding names for them—"completely collapsed objects" was common. "Well, after you get through saying 'completely collapsed objects' six times, you look for a shortcut," Wheeler later recalled. "And that's when I found myself using the phrase 'black hole.'"[27]

"It was a stroke of genius: The name ensured that black holes entered the mythology of science fiction," Hawking told an audience at Berkeley in 1988. "It stimulated scientific research by providing a definite name for

---

* In fact, Jane's justification for the Freudian concept occupies a paragraph at the beginning of the fourteen-page document and presents neither a deep nor a rigorous argument. Jane Hawking, "Madre Celestina," *Annali—Sezione Romana* 9, no. 2 (1967): 177.

something that previously had not had a satisfactory title. The importance in science of a good name should not be underestimated."[28]

It was turning out to be a banner year for black holes. Just a few months before Wheeler came up with the term for black hole, a South African physicist, Werner Israel, had taken the first big step toward what are now called (in another Wheeler-catalyzed coinage) the no-hair theorems.

To understand a black hole's behavior—what it looks like to a nearby particle, how it can move, how it can interact with the environment around it—physicists need to know how a black hole shapes the fabric of space and time nearby. And that's easier said than done. Despite thinking about spacetime around black holes for decades, by the early 1960s physicists had only really gotten a handle on the shape of spacetime around the very simplest possible black-hole system—the neighborhood around totally motionless, perfectly spherical black holes. But though they could solve the Einstein field equations for that idealized situation, they feared that actual, astrophysical, black holes would be quite a bit more complicated. They might rotate. They might be asymmetric and have hills and valleys, or regions of greater or lesser density. They might pulsate or change shape. And it was one thing to solve equations regarding idealized, simple Platonic objects; doing the same thing with complex, messy astrophysical black holes, rather than mere models, was incredibly difficult. It was only in 1963 that an Australian scientist, Roy Kerr, had come up with a solution to the Einstein equations to describe the behavior of spacetime around a spinning black hole. The Kerr black hole was still an oversimplified object, as it was symmetric and uniform, but it was at least one step more realistic than before. His discovery marked the beginning of the Golden Age of Black Holes—a rapid series of discoveries about these collapsed stars that turned them, step by step, from a cartoonish theoretical idea into a real astrophysical object worthy of study.

Within two years of Kerr's discovery, Ezra Newman, a physicist at the University of Pittsburgh, was able to take Kerr's twirling black hole and add one more layer of complexity: charge. Now, scientists could start examining spinning black holes with or without electric charge, and begin deducing their properties. Brandon Carter at Cambridge, for example, quickly became an expert in the Kerr-Newman solutions to the Einstein field equations, and he soon created a stir by finding completely unexpected new physics in the way objects in orbit around a Kerr-Newman black hole move about.

So, by the mid-1960s, scientists could handle black holes that were spinning, and that carried an electric charge. But that was still an oversimplification. What about asymmetries? What if the black hole were lumpy rather than perfectly spherical? In 1967, Werner Israel proved that it didn't matter how asymmetric or inhomogeneous a collapsing star was. The end result was always the same: a perfectly symmetrical, spherical black hole. No matter how bizarrely the matter might have been distributed, the final black hole would always be an unblemished sphere. Over the next few years, Carter and a number of other scientists extended Israel's proof, putting it on more and more solid theoretical ground and exploring its implications. They figured out, for example, that any asymmetries or pulsations in the collapsing star would wind up being radiated away as gravitational waves—only once the black hole was as quiescent and featureless as a cue ball would it stop radiating any asymmetries away. Even though it might be born in complexity, a black hole always winds up extremely symmetric and simple. It has no hair.

The no-hair theorems were great news for physicists; no longer did they have to worry about solving equations for messy, asymmetric, pulsating matter. When it came to black holes, that asymmetry, that mess, was all lost—what had been a complicated lump of material had become simpler as it collapsed. Indeed, these theorems implied that a black hole was the simplest macroscopic object in the universe: there's nothing more to learn—nothing more that you *can* learn—about a black hole other than its mass, rate of spin, and charge. A black hole is an almost featureless object.

It truly was a Golden Age. By the late 1960s, physicists had a whole suite of powerful new tools: the Kerr-Newman solution, the no-hair theorems, and Penrose's black-hole singularity theorem. However, despite being "tops" in singularities, Hawking wasn't really spending much time thinking about black holes. At the time, with a few exceptions, his work was on cosmology, on trying to understand the conditions (and the singularity) at the beginning of the universe, rather than astrophysical objects like black holes (and their singularities). So Hawking had not yet played a major role in the Golden Age. That, of course, would change within a few years.

However, in 1967, Hawking was occupied with other matters. Around the time Robert was born, he ceased being able to walk to the department on his own. Typically, he would hitch a ride with George Ellis, who would bring him back and forth from work, or he would ride in a three-wheeled battery-powered car designed for people with disabilities, occasionally bringing a passenger along for the ride. ("I found this rather scary because I thought he drove faster than was safe," wrote one of Hawking's early students.) Hawking managed to get his research fellowship at Gonville and Caius renewed for another two years, but the rules prevented another renewal. And as he struggled to find another idea as powerful as his singularity theorem, Hawking was no doubt acutely aware that some of his younger colleagues were more than giving him a run for his money, research-wise. Not only was Brandon Carter making a name for himself with black holes, but Martin Rees, also a student of Dennis Sciama's, was turning out to be a powerhouse; he was coming up on a dozen publications in *Nature* on a variety of topics, including X-ray astronomy, radio astronomy, and cosmic rays. (Rees was skyrocketing up the academic ladder; in 1972, he would land a professorship, and in 1973, he would be named to the Plumian chair in astrophysics—like the Lucasian chair, a centuries-old position of high honor.) When 1969 rolled around and Hawking's contract expired once more, Dennis Sciama helped wangle a "special category" of research fellowship for his young protégé.[29]

At the time, there was a little-used (in fact, perhaps never-before used) provision in the university statutes allowing for six-year fellowships for "persons of exceptional distinction in science, literature, or art." It probably wasn't intended for someone so young and inexperienced, but it allowed the university to throw Hawking a lifeline. According to the terms of the fellowship, Hawking would spend half of his time at the Department of Applied Mathematics and Physics, and half at the newly created Institute for Theoretical Astronomy, run by Fred Hoyle. Simon Mitton, the institute's departmental secretary at the time, recalls afternoons filled with mundane duties tending to Hawking's needs. "They included getting him out of his three-wheel invalid car and putting him in the wheelchair which we kept at the Institute of Astronomy. I used to have to give him medication, you know, a couple of pills, and I remember having to get fixed up for his office, when he first joined, a pushbutton telephone system in which something like twenty lines were preloaded. It's hard to imagine this today, but this was a massive procedure, involving telephone engineers of what was back then a nationalized industry having to come along and fix all of this stuff up by hardwiring it."[30]

Cambridge was doing its best to look after Hawking. The position they gave him was a renewable six-year appointment that took the pressure off. And, as internal documents made clear, the university had committed to "accept[ing] the obligation of wholly maintaining him as long as he can do his scientific work." Nevertheless, at the time, Hawking was no closer to winning a professorship than he had been in 1966. He needed another big idea, and time was not on his side.[31]

Before Hawking's attention turned squarely to black holes, he even flirted with becoming an experimentalist. In 1969, a set of experiments by University of Maryland physicist Joseph Weber was the talk of the relativity community. Using a set of refrigerated metal cylinders, Weber claimed to have been detecting gravity waves; the idea was that the distinctive stretch-and-squash distortion of a passing gravity wave would set these specially sized cylinders ringing like a bell. And Weber was claiming to have seen dozens of gravity waves, many of which seemed to be coming from the center of the Milky Way. It sounded almost too good to be true.

It *was* too good to be true. Within a few years, other physicists demolished Weber's claims; he hadn't been careful enough with his measurements and had convinced himself that he was seeing gravitational radiation when it was mere noise. But in 1969 and 1970, Weber's results were being taken very seriously indeed. Hawking flew out to see Weber's equipment and was soon at work with one of his graduate students thinking about how to improve Weber's results. By November 1970, the two had come up with a design that would be more sensitive by a factor of ten—or even more—and Hawking applied for a grant to build their own gravity-wave detectors.[32]

Luckily, the pair eventually decided not to pursue the idea. "That was a narrow escape! My increasing disability would have made me hopeless as an experimenter," Hawking later wrote. "One is often only part of a large team, doing an experiment that takes years. On the other hand, a theorist can have an idea in a single afternoon, or, in my case, while getting into bed, and write a paper on one's own or with one or two colleagues to make one's name."[33]

When the next big idea came in 1970—the area theorem—and then the huge one after that in 1974—Hawking radiation—Hawking would indeed make his name. And he would do it alone.

• • •

In September 1969, Jane, pregnant with her second child, awoke to the sight of her first, Robert, covered with sticky pink liquid. With a start, she realized that the toddler had used a chair to get at the collection of medicine bottles on the top of the refrigerator, and had drunk the lot down. Among the bottles was a "stimulant" that Jane had been prescribed to "pep [her] up." She immediately ran with Robert to the doctor, who sent them off to the hospital. As Robert began convulsing, the nurses held the little boy down and began pumping his stomach.[34]

Jane watched helplessly as her child, strapped down on the cot to avoid hurting himself, stopped flailing and sank into a coma. After some hours, his condition stabilized. "His state was not hopeful; all that could be said was that it was not deteriorating further," Jane wrote in her memoir. "Coming to my senses at this slightest of changes, I was shocked to remember that I had left Stephen alone in the house, scarcely able to look after himself." She had to make a terrible decision: stay by the side of her comatose baby, or rush home to make sure that Stephen was all right. She ran home to check on her husband.[35]

Stephen was fine; he hadn't been alone. George Ellis had come in the morning and gotten Stephen to work. Jane returned to the hospital, not knowing whether her son was dead or alive. When the nurse told her that Robert had emerged from the coma, Jane could do nothing more than weep. And while the story had a happy ending, the deep scar from having to choose between caring for her son and caring for her husband never fully healed: "That day," she later wrote, "Robert survived, but a little piece of me died."[36]

CHAPTER 15

SINGULARITY (1962–1966)

I n the late spring of 1962, Stephen Hawking was facing the most important academic examination of his life. Most undergraduates at Oxford
didn't have to go through the ordeal of the *viva*, an oral exam akin to a
thesis defense—it was reserved for special or difficult cases. Stephen Hawking happened to be a special and difficult case.

The three undergraduate years at Oxford were structured around the
"tutorial." Each student would be assigned to a tutor—a faculty member in
a relevant field who would meet with a small group of undergraduates once
or twice a week. The undergrads would have to go to lectures and complete
weekly assignments, but those assignments were for the student's own benefit. All that really mattered, gradewise, were the final exams.

Hawking's tutor, Robert Berman, was a physicist—a thermodynamicist
who studied how heat flowed through exotic solids. Berman recognized that
Hawking was brilliant, "completely different from his contemporaries." But
for all his brilliance, Hawking had been an indifferent student—even lazy.
He thrived in theoretical physics, but he found it easy; he had to expend
little effort to get through the problem sets even as his fellow undergrads
struggled. "He could do any problem put before him without even trying,"
Berman told a reporter in the 1990s. But Hawking had little patience for
anything else; experimental physics, much less anything having to do with
the history of science, held no interest.[1]

During one summer, Hawking worked at the Royal Greenwich Observatory helping the Astronomer Royal make measurements on double stars

and found the whole exercise dreadfully disappointing. Nothing at Oxford excited any passion in him, and as a consequence, Hawking spent as little effort on his studies as he possibly could. "At that time, the physics course at Oxford was arranged in a way that made it particularly easy to avoid work," Hawking later wrote, and he avoided work with vigor, spending a mere one thousand hours or so (by his own estimate) on his studies during his entire time as an undergraduate. Indeed, working hard would have been a source of shame rather than of pride, the mark of a "gray man" rather than a lively fellow. Underneath the bravado, though, there was a creeping ennui, "an attitude of complete boredom and feeling that nothing was worth making an effort for."[2]

Because Hawking didn't feel challenged by the work his tutor gave him, he didn't exercise his mathematical muscles very well; he didn't know it yet, but he was actually a bit weak in the mathematical techniques he would need to succeed in his chosen profession. For Hawking had decided to be a cosmologist. But he had to get through the finals first, and he found it more difficult than he had expected.

"Because of my lack of work, I had planned to get through the final exam by doing problems in theoretical physics and avoiding questions that required factual knowledge," Hawking said in the *Brief History of Time* documentary. "I didn't do very well. I was on the borderline between a first- and second-class degree, and I had to be interviewed to determine which I should get." This was the viva.[3]

Hawking's future was resting on the outcome of the viva: he had been accepted to Cambridge University, but that acceptance was contingent upon getting a first-class degree. Without it, his career path was very much in doubt—at one point, he even interviewed for a position with the Ministry of Works, but then forgot to show up for the civil service examination.[4]

The professors at the viva knew how much was at stake for the young Hawking. "They asked me about my future plans," Hawking recalled. "I replied, if they gave me a first, I would go to Cambridge. If I only got a second, I would stay at Oxford. They gave me a first."[5]

• • •

In his last year at Oxford, it was not at all clear that Hawking would ever amount to anything. He was brilliant, to be sure—he had an extraordinary gift for physics, and he could instantly see things that were obscure to others who had expended much more effort. But expending effort didn't much appeal to him.

He would soon be a different man.

He would be diagnosed with an incurable disease and be given only a few dozen months to live. He would find love. He would find a purpose. And he would take his first few steps down a singular path.

•••

Stephen Hawking wanted to go to Cambridge to study with Fred Hoyle, the most famous astrophysicist in Britain, if not the world.

Hoyle started out his career trying to understand how stars form and evolve—how gas clouds accumulate, ignite into a stellar furnace, age, and die. After a hiatus to work on naval radars for the war effort, Hoyle realized that he could explain the abundance of heavy elements by looking at the reactions in stars. At the very core of a sun, hydrogen was fusing to produce helium and lithium and a few heavier elements. And when a star explodes, the violence of the explosion created elements that are heavier still. This was Hoyle's main scientific insight, and what earned him his place at the forefront of astrophysics. But it was not what Hoyle was best known for.

Hoyle, a cosmologist as well as an astrophysicist, was the principal architect of what became known as the steady-state theory. In the late 1920s, astronomer Edwin Hubble used painstaking measurements of distances to nearby galaxies to deduce that the universe was expanding—that galaxies were all rushing away from each other at a rapid rate.* This meant that the fabric of spacetime was constantly stretching—or, if you reverse the arrow of time, looking into the past means seeing all of spacetime shrinking into a point. This implied that the universe had to have been born at a single violent moment, a cosmic explosion that set all the fabric of spacetime in motion. Hoyle rejected this idea, dismissively dubbing it the "Big Bang." (The name stuck.) He preferred a universe that had no beginning and no ending, an eternal cosmos.

Unfortunately, the Einstein equations wouldn't permit such a solution, so in the late 1940s, Hoyle tweaked the equations to allow for a *creation field*—a mysterious force that continuously created matter out of nothing, forcing the fabric of the universe to expand. Where Big Bang theory created all the matter and energy in the universe in a single instant, Hoyle's steady-state theory created matter and energy in a slow, everlasting process that needed no moment of creation or, heaven forbid, destruction. Hoyle's

---

* For a deeper explanation of Hubble's work, see my *Alpha and Omega*, chap. 3.

colleagues Hermann Bondi and Thomas Gold, also at Cambridge, came up with a slight variant on the same theme, but the idea was the same: the universe wasn't born and will never die. It just is.

Big Bang versus steady-state quickly became the cosmological controversy of the decade. Of several decades, in fact, because there wasn't a heck of a lot else going on in cosmology at the time. Scientists were trying to improve on Hubble's estimates of the expansion of the universe, and performing a number of increasingly precise observations, but there was only so much one could do without a major new astrophysical discovery or a new mathematical technique brought to bear on the old theory. In fact, steady-state theory was never embraced by the cosmological community. From the moment Hoyle proposed the idea, astronomers were able to poke small holes in it—finding observations that didn't quite match Hoyle's predictions. These problems were never enough to kill the theory outright, but true believers were relatively uncommon. Most of the diehards were in the south of England: Hoyle at Cambridge, Bondi at the University of London, and Gold in Sussex (until he moved to the States in the late 1950s). By the early 1960s, when Hawking was applying to study under Hoyle at Cambridge, steady-state theory was badly wounded, but still alive—the death stroke would only come with the discovery of cosmic microwave background radiation in 1965.

However, Hawking probably didn't learn about Hoyle in a class at Oxford, or by reading physics texts. He first heard of Hoyle, almost certainly, by listening to the radio. Hoyle was the world's premier popularizer of astronomy and cosmology.* He got his start speaking to the public when he adapted a few of his talks for BBC Radio in the late 1940s and early 1950s—explaining such topics as sunspots to the audience in an earthy Yorkshire accent that was only slightly marred by Oxbridge posh. Hoyle had a great talent for packaging ideas to make them understandable to an untutored listener; indeed, the term "Big Bang" was one of his homely little radio coinages intended to conjure a powerful image in the minds of his audience. Hoyle enjoyed the limelight; not only did his radio work (and the books that followed) provide a much-needed boost to his finances, but it gave him a platform for his favorite theories—theories that became

---

* George Gamow and Isaac Asimov were the only people, at least in the English language, who could even come close, at least until Carl Sagan claimed the title in the late 1970s.

increasingly barmy as the astrophysicist aged.* For Hoyle lived by the dictum "It is better to be interesting and wrong than boring and right."[6]

Hawking was drawn to Hoyle. He probably didn't know that cosmology was a stagnant field, with little real progress since the 1930s. He was also unaware that he was woefully underprepared to use the mathematical techniques that general relativists use to analyze the shape of spacetime—a field known as differential geometry, populated with exotic creatures such as tensors and manifolds and simplices that Hawking had yet the occasion to tangle with. Apart from the heavy reliance on theory, cosmology wasn't an obvious fit for the young wanna-be physicist. But Hawking knew without a doubt that he wanted to study under Hoyle.

When Hawking arrived at Cambridge, Hoyle turned him down. Instead, Hawking was assigned to a relatively new faculty member, Dennis Sciama. It was a major disappointment.

• • •

Shortly after birth, a newborn baby begins trying to exert control over its little body. The struggle goes from top to bottom. At first, she can't even hold her head upright; when her father scoops her up, he must take care to cradle the head so that it doesn't hang limply. Within a few weeks, the infant is able to hold her head up, and then turn over in bed. A month or two later, the baby can sit up, and, once she has control over her larynx, begins to babble rather than merely cry. Soon, she learns to crawl, and the new parents realize how much more effort it takes to care for a mobile child than an immobile one. A few more weeks, and the baby is pulling herself upright. All too soon, she is able to balance herself on her feet and begins to take a few unsteady steps, trying to keep her oversized head and body upright supported only by a pair of absurdly stumpy legs. And then, a year and change after the process has begun, the child—no longer an infant, but a toddler—has mastered the ability to balance, to walk, to run. From head to toe, she has tamed her muscles and taught them to obey.

To many patients with amyotrophic lateral sclerosis—or motor neurone disease, as it's called in the United Kingdom—the disease is an unraveling, a gradual loss of obedience of the muscles tamed in the first

---

* Steady-state theory was on the respectable end of the spectrum, but Hoyle clung to it long after it had given up the ghost. On the wackier side of the spectrum was the idea that oil had not come from dead organic material but was being created continuously in the Earth's core; another was that the sunspot cycle was linked to flu pandemics, because influenza viruses came from outer space.

year of life. For reasons yet poorly understood, the long nerve cells that transfer electrochemical messages from the brain to the spine and the spine to the muscles wither and die. The legs, the arms, the fingers, the larynx, all these are cut off from the brain's command, and, by default, disobey the orders that they never receive. The denervated muscles slowly atrophy as they wait in vain to be put to use, and the patient gradually wastes away.

By the end of his life, ALS had claimed almost all of Hawking's muscular control; he could not even move his fingers enough to click a button, but instead had to control his computer by twitching his cheek. His head lolled uncomfortably to the side unless propped up by a nurse, and a rivulet of saliva typically rolled down his chin. Play the progress of his disease backward through time and it's remarkably similar to the toddler gaining control, painstakingly re-created over the course of half a century. He gains control over his limbs and is able to operate his computer and his wheelchair; he regains his voice and speaks more and more intelligibly; his fingers uncurl and he regains the ability to scrawl his name, to write, to walk with crutches. And in 1962, at the beginning of the process, the only sign of the disease was a bit of clumsiness, an unexpected fall here and there, and a bit of trouble tying shoelaces.

In his last term at Oxford, Hawking was coming down a spiral staircase when he slipped. "He kind of bounced all the way to the bottom. I don't know whether he lost consciousness, but he lost his memory," said fellow Oxford undergraduate Gordon Berry a number of years later. At first he was a total amnesiac—unable to remember who he was. Over the course of a couple of hours, his memory came back. Worried that he had damaged his intellect, Hawking decided to take the test for Mensa, the "genius" society. There was apparently no lasting damage. "He came back delighted that he was able to get into Mensa. Absolutely delighted."[7]

Stephen didn't tell his mother, Isobel, about his increasing clumsiness: "I think he began to notice that his hands were less useful than they had been, but he didn't tell us," she later said. But he couldn't keep the secret for long. During the Christmas break home from his first term at Cambridge, Stephen went skating with his family. Stephen was skating about, and then Isobel saw him tumble helplessly to the ice. "He fell and he couldn't get up. So I took him to a café to warm up, and he told me all about it." Isobel took Stephen to see the family doctor, who referred Stephen to a specialist. In the meantime, it was life as usual.[8]

New Year's Day, 1963. Stephen's friend Basil King threw a little party, and Stephen was there in a black velvet jacket and bow tie, joking with his friends. Basil's younger sister, Diana, had invited her friend Jane

Wilde. Stephen and the eighteen-year-old Jane hit it off—she found his self-deprecating humor delightful and decided she could overlook his long fingernails and mop of floppy hair. And there was something about his eyes—soulful gray eyes—and his broad smile that won her over. They exchanged names and addresses, and Jane soon received an invitation to Stephen's twenty-first birthday party, on January 8.[9]

Stephen's birthday was a fairly dreadful affair. Even though she knew many of the people in attendance—Diana was there as well—Jane was uncomfortable. The house was cold, the conversation stilted, and much of the evening was spent with the guests challenging each other with brain-teaser riddles. At that point, Jane didn't see much of a future with Stephen—and vice versa.[10]

Soon after his birthday, Stephen was admitted to St. Bartholomew's Hospital in London—where his father, Frank, had gone to medical school— to undergo a battery of neurological tests. His frugal parents were willing to spend for a private room, but Stephen refused; his "socialist principles" wouldn't allow him to use capital for that purpose. He was there for two weeks, undergoing a series of increasingly unpleasant examinations to determine what was wrong with him. Doctors took a muscle biopsy to see if the problem might lie with the muscles themselves rather than the nerves. They did an electromyographic study, sending shocks of electricity down his neurons to test how well they conducted signals and transmitted them to the muscles. And they did a procedure known as a myelogram to look at the spine and spinal canal: as Hawking described it, doctors "injected some radio-opaque fluid into my spine and with x-rays watched it go up and down as they tilted the bed."[11]

At the end of the ordeal, the doctors didn't give Hawking a diagnosis, telling him instead that he was an atypical case, and that his problems would continue to get worse. Hawking wrote that he didn't know that the culprit was ALS until a few years later,* but he knew he had been struck with an untreatable, progressive ailment—a terminal disease. "The doctors offered no cure and gave me two and a half years to live," he later said. All they could suggest is that he go back to Cambridge and try to finish his PhD.[12]

. . .

---

* If this is true, it was almost certainly willful ignorance; there's little chance that Frank would have been content until he had an affirmative diagnosis in hand. As Stephen admitted in his autobiography, he didn't press the doctors for details at the time because he knew the details were bad.

Hawking quite naturally fell into a depression. "[I] thought I would be dead in a few years. There didn't seem to be any point in carrying on," he told one early interviewer. Instead, he spent most of his time in his room listening to Wagner and "drinking a fair amount"—an account Hawking tried to distance himself from later on. When a reporter from *Playboy* asked him a number of years later about his drinking binge, Hawking responded, "It's a good story, but it's not true. . . . I took to listening to Wagner, but the reports that I drank heavily are an exaggeration. The trouble is, once one article said it, others copied it, because it made a good story. Anything that appeared in print so many times has to be true."[13]

On top of that, Hawking's studies weren't going well. Even before his diagnosis, he was floundering and in danger of failing out. "At first, I was doing very little work. I had very little mathematical background, so that made it difficult to make any progress," Hawking told a journalist two decades later. His condition was getting worse day by day. "It seemed to be developing very rapidly at first," he said, "and I was very depressed. I didn't think there was any point in doing any research, because . . . I didn't feel I would live long enough to get my Ph.D."[14]

Dennis Sciama watched his student with concern, but it would be up to Stephen to pull himself out of his rut. Frank Hawking paid a visit to Sciama to pressure him to let Stephen graduate early—presumably on humanitarian grounds—but Sciama refused to bend the rules. If Stephen were to earn a PhD, it wouldn't be due to any special favors; he would have to earn it.[15]

In fact, despite Hawking's disappointment at not being apprenticed to Hoyle, Dennis Sciama was a boon for the young student. Like Hoyle, Sciama had gotten his PhD from Paul Dirac, and he, too, was an advocate of steady-state theory. But Sciama was less ambitious—and less fractious—than the irascible Hoyle. He devoted most of his working hours to his students, and like John Archibald Wheeler, he spent an enormous amount of time and effort helping them launch their careers. Even if Sciama had been able to bend the rules for a dying student, he probably sensed that Stephen needed to succeed despite—not because of—his disease. Otherwise, any triumph he might achieve would be hollow. But in early 1963, that hardly seemed a kindness.

The one thing that was going well for Stephen was his love life. In February, he ran into Jane Wilde on a train platform. She had heard about his diagnosis, but he refused to talk about the matter. So the two discussed more pleasant things, and she accepted his invitation for a date. The two were soon courting, which meant bopping back and forth variously to Cambridge or to the outskirts of London on weekends. Usually by train. When they went by car, Jane hung on for dear life; Stephen apparently drove like a man with nothing to lose.* By the end of the spring semester, it seemed like they were a couple.[16]

However, Stephen's courtship, like his disease, progressed in fits and starts. Jane went to Spain for the summer, and upon her return, Stephen didn't make much of an effort to track her down; he only contacted her in November when he had come down to London for other reasons. And when Jane left for a term abroad in Spain in the spring of 1964, Stephen didn't bother to answer any of her letters. Upon her return, she found him "terse and uncommunicative." On occasion, he exhibited such "hostility and frustration" that Jane suspected he was "deliberately trying to deter me from further involvement with him. It was too late. I was already so deeply involved with him that there was no easy or obvious way out." She was almost relieved when the two decamped to different countries for the summer holiday.[17]

· · ·

In the middle of 1964, Jane perceived a young man dogged by depression, cynical and disheveled. However, in many respects, Stephen's situation had brightened considerably since the dark winter of his diagnosis two years earlier. After a steep decline, the disease loosened its grip, and his condition stabilized; he now had to walk with a cane, but it seemed quite certain that he would defy the dire two-and-a-half-year sentence the doctors had passed upon him. And after two years of flailing about, trying to master the mathematics of the Einstein field equations, he finally got it— and he had a distinct talent for the work. He had chosen his field wisely. It was around that time that Dennis Sciama tried to get his young student's brain into gear, saying something to the effect of: "Well, you're not dead yet. So, are you ready to work on that problem I suggested?"[18]

He was. Hawking was primed to attack a thesis-worthy problem in cosmology, but finding that problem was itself a major task. None of the problems Sciama was suggesting were sticking. "It's very difficult in these advanced fields to find a good thesis topic. Cosmology at that time was a bit fragmentary. It was not easy to say, 'Well, here's a problem. It will take you three years; now go on and do it,'" Sciama said two decades later. "So

---

* Even long after Hawking himself lost the ability to be behind the wheel, this reckless-driving habit continued whenever he had any degree of control over a car. Marika Taylor remembers driving her adviser around California in a minivan in the late 1990s: "Stephen told me to do a U-turn, and I'm like, 'But it says no U-turn!' So he's like, 'No, you're doing a U-turn.' So I do a U-turn." Naturally, the rearview mirror filled with flashing red and blue lights, and the police pulled the van over. Taylor adds, "They open up the back of the van and say, 'Oh, hello, Dr. Hawking! It's you again.'"

indeed [Hawking] looked around for quite a bit. While he did quite interesting things, the real measure of his ability was hardly emerging yet, and
I can imagine it was a bit frustrating for him." And Hawking, for his part,
didn't think much of Sciama's proposed topics for a thesis. "He was always
stimulating, but I didn't agree much with his ideas," Hawking later recalled.
But once out of his depression, Hawking's mind was primed, just awaiting
the tinder to set it ablaze.[19]

The first sparks came in early 1964. Hawking shared an office at Cambridge with a young postdoc—Jayant Narlikar—who had studied under
Hoyle and was busy hammering out some steady-state-related theories
with his former adviser. Specifically, Narlikar and Hoyle were trying to
create a new mathematical description of gravity in hopes that their version would help them support the idea of an eternal universe. Hawking,
quite naturally, was interested in what his officemate was working on—and
Narlikar was more than willing to share. Narlikar even provided an early
draft of a key paper that he and Hoyle were about to submit to peer review.
Hawking played around with the equations on his own and found a flaw
in their theory. He realized that their idea wouldn't work because certain
key values "diverged"—they went off to infinity, rendering the whole calculation meaningless. But rather than telling Narlikar about the problem,
Hawking attended a meeting of the Royal Society in London where Hoyle
was presenting the idea to the scientific community for the first time.[20]

There were about a hundred scientists in attendance, and when Hoyle
finished his talk, he asked if there were any questions. Hawking rose slowly
to his feet and balanced himself with his cane. A hush fell over the crowd
as the unknown twenty-two-year-old graduate student had the temerity to
challenge the nation's most famous astrophysicist:

> "The quantity you're talking about diverges," [Hawking] said.
>
> Subdued murmurs passed around the audience. The gathered
> scientists saw immediately that, if Hawking's assertion were correct,
> Hoyle's latest offering would be shown to be false.
>
> "Of course it doesn't diverge," Hoyle said.
>
> "It does," came Hawking's defiant reply.
>
> Hoyle paused and surveyed the room for a moment. The audience
> was absolutely silent. "How do you know?" he snapped.
>
> "Because I worked it out," Hawking said slowly.[21]

Since this was the first time that Hoyle had presented his theory to the
public, everyone assumed that Hawking had done the calculations in his
head, on the spot during the lecture. Nobody suspected that he had seen

a draft of the paper beforehand, and had come to the meeting forearmed with what would be the fatal blow to Hoyle-Narlikar gravitation. "Hoyle was furious," Hawking later wrote.[22]

Hoyle's fury was irrelevant; Hawking was right. The values diverged. The young upstart had gone head-to-head with the most famous astrophysicist in Britain and come out on top. Not just that, he cultivated the impression that he had done it off the cuff—that the eccentric-looking young gentleman leaning on a cane harbored a mind so vast that it could do calculations in minutes that Hoyle and his graduate student couldn't figure out how to do in months. It was a Feynman-quality performance. And, as it turned out, it was Hawking's first publishable-level research.

Knocking down someone else's theory—especially a brand-new one that hadn't received much scrutiny—isn't itself a huge feat, but it is real physics nonetheless. Hoyle and Narlikar had submitted their calculations to the *Proceedings of the Royal Society A*, where they were published at the beginning of the summer; Hawking's counterstrike, proving that the Hoyle-Narlikar calculations were a dead end, was almost guaranteed a spot in the same journal. But as with many other society journals, Hawking needed to find a sponsor—in this case, a member of the Royal Society— who was willing to transmit the paper to the editors of the journal. So he approached Hoyle collaborator Hermann Bondi, whose lectures Hawking had attended a few times. Bondi said yes; he probably felt honor-bound to help air any substantive critiques of steady-state-related work—but the whole affair probably didn't leave Bondi with a great love for Hawking. That didn't matter . . . at the moment. For in October 1964, Hawking submitted his first peer-reviewed paper to a prestigious journal; it would be published in the summer of the following year.

The takedown of the Hoyle-Narlikar theory also became Chapter 1 of Hawking's PhD thesis. By itself, it wasn't enough to merit a degree. But it was a solid little piece of research, and Hawking was finally producing something of merit. Despite the death sentence hanging over him, he had transformed himself into a physicist.

...

Jane likes to think she broke Stephen out of his depression and gave him the will to live. "We were going to defy the disease. We were going to defy the doctors," she told an interviewer in 2013. "And we were going to challenge the future." Stephen, too, cleaved to this narrative, giving Jane credit for his renewed sense of purpose. "The disease wasn't progressing so rapidly, and I got engaged to be married. So that was really the turning point," he told an interviewer. "I realized that if I was going to get married,

I would have to do some work—I'd have to get a job. About that time, I began to understand what I was doing as a mathematician."[23]

However, in mid-1964, when Hawking was busy demolishing the Hoyle-Narlikar theory, it wasn't at all clear that the romance would survive. Their parting at the end of the spring term left Jane wondering if they would see each other when they both returned to Britain at the end of summer break. And then, as Jane was traveling through Italy with her family, she received a postcard from Stephen. "I was overjoyed to receive such an unexpected piece of correspondence," she later wrote.[24]

"Could Stephen really have been thinking of me as I had been thinking of him?" she wondered. "It gave me grounds for daring to hope that he was looking forward to seeing me at the end of the summer." He was, in fact. Stephen proposed to Jane in October 1964, at almost the same moment that Hermann Bondi transmitted Hawking's paper on Hoyle-Narlikar gravitation to the *Proceedings of the Royal Society*.

Jane said yes.

She would wed a young man who wasn't expected to live to the end of the decade. As Jane put it in 2015: "[I] loved Stephen and wanted to do my best for him. So I thought that I could easily devote two years of my life to help somebody I loved—someone who had so much potential—achieve his ambitions." Neither she nor Stephen dreamed at the time that it would not be two years of devotion, but twenty-five.[25]

Even early on, Jane got the sense that it wasn't going to be easy to live with Stephen. He had a tendency to assert dominance in a crowd by playing the contrarian—picking fights with innocent bystanders. Jane herself was by no means spared: "With me he would argue that artificial flowers were in every way to be preferred to the real thing and that Brahms, my favourite composer, was second-rate because he was such a poor orchestrator." He even embarrassed her in front of Alan Deyermond, her PhD adviser, with a tirade about the uselessness of studying medieval literature. "You shouldn't take it personally," he told her when she took offense. Yet she loved him, and she was committed to helping him for the rest of his life, tragically foreshortened though it might be.[26]

The upcoming nuptials forced Stephen to think more carefully about his future, at least in the near to medium term. With a wife who wanted to continue her studies, he was going to have to be the breadwinner, and that meant finishing up his PhD and getting a fellowship of some sort. Luckily, his research was finally going well; in late 1964 and early 1965, Sciama and his group of acolytes—Stephen often among them—regularly took the train to physics lectures in London. The Penrose talk in January, faithfully recounted by Brandon Carter, provided that final spark, that one big idea, worthy of a PhD thesis and more. The singularity theorem he had in mind was good enough to fuel the start of his career. But he still had to find a job.

At the start of a chilly weekend in February 1965, Hawking eagerly awaited Jane's arrival. When she came through the door, left wrist in a cast, he was horrified—horrified because he needed her to type up his fellowship application. "I must admit that I was less sympathetic than I should have been," Hawking noted in his autobiography. Jane spent the entire weekend writing it out longhand.[27]

It almost came to nothing. The application required two recommendations, and Hawking had chosen to approach Hermann Bondi. "After a lecture in Cambridge, I asked him about providing a reference, and he looked at me, and in a vague way said yes, he would. Obviously, he didn't remember me, for when the college wrote to him for a reference, he replied that he had not heard of me." Sciama immediately got on the phone. Hawking soon had a glowing recommendation.[28]

He got the fellowship. It would start that fall.

Jane and Stephen were married on July 15, a mere five days after the end of the International Conference on General Relativity. At that conference, Hawking presented his singularity theorem to some of the most important relativists of the day, including John Archibald Wheeler and his students Kip Thorne and Charles Misner. Hawking made quite an impression on them, and he wound up fast friends with Thorne and Misner (who would be the godfather to his son Robert two years later). And then it was off to an all-too-brief stay in Suffolk before packing off to the Cornell summer school a week later.

Though Jane was now married to Stephen, it was beginning to dawn on her that Stephen was now married to physics.

CHAPTER 16

YLEM (1942–1962)

―――――――

S cenes from a youth. Photos reprinted in books, images preserved online—brief glimpses of the days before Hawking became Hawking. Nearly seventy years on, little else remains: a few snippets of memory from those old enough to remember, and mute images of a boy yet to know fame.

...

I n front of a stone wall, eroded with the passage of time, some three dozen members of the Oxford University boat club goof around as they pose for a photograph. Four chaps sit on a sofa while a fifth stands on his head, supported by grinning boys surrounding him. Several sit on or dangle from the wall in various states of precariousness. One holds a hammer, seemingly ready to strike the unsuspecting student in front of him; another proudly displays a large letter "Y," possibly the trophy of some sort of illicit raid of a rival college. Some stare at the camera, some look away. Some are in tidy suits, some are disheveled, one is entirely shirtless, and three do their best Lawrence of Arabia imitation—keffiyahs and all. It's a mishmash of chaotic action.[1]

But one young man on the far right, near the edge of the frame, dominates the tableau. He stands as rigid as a statue, left arm clutched to his chest and right arm upraised with a kerchief in his hand—head thrown back, grimacing in mock solemnity. That young man is Stephen Hawking.

Hawking seems younger than most of the others in the boat club; many of the young men had been conscripted into the military before coming to Oxford. Hawking was fortunate in that compulsory military service was abolished in the United Kingdom in 1960, just as he would have been called up.

The boat club offered some comradeship to the young Hawking, who was not only bored at Oxford, but lonely. Not being the brawny type, he was drafted to be a coxswain. Hawking would sit in the front of the boat in a white straw hat, telling the eight rowers in the crew when to stroke and where to steer. He was good at the job, but Norman Dix, the Oxford boatman—the official in charge of the crews—was torn about whether or not to put Hawking in charge of the best crew on the river. "The question always with Stephen was should we make him coxswain of the first eight or the second eight," Dix later said. "Coxes can be adventurous. Some coxes can be very steady people, you see. He was an adventurous type. You never knew quite what he was going to do when he went out with a crew."[2]

· · ·

Oxford University, night. The river Thames flows silently in from the northwest and, joining the Cherwell just south of campus, streams away to the south and east. All's quiet. On one of the footbridges spanning the river, Hawking and a friend try not to snicker as they use a rope to suspend a plank from the bridge. They climb down with paintbrushes:

> A few minutes later, just visible in the dark, were the words VOTE LIBERAL in foot-high letters along the side of the bridge and clear to anyone on the river when daylight broke.
>
> Then disaster struck. Just as Hawking was finishing off the last letter, the beam of a flashlight shone down on them from the bridge and an angry voice called out, "And what do you think you're up to then?" It was a local policeman.[3]

Hawking's friend jumped off the plank, made it to the riverbank, and scurried away. But Hawking was nicked. He was hauled off to the local constabulary and given a good talking-to, but nothing more came of the incident.

Given the potent mix of boredom and beer—spiced with more than a touch of arrogance—it's amazing that Hawking didn't get into more trouble than that. "At Oxford, where he spent a lot of time drinking with oarsmen,"

wrote an old friend of his, "he seemed angry and frustrated in an unfo-
cused way, and his manner was provokingly raffish." He was lucky enough
to have a tutor, thermodynamicist Bobby Berman, who wasn't offended
by the young man's affect, even when his peers were aghast at Hawking's
attitude. "He used to produce his work every week for tutorial, and, as he
never kept any notes or papers or that sort of thing, on leaving my room, he
would normally throw it in my wastepaper basket," Berman told filmmaker
Errol Morris. "And when he was with other undergraduates at the tutorial,
and they saw this happen, they were absolutely horrified. They thought he
did this work in probably half an hour. If they could have done it in a year,
they wouldn't have thrown it in the wastepaper basket, they would have put
it in a frame up on their walls."[4]

But Berman couldn't engage Hawking, who spent most of his time
coxing on the river, playing bridge or poker or darts, and drinking port
or beer—and as little time as possible on physics. Even so, he quickly out-
shined the three other undergraduates at Oxford in his year. One of them,
Derek Powney, told of one particularly hard problem set—thirteen problems in electricity and magnetism—all of which seemed almost totally
intractable. "I discovered very rapidly that I couldn't do any of them,"
Powney said. He teamed up with one of the students, and by the end of
the week, as the problem set was coming due, the two of them together
had completed one and a half problems. The third student, working alone,
managed to finish one. "Stephen as always hadn't even started." Hawking
then went to his room to do the assignment, and emerged about three
hours later. "'Ah, Hawking,' I said, 'how many have you managed to do,
then?'" Powney recalled. "'Well,' he said, 'I've only had the time to do the
first ten.' I think at that point we realized that it's not just that we weren't on
the same street, we weren't on the same planet." He added, "It's quite difficult to live all the time with people who are a lot stupider than you are. I
think that therefore you tend to become a very private person, and to build
almost a caricature of yourself as a defense."[5]

. . .

Frank Hawking ponders a book, his head on his hand. Despite the round-
rimmed glasses, tweedy jacket, and mop of wispy gray hair—the very
picture of an academic type—there's something steely about Frank. Where
his son is lanky, Frank is broad and robust, clearly used to the outdoors.
There's more severity than warmth in his visage.[6]

Another photo, from a few years earlier, shows him on a rocky shoreline
in a khaki shirt and toting some sort of field kit over his shoulder. He gives
a half smile to the camera as he cups something indistinguishable—some

specimen, perhaps—in his right hand. One would guess him to be an archaeologist rather than a medical doctor.[7]

Frank's alma mater was Oxford, and he desperately wanted Stephen to attend as well. But Stephen's performance at St. Albans School was far from stellar—despite his genius, Hawking always wound up in the bottom half of his class. ("It was a very bright class," Hawking liked to joke.) This made Oxford very unlikely for Stephen, even without the additional worry of obtaining a scholarship. So Frank decided to take matters into his own hands.[8]

In the early spring of 1959, just before Stephen was to take the entrance examinations for Oxford, Frank and Stephen paid a visit to Bob Berman, Stephen's prospective tutor. Frank reportedly applied pressure, enough that Berman would ordinarily have been ill disposed to the candidate. But Stephen scored so high on the exams that his father's meddling—and Stephen's mediocre high-school grades—were moot. Hawking would go to Oxford to study physics.[9]

. . .

Four young men are gathered around a pinball-machine-sized contraption full of wires and switches. All four are smiling broadly. The one on the left, the shortest one, is Stephen Hawking.[10]

It was 1958, and the four boys, nearing the end of their high-school career, were posing in front of their home-built computer, which they dubbed LUCE—the Logical Uniselector Computing Engine. It was incredibly primitive—old wires and electromechanical switches recycled from a defunct telephone exchange soldered together—but it could solve some very basic logical problems. It wasn't really a computer, or even an adding machine, but it was a first step along the way, and it was impressive enough to make it into the local paper.[11]

LUCE was the brainchild of Dikran Tahta, a new math teacher at St. Albans. Tahta got as close as a teacher could get to inspiring the disinterested young man. "Many teachers were boring. Not Mr. Tahta," Hawking later said. "His classes were lively and exciting." By the end of high school, Stephen wanted to study mathematics and physics. "But my father was very much against it, because he thought there wouldn't be any jobs for mathematicians except as teachers," Hawking wrote. "He would really have liked me to do medicine, but I showed no interest in biology. . . . [H]e made me do chemistry and only a small amount of mathematics. He felt this would keep my scientific options open."[12]

Stephen's younger sister Mary would become a physician, but to Frank's disappointment, Stephen never gained any interest in medicine or biology. And even with the inspiring classes of Mr. Tahta, Stephen never rose above the middle of the pack in his class. That, too, must have disappointed Frank.

As Stephen studied for his A-level exams in physics, the Hawking family spent the year in northern India, so Frank could do some research at the Central Drug Research Institute in Lucknow. During the academic year, Stephen stayed behind with a colleague of his father's, but he managed to come down to rejoin the rest of the family in the summer, just in time for the monsoon. He was disappointed to find Frank's India a little less exotic than he had hoped: "My father refused to eat Indian food during his time there, so he hired an ex–British Indian Army cook and bearer to prepare and serve English food. I would have preferred something more exciting."[13]

· · ·

A drawing of a man in profile, standing, facing to the right. Apart from a slight frown, he looks perfectly fine above the waist. But shortly below the knee, his right calf balloons outward grotesquely. The excess flesh seems to sink down the leg, dripping pendulously toward the ankle and the swollen, oversized foot.[14]

It was July 1958, and Frank Hawking had written a large feature article in *Scientific American* about elephantiasis, a parasitic disease endemic to China, India, and the South Seas that caused monstrous deformities in its victims. Frank and his colleagues had been studying the parasites—threadlike worms known as filaria—that caused the disease, and were making headway in understanding how it behaved and how to control it. Since 1951, Frank had been testing the use of a drug called diethylcarbamazine (DEC) that seemed to kill most of the parasites. It even offered the hope of eradicating the disease altogether, if only large populations of people could be dosed with it on a regular basis.

With his *Scientific American* article, Frank was bringing his scientific ideas to the largest audience he had ever reached. For years, he had been publishing scientific articles in peer-reviewed journals, including *Nature*. But reaching a popular audience was another thing entirely, and Frank must have been beaming with pride.

Even Stephen must have been impressed. Truth be told, his father's research gave him the willies. While he enjoyed visiting his father's lab and looking through the microscope, he hated visiting the hothouse where his father kept mosquitoes infected with various tropical diseases. "This worried me, because there always seemed to be a few mosquitoes flying around loose," he later wrote.[15]

Scientific American articles were aimed at the layman, and no doubt the sixteen-year-old Stephen read his father's piece with great interest. Not only could he finally get a sense of what his father's research was all about—and understand why Frank spent so much time away from home—but he could see just how important his father really was. Simply by virtue of having been published in as prestigious a magazine as *Scientific American*, Frank Hawking must have seemed a superstar.

In fact, Frank was a superstar. He advocated adding that anti-filaria drug, DEC, to table salt in regions where the disease was endemic, and proved its effectiveness by testing it out in Brazil. In the mid-1960s, he traveled to China, where he likely proposed fortifying salt with DEC in affected areas; at any rate, the Chinese government began doing so shortly thereafter. In the 1950s, roughly thirty million Chinese citizens were afflicted with filariasis. By 1994, thanks to DEC-fortified salt, the disease was officially eradicated in China. Frank was primarily responsible for preventing the disability and disfigurement of millions upon millions of people. To young Stephen, Frank must have cast a very long shadow.[16]

. . .

A large Victorian house: a small handful of narrow windows and a single large bay window can't soften the silhouette of the imposing brick walls. Everything's covered in a deep blanket of snow—probably what spurred one of the Hawkings to venture out into the yard to take the photo in the first place.[17]

Frank and Isobel had bought the house in St. Albans—a small town on the northern outskirts of London—in the early 1950s. After an initial effort to get it fixed up for the move-in, Frank and Isobel, in their frugality, neglected the house and it slowly fell into a state of clutter and disrepair. It wasn't heated, and as broken windows went unreplaced, it got colder and wetter as the years went by. Stephen's younger brother, Edward, described the home as a "very large, dark house . . . really rather spooky, rather like a nightmare." To an outsider, it could seem to be an intellectual version of the Addams Family house.[18]

Frank was the local beekeeper, and he kept beehives in the basement. Grandma Walker—Isobel's mother—lived in the attic, emerging occasionally to join the family or to play the piano at the local folk dances. There were four Hawking children: Stephen, the eldest; Mary, eighteen months younger; Philippa, born when Stephen was five; and Edward, who was adopted when Stephen was a teenager. (Stephen described him as "completely non-academic and non-intellectual, which was probably good for us.")[19]

"It was the sort of place where if you were invited to stay for supper, you might be allowed to have your conversation with Stephen, but the rest of the family would be sitting at the table reading a book," recalled childhood friend Basil King, "a behavior which was not really approved of in my circle, but which was tolerated from the Hawkings. Because they were recognized to be very eccentric, highly intelligent, very clever people, but still a bit odd."[20]

...

An old Romani wagon—the type that make up "Gypsy" caravans in the movies—sits against a copse of trees. It is painted green in a half-hearted attempt at camouflage, but it's a fairly prominent eyesore, especially with an army-surplus tent propped open right next to it. The three Hawking children (it's before Edward joined the family) sit on the ladder-cum-stairway leading to the door of the wagon, amid a collection of pots and pans and assorted junk that had accumulated over the years.[21]

Frank had somehow acquired the old wagon, plunked it down in a town on the southwest coast of England, and left it there as a ramshackle summer home. For years, the Hawkings spent their summer vacations in the wagon and the tent. In 1958, no doubt much to the locals' relief, the county council managed to get the thing removed.[22]

...

Hair slicked back and in a dapper slate-gray suit, Billy Graham addresses a sellout crowd at Harringay Arena in London, kicking off a three-month evangelical "crusade." "It's been a long time since evangelism and revival and Christ and God was front-page news around the world," he exclaimed. "And we thank God for it!" In 1954, Graham's visit, full of American-style evangelism, backed with a chorus of two thousand singers, was guaranteed to set British tongues a-wagging.[23]

It even landed a convert or two. One of the boys in Hawking's circle of friends caught the spirit from Graham, and soon the boys of St. Albans were chatting about matters theological.[24] Hawking watched his schoolmates' sudden zeal with some bemusement. As Michael Church, a boyhood friend, described it:

> We began talking about life and philosophy and so on, which I thought I was very hot on at the time, and so I held forth.
>
> Suddenly I was aware that he was egging me on, leading me to make a fool of myself. It was an unnerving moment. I felt looked

down upon from a great height. I felt that he was watching, amused and distant. . . . And there was some overarching arrogance, if you like, some overarching sense of what the world was about.[25]

It must have irked his peers when Stephen won his school's divinity prize one year. As his mother, Isobel, put it, "It's not surprising because his father used to read him Bible stories from a very early age and he knew them all very well."[26]

For a brief moment, Hawking got caught up in an ESP craze, boosted, in part, by the work of J. B. Rhine, a parapsychologist at Duke University who wrote a few popular books on the subject, including one that came out in 1953. Stephen scrutinized the Duke research and quickly came to the conclusion that the work was bunk. "Whenever the experiments got results, the experimental techniques were faulty. Whenever the experimental techniques were sound, the results were no good," he later told a journalist. "People taking it seriously are at the stage I was when I was a teenager."[27]

Science held more interest to Stephen than theology or parapsychology, and his opinions were equally strong. When he heard that the universe was expanding, he refused to believe it. "A static universe seemed much more natural," he said. "It could have existed and could continue to exist forever." He would be a PhD student before he realized that he was wrong and that the universe was, indeed, expanding from a Big Bang.[28]

Even though he was obviously bright, Stephen refused to apply himself in school. Isobel remembered confronting him about being third from the bottom in his class: "So I said, 'Well, Stephen, do you really have to be as far down as that?' and he said, 'Well, not a lot of people [did] much better.' He was quite unconcerned."[29]

• • •

Right next to Westminster Abbey in downtown London is Westminster School—an imposing Gothic compound with pointed arches and passageways with vaulted ceilings and crenellated battlements. It is one of Britain's elite "public" schools (confusingly named, as these schools charge large sums for tuition), and for more than five centuries it had been preparing young scions for Oxford and Cambridge and from there to further the glory of the Empire. By the early 1950s, that Empire was greatly reduced, but Frank well knew that the future would still be brighter for a public schooler than it would for an ordinary grammar-school boy. "He had a bit of a chip on his shoulder because he felt that other people who were not as good but had the right connections had got ahead of him," Stephen later wrote. But Westminster would be out of the question for Stephen if he couldn't win a scholarship.[30]

Alas, the day of the scholarship examination, Stephen was struck ill and had to stay home in bed. Thus was Frank's dream of an elite public-school education for his son dashed forever. Writes Stephen: "Instead, I stayed at St. Albans School, where I got an education as good as, if not better than, the one I would have had at Westminster."[31]

About that time, Stephen was suffering from a strange low, recurring fever that caused him to miss quite a bit of school. It was diagnosed as some sort of "glandular fever"; his mother halfway suspected that this condi-tion—which disappeared as mysteriously as it arrived—had something to do with his later struggle with ALS.[32]

It's doubtful that Stephen would have felt much loss at staying home from school. Attending St. Albans was often a trial for the geeky and awk-ward young man. "He was by turns the classroom clown and the defenseless butt for violence in the showers," his friend Michael Church later wrote. And even without the mysterious fevers, given how poorly Stephen did at St. Albans when he was there, most of his intellectual development had to have happened at home—nose in book—or with his friends. In the after-noons and on the weekends, he would spend hours with his pals debating various topics, building things like model airplanes and boats, listening to classical music on the radio, and inventing board games with elaborate and Byzantine rules. One was a complicated war game, another about logistics—manufacturing and transport and getting goods to market. "He ended up with a fearful game named 'Dynasty,'" his sister Mary recalled. "It went on forever because there was no way of ending it." Isobel, too, remem-bered that particular game: "It took hours and hours and hours. I told him it was a terrible game, and I couldn't imagine anybody being taken up with it. But Stephen always had a complicated mind, and I felt, as much as any-thing, it was the complication of it that appealed to him."[33]

"I think these games, as well as the trains, boats, and airplanes, came from an urge to know how systems work and how to control them," Ste-phen later wrote. "If you understand how the universe works, you control it, in a way."[34]

• • •

Two young boys sitting on the edge of a dock; behind them, the clear water revealing no color but that of the green hills in the distance. The smaller of the two squints in the Mediterranean sun, all buck teeth and splayed ears and grinning with boyish excitement. That's the eight-year-old Stephen. Next to him, a little larger, a little older, and a little cooler, William Graves scowls at the camera, hiding behind a pair of sunglasses.[35]

It was the early spring of 1950, and Frank was away, traveling, as always. One year it was India, another China, another Africa, another to the swamps of the Americas. Stephen's sister Mary remembers thinking that her father—and, by extension, fathers in general—were "like migratory birds. They were there for Christmas and then they vanished until the weather got warm." This time, when Frank disappeared for the season, Isobel decided to take the rest of the family to visit her friend Beryl, on the island of Majorca.[36]

Isobel and Beryl had met when the two were studying at Oxford—an unusual thing for women to do in those days—and both were ardent leftists (Isobel had joined the Young Communist League). Beryl had just married the writer Robert Graves after a complicated love affair.[37]

While Stephen and the family had a "wonderful time," he didn't quite fit into the Graves household. He sparred with his and William's tutor—the poet W. S. Merwin—who insisted that, each day, the boys were to read a chapter of the Bible and write an essay about what they had read. Stephen realized that it was busy work, as his tutor "was more interested in writing a play for the Edinburgh Festival than in teaching us." At the time, Graves was deep into a mystical phase, busy deconstructing Greek and biblical myths in search of the ur-stories hidden beneath. "So there was no one to appeal to," Hawking lamented.[38]

William was put off by the young visitor. "[Stephen's] arrival rather upset my routine," William told an interviewer in 2018. "He was a year and a half younger than me, which is an age for small boys. Also, my friends were the village children and I was used to speaking to them in Mallorquin, which Stephen didn't speak—of course." But the one who was most upset by Stephen's presence had to be Robert Graves himself.[39]

Robert had been badly injured in World War I, nearly succumbing to a chest wound—one that he never truly recovered from. He suffered, too, from what's now called post-traumatic stress disorder. As he described it in his memoir, *Goodbye To All That*:

> Since 1916, the fear of gas had obsessed me: any unusual smell, even a sudden strong smell of flowers in a garden, was enough to send me trembling. And I couldn't face the sound of heavy shelling now; the noise of a car back-firing would send me flat on my face, or running for cover.[40]

Unfortunately, recalled William, Stephen the prankster had come to Majorca armed with a goodly supply of stink bombs and home-made firecrackers. "My father didn't appreciate this."[41]

• • •

Stephen, maybe four years old, sits in front of a model train set—a sleek steam engine with four cars—sitting on a small wooden track. He has a toddler's face, the cheeks of a baby, but there's a depth to the expression in his eyes that's a bit startling in a boy that young.[42]

The train set was a gift from his father, the spoils of an overseas trip to the Americas right after the end of the war. The family had lived in north London and suffered the privations of rationing and random bombing from German rockets, and now that the war was over, they began to settle into the new normality of peacetime.

Frank and Isobel enrolled Stephen in the Byron House School. In his memoir, Stephen recalled complaining to his parents that he wasn't learning anything. He had a point—he didn't learn to read until the age of eight. "We were more concerned with my husband's brilliance rather than Stephen's," Isobel later said. "Still, Stephen was a self-educator from the start, and if he didn't want to learn things, it's probably because he didn't need to."[43]

The postwar years also saw the introduction of a spate of modern drugs and compounds that could be used against tropical diseases. Frank investigated how safe and effective some of these compounds were against parasitic ailments like malaria, and he studied the life cycles of disease parasites in laboratory animals. He had a spate of publications in the 1940s—including in esteemed journals like *Nature* and the *British Medical Journal*—and was soon promoted to head up the division of parasitology at the National Institute for Medical Research. The family moved to St. Albans, a little farther north, to reduce Frank's commute. Unlike the London neighborhood where Stephen had spent his first few years, St. Albans wasn't populated with academics and intellectuals. He, like the rest of the Hawkings, would find it a bit more difficult to fit in.[44]

• • •

An infant Stephen, swaddled in a white knit blanket. His father holds him and gazes at him intently through round-rimmed glasses. Stephen peers back at him just as intently, studying his father with intense gray eyes.[45]

• • •

Stephen William Hawking entered the world on January 8, 1942. It was three hundred years to the day after the death of Galileo, a coincidence that he would later find quite amusing. He would depart the world some 76 years and change later, on March 14, 2018. It was 139 years to the day after the birth of Einstein. That—that he would have found hilarious.

CHAPTER 17

ON THE SHOULDERS OF GIANTS

———————————

n the early 1930s, so the story goes, Albert Einstein was in Hollywood, entertaining a visit by a friend, the comedian Charlie Chaplin. They were enjoying some tarts baked by Elsa Einstein and idly chatting when Einstein's son turned to Chaplin. "You are popular," he said, "because you are understood by the masses. On the other hand, the professor's popularity with the masses is because he is not understood."[1]

In 1919, Albert Einstein underwent a rapid metamorphosis, transforming almost overnight into an international celebrity—from a mere theoretical physicist into the archetype of scientific genius. From the start, that transformation was founded largely upon unintelligibility. In introducing the public to the theory of relativity, the *New York Times* took pains to convince readers of its inaccessibility. Einstein's book about relativity was touted as "A Book for 12 Wise Men": "When he offered his last important work to the publishers he warned them there were not more than twelve persons in the whole world who would understand it, but the publishers took the risk." That legend—and that number, twelve—was repeated over and over as Einstein's fame grew, even as the man himself denied it. "That the story of the twelve men never died out, even though Einstein himself and others offered disclaimers, shows the hold that this phrase had on the public mind," one historian writes. "It was probably the most important factor in the growing fame of the theory of relativity."[2]

Surrounded by his disciples, Einstein had become the bearer of the new physics, superseding the old Newtonian testament. That the physics was written in a mathematical language that few members of the public could fully understand enhanced the awe surrounding Einstein rather than diminishing it. People were less interested in the theory of relativity than what it represented, just as they were less interested in the flesh-and-blood Einstein than they were in him as a symbol of the triumph of the human mind.

It was a symbol so powerful that members of the public would fight to get a glimpse of it. In 1930, a crowd of more than four thousand people trying to see a film about Einstein rioted through the American Museum of Natural History. The riot began not because a group of New Yorkers were desperate to learn more physics; it was because they wanted to share in the vision of someone destined to become immortal, someone whose fame would outlast even that of the great Charlie Chaplin.

A scientist with that much hold over the public imagination is born perhaps once a century, or even more rarely. Since the death of Einstein, there is only one scientist who has come close to achieving that level of fame: Stephen Hawking.

．．．

It's extremely rare for a scientist to become a celebrity, much less achieve public immortality. At any given moment, perhaps a dozen or two in the world might have enough fans to be recognized on the street once in a while. Those who do almost never gain their fans because of the science they've done; more often than not, a scientist's celebrity has little to do with the quality of his or her research.

Each year, the names of the new Nobel laureates are forgotten within days—if those names ever register at all. Even the extraordinarily rare double Nobelists, like John Bardeen, don't necessarily get a share of fame. The list of former Lucasian chairs alone is filled with names that are instantly recognizable to mathematicians and physicists—Paul Dirac of the Dirac equation, Joseph Larmor of the Larmor frequency, Gabriel Stokes of Stokes' theorem—but who never received much attention from the public while they were alive, and are now almost totally unheard of outside of the scientific community. Even scientists who had an absolutely pivotal insight don't automatically achieve fame. James Clerk Maxwell, Erwin Schrödinger, Murray Gell-Mann, Ludwig Boltzmann, Niels Bohr, Ernest Rutherford . . . the names of these physicists might trigger a glimmer of recognition in the sort interested in science, but even when they were alive, few people on the street would have known who they were or what they had accomplished. No, it's not scientific accomplishment that turns a physicist into a celebrity, but something else.

The most common form of scientific celebrity comes from popularization. In some sense, bringing the ideas of science to the public is an act of translation, a clever use of metaphor and other literary tricks to make work written in the language of mathematics accessible in the vernacular. A popularizer is an intermediary between science and the public, a priest who helps the laity commune with a higher knowledge. Generally speaking, scientists are not expected to take on the mantle of this priesthood.*

Even so, there is a steady flow of scientists who attempt to popularize scientific advances, including, occasionally, ones at the very pinnacle of their field. Of these, a few each generation achieve some degree of celebrity. But it's usually not the sort of celebrity that lasts. Fred Hoyle was a household name in the 1950s, but few today not interested in cosmology can recall his name, let alone recognize his face. In the 1970s and 1980s, Carl Sagan was arguably the most famous living scientist, yet he's probably better known today for his novel, *Contact*, than he is for anything he did having to do with science or science popularization. Even high priests come and go, and soon fade into obscurity.

• • •

Einstein was different.
Albert Einstein was not a priest of science—he was a prophet.

High priests might be admitted into the presence of the ultimate secret of nature, but only a prophet is thought to be able to gaze directly at it.† Priests, even high priests, work collectively to serve their deity, toiling for years within the hierarchy of the temple. A prophet might be alone in the wilderness yet is suddenly elevated above all others—including the priesthood—to serve as the mouthpiece of God. Priests represent an establishment; prophets bring their message, their Truth, directly to the people. And a prophet is almost always astonished, bewildered even, that he has been chosen to serve as a conduit of the Lord's truth.

Einstein was as close as one gets to a secular prophet. He was not just at the very forefront of physics, but he, and he alone, seemed to be responsible

---

* Sometimes, their reputations even suffer if they try. In 1992, Carl Sagan was nominated to the National Academy of Sciences. He was rejected. "If he had not done television, he probably would be in the academy," one academy member who was present at the vote later commented. More often, though, a scientist's attempt to commune with the public is met with mild scorn or mere indifference from his or her peers.

† Or at least its backside. See Exodus 33:21–23.

for a brand-new fundamental insight into the workings of the universe.* The legend was almost too perfect: he came from the wilderness of the patent office to overturn the tables in academic physics. His work was so ineffable and so profound that only twelve others could understand it. He was a singular genius. That singularity is crucial to the legend.

Quantum theory is just as important and just as difficult to understand as relativity theory, but it was a shared triumph; quantum theory was not associated solely with Planck or Schrödinger or Bohr or Heisenberg or Dirac or, yes, Einstein. Einstein's legend should rest as much on his contribution to quantum mechanics as it does on relativity, yet it is relativity that establishes his stature. A prophet can only be born from a superlatively esoteric idea that has—or is perceived to have—a single parent.

Yet, like any true prophet, Einstein was a man of the people. Even as he pondered the fundamental laws of nature, he brought his message directly to the public. His book, *Relativity*, was specifically aimed at those who didn't have a strong mathematical background; even though few read it, the attempt was admired. More appreciated were his frequent appearances in the popular press. Einstein initially hated speaking to journalists, but he quickly learned to make an effort to answer all their questions and to pose, smiling, for their photographs. Even when they asked stupid questions, Einstein would give a polite answer. "On such occasions, he usually said something that was not a direct answer to the question, but was still rather interesting, and which when printed conveyed to the reader a reasonable idea or at least gave him something to laugh about."[3]

Einstein's warmth and sharp sense of humor showed through, as did his vulnerability. As one Einstein scholar put it:

> Besides his easy-going manner, which so surprised everyone, there was his actual physical appearance, which was also reassuring. He did not look like a haughty European scientist. The first photographs that appeared of him in newspapers in April of 1921 showed him standing in an ill-fitting coat, holding a violin, and smiling in a bemused way. It was not at all a frightening picture.[4]

Einstein had tapped unfathomable forces of nature—indeed, he had helped set in motion the project to build a weapon that could kill hundreds of

---

* Even Einstein's relativity—which was very much his brainchild—wasn't devised in a vacuum. Marcel Grossman, for example, was an important collaborator who helped Einstein devise the mathematics necessary for the general theory. In addition, mathematicians like Henri Poincaré and David Hilbert were also having insights close to full-on discovery of the theory.

thousands of people in a fraction of a second. However, to the public, he was always an avuncular, funny little man with fuzzy white hair, rather than a destroyer of cities. He was not to be feared, but adored, thought of as someone who—prophet-like—tries to save us from nuclear doom rather than being one of its architects. Nor was there any hubris about him; he was modest, even self-effacing. Even as a secular saint, and himself a non-religious man, Einstein did not attempt to displace God.

Einstein was somewhat bewildered at his own fame. He had no illusions about how important his work was, but the degree of his celebrity— especially over that of all other scientists of the day—wasn't rational. At one point, he told a reporter that it was a psychopathological question about why people otherwise indifferent to science were so fascinated with him and with relativity. But they were, even though few of his fans had anything more than an inkling of what Einstein's scientific contribution truly was.[5]

Not only had Einstein, seemingly alone, laid claim to an important and arcane idea at the very edge of human knowledge, but he did it in a way that didn't feel threatening. His disarming affect, his sense of humor, his willingness to share his ideas with the public—and his inability to do so in a way that could truly be understood by most—assured people that he walked among us even as he inhabited a different plane.

It was a potent combination. Einstein had become the prophet of science. And Albert Einstein as prophet was the product of the modern press more than it was that of modern physics. As a biographer and friend of his wrote, "Einstein, creator of some of the best science of all time, is himself a creation of the media in so far as he is and remains a public figure." For all of Einstein's natural talents and abilities—his scientific brilliance, his playful sense of humor, his endearing imperfections—he wouldn't have become the public face of science had not the media bought into the narrative. They did so in part because Einstein helped them to. Einstein carefully cultivated his image, up to and including his physical appearance; whenever he saw photographers coming his way, he reportedly "mussed up his hair with both hands to freshen up his typical Einstein look." Einstein became the preeminent mind of our time not just because he had a genius for physics, but also because he had a genius for packaging himself for the media.[6]

It was the media who took Einstein and anointed him prophet. In the process, the complexities of Einstein the human were washed away and replaced with the simplicity of Einstein the symbol.

...

The symbol of Stephen Hawking in the second half of the twentieth century was just as much a product of the media as Albert Einstein's in the first half. In Hawking, the press had seemingly found someone who could match Einstein—overmatch him, even—on all fronts.

Hawking's science extended directly from Einstein's, making him the natural successor to the old prophet. This armed Hawking, like his predecessor, to be one of the few who could gaze directly at the most destructive forces in the universe; he descended, alone, into the maw of a black hole and came back with esoteric knowledge that nobody else could have retrieved. And he was purportedly on the path to complete Einstein's unfinished quest for a theory of everything: even though this never formed a serious thread in Hawking's scientific research (and he lagged far beyond the string theorists and others who were working on unification), this quest was a central part of the Hawking legend.

Hawking, like Einstein, was a man of the people—someone to be adored rather than feared. Hawking's sense of humor was wicked, and every bit as self-effacing as Einstein. Einstein had tried to popularize his science. Hawking had outdone him, writing a best-seller about his physics. And whereas the public was disarmed by Einstein's charming eccentricities, it was transfixed by Hawking's infirmity, finding it heroic as well as tragic. In his shriveled body, he was a pure mind: a Tiresias whose crushing physical disability was more than repaid by a divine gift of insight.

Little wonder that, like Einstein, Hawking was anointed prophet. But in becoming a prophet, he would become a symbol. Unlike humans, symbols can achieve immortality.

It took charm and discipline to maintain the simplicity of that image, even as the complexity of Hawking the human being was visible just beneath the surface. Hawking the symbol was arguably the most celebrated popularizer of science in our time. Yet he was a man who had such difficulty communicating that he had little choice but to save time and effort by populating his speech and his writing with recycled phrases—when, in fact, he himself was writing the prose that he set his name to. Hawking the symbol was a secular saint, kindly and humble and, above all, harmless. Yet in person, he could be stubborn, peevish, and proud. Divorced twice and with periods of alienation from his children, even his family knew that he could be a hard person to be around.

Hawking the symbol was a stoic who transcends life; his perseverance was inspiring, even to—especially to—the people who knew him best. "I was with him in the hospital a few times," Kip Thorne recalls. "Once, he was in a fairly bad shape, and the only way for him to communicate was by using cards his carers would put up, and he basically indicated yes or no for whether some letter on a card or some symbol on the card was what he was trying to focus on," Thorne says. "Even then, he didn't show particularly great frustration. It was quite incredible, really." He never showed any self-pity, and seldom even any vulnerability, to his family and friends, much less to the public. At the same time, Hawking the person was dependent upon the people around him to sacrifice themselves for his survival. And for his hedonism. Hawking had a strong appetite not just for food and wine, but for women—for extracting the pleasures he could from a life he thought could be snatched away at any moment.[7]

As a scientist, Hawking the symbol was supposed to be a lone genius who towered over his peers from his wheelchair, working to fulfill Einstein's dream of a theory of everything. The authentic Hawking was just one bright star in a constellation of scientists in the United Kingdom, the United States, and the Soviet Union who were transforming general relativity and cosmology in the 1960s and 1970s; some of Hawking's most important contributions to physics were collaborative rather than solo achievements. Hawking's cardinal scientific achievement—the discovery of black-hole radiation—was a triumph that used both quantum theory and relativity in a physical regime where the two must give way to an overarching theory of everything. Yet his work brought the world no closer to that theory, and by the time he became an international celebrity in the late 1980s, his most important scientific contributions were well behind him.

Hawking the scientist, who didn't find the theory of everything, died in 2018; but Hawking the symbol, who brought us *The Theory of Everything*, lives on. Just as Hawking worked to maintain his image during his lifetime, the nonprofit set up by his family and friends—the Hawking Foundation—is continuing to do so after his death.

More than that of any other human on the face of the planet, Stephen Hawking had a life that was preserved in a digital medium. Every utterance for the last few decades existed only through the mediation of a computer—a computer that could, and did, store his conversations. Sitting quiescent in those digital memory banks are, no doubt, traces of all the hidden collaborations, the worries about money, the petty jealousies, his uncertainties about his scientific legacy: all the rough edges of his humanity. But the foundation has not made even Hawking's paper archives available to those who wish to see the human behind the symbol—despite a supposed plan to donate at least some of them to public collections in order to avoid the inheritance tax. Even a small cache of papers that were housed at a library at Cambridge University were reclaimed and removed from display. Only a biographer selected by the foundation is granted even limited access to consult not just Stephen's personal archives, but to get access to Jane, Lucy, and the rest of the Hawking family.[8]

But it is in the messy humanity of the person rather than the pristine symbolism of the image where the real tragedy and triumph of Stephen Hawking are to be found.

The tragedy of Hawking's life was not being diagnosed with a terminal disease; nor was his triumph in overcoming that disease to become the greatest physicist since Einstein. That is the story of the symbol. The human's story is much more complex.

To be sure, Hawking's persistence and good humor in the face of his deteriorating condition was inspiring, but to Hawking, that was hardly any sort of triumph; it was merely survival. Nor was it false modesty when Hawking rubbished the idea of his being a latter-day Newton, even though there's little doubt that he was a physicist of the first rank and made major contributions to general relativity and cosmology. Hawking's was not the best mind since Newton and Einstein and Galileo, as he himself repeatedly insisted.

Hawking's triumph was to discover something new that changed our understanding of the universe in important ways—and to do so multiple times. More than that, Hawking's triumph was also to act as a catalyst for important ideas that were brewing in his field, and to inspire a generation of scientists who wished to follow in his footsteps.

The central tragedy of Hawking the human being is also quite different from that of Hawking the symbol. Hawking didn't retreat into his mind as a result of his disease. Since childhood, Hawking had been cerebral to the extreme. Even when it wasn't clear whether he would fail out of school, the core of Hawking's identity, of his self-worth, was the superiority of his brain. It was what he always wanted to be known for.

Yet even as he achieved this goal, he was denied it, at least in his own mind. Hawking always suspected that, to a large degree, his celebrity—and even to some degree his academic success—rewarded his infirmity rather than his brain.

One of the curious facts about Hawking was that he was unusually fond of—even devoted to—a celebrity that, at first glance, seems to be his polar opposite. Stephen Hawking absolutely adored Marilyn Monroe.

Errol Morris, the film director, was chatting with Hawking during the filming of *A Brief History of Time* when the subject of Monroe came up. As Morris put it:

Finally, I said, "I figured it out, why you have all these pictures of Marilyn Monroe on the wall. Like you, she was a person appreciated for her body and not necessarily her mind."

And he gave me this really crazy look, like, "What the fuck are you saying, Mr. Morris?" He gave me this crazy look, and then finally, there's a click, and he says, "YES."[9]

Of all the paradoxes in Hawking's life, this one may have been the most profound.

ACKNOWLEDGMENTS

For the first time in my career, I am struggling to write an acknowledgments section. For this was not an easy book to write—and I am indebted to more people, more deeply, than I have ever been before for helping make it happen.

Any unauthorized biography is going to have its unique difficulties; to me, the virtue of being able to write an independent history unencumbered by the expectations of the subject or his estate far outweighs the problems caused by the lack of access to the subject's papers, family, and friends.

Which is why my gratitude is all the greater for all of those who knew Stephen Hawking best and decided that the best way to honor his legacy was to help paint as accurate a picture as possible about the man and his science. So I wish to give my most heartfelt thanks to the physicists who were so generous with their time: Raphael Bousso, Peter D'Eath, Daniel Freedman, Christophe Galfard, Jim Hartle, Ray Laflamme, Andrei Linde, Simon Mitton, Roger Penrose, John Preskill, Andy Strominger, Marika Taylor, and Neil Turok. I am also indebted to many others who were a part of his life—physicists, writers, directors, editors, agents, and aides—people like Andy Albrecht, John Gribbin, Peter Guzzardi, Leonard Mlodinow, Errol Morris, Lenny Susskind, and Al Zuckerman, among others. And finally, a special thanks to those who did not want to be named in this book, but still spent time speaking to me about Hawking in hopes of helping me understand the real person behind the myth.

I'd also like to thank New York University, not just for allowing me to take a semester of sabbatical to work on this book, but for the opportunity it has given me to work with such wonderful colleagues. Special thanks go to James McBride for taking the time—at a moment when he didn't have much time to spare—to give me the benefit of his immense talent as a writer and a critic. Laura Helmuth also was kind enough to share her wisdom and skill even as she was transitioning to a new job. Thanks, too, to the librarians, scholars, and various officials from around the world who were more than willing to assist me in finding little scraps of information about Hawking. Of particular note were the Washington University of St. Louis Libraries; the California Institute of Technology Archives; the American Philosophical Society Library; the St. Johns College, Churchill College, and Cambridge University Libraries; and the Cambridge County and Family Court. I'd also like to thank my editor, T. J. Kelleher; my copyeditor, Katherine Streckfus; and my agent, Katinka Matson, for their wonderful work on my behalf.

This research was also made possible with a grant from the Alfred P. Sloan Foundation; my thanks to the foundation generally and to Doron Weber in particular for their assistance and their flexibility when a global pandemic upended the interview and research plans for the book.

Great thanks are also due to the essential workers of New York City— the grocery store workers, the package delivery people, the doormen of our building, and others too numerous to name—people who bore the heaviest risk so that everyone else would be able to stay safe in their homes.

My greatest thanks are to my wife, Meridith, and my children, Eliza and Daniel. Locked together in a three-bedroom apartment, all trying to give me space to write while simultaneously trying to work and home-school—it was far from easy. I thank Meridith for her patience with me and my children for their bravery and forbearance during what was undoubtedly the scariest time in their young lives.

Finally, I want to thank my mother, Tama, and to give a last thank you and a farewell to my father, Burt—you've always been my most important audience.

NOTES

PROLOGUE

1. Jacqui Deevoy, "Has Stephen Hawking Been Replaced with a 'Puppet'?," *Daily Mail*, January 12, 2018, www.dailymail.co.uk/femail/article-5261939/Has-Stephen-Hawking-replaced-puppet.html.

2. Leonard Susskind, personal communication with author.

CHAPTER 1: NEXT TO NEWTON

1. See "Remarks by President Clinton and Q&As at Hawking Lecture," White House Millennium Council, March 6, 2000, https://clintonwhitehouse4.archives.gov/Initiatives/Millennium/19980309-22774.html.

2. Stephen Hawking, *A Brief History of Time* (New York: Bantam Books, 1998), 141.

3. John Preskill, personal communication with author.

4. Bob Sipchen, "The Heroism of Stephen Hawking—Famous Physicist Is Mandela for Disabled," *Los Angeles Times*, June 7, 1990, reprinted at *Seattle Times*, http://community.seattletimes.nwsource.com/archive/?date=19900607&slug=1075967.

5. Jane Fonda, "My Meeting with Stephen Hawking," Jane Fonda (blog), February 3, 2011, www.janefonda.com/2011/02/my-meeting-with-stephen-hawking.

6. Christophe Galfard, personal communication with author; Sipchen, "Heroism of Stephen Hawking."

7. "Stephen Hawking's Ashes Buried at Westminster Abbey," ABC News, June 15, 2018, https://abcnews.go.com/International/stephen-hawkings-ashes-buried-westminster-abbey/story?id=55922397; "The CD: The Stephen Hawking Tribute CD," Stephen Hawking Foundation, https://stephenhawkingfoundation.org/the-cd.

8. Leonard Susskind, personal communication with author.

9. Ray Laflamme, personal communication with author.

CHAPTER 2: RIPPLES

1. Maggie Fox, "Gravitational Wave Work Wins Physics Nobel Prize," NBC News, October 3, 2017, www.nbcnews.com/science/science-news/gravitational-wave-works-wins-physics-nobel-prize-n807081.

2. Sarah Lewin, "Nobel Prize for Physics: Einstein Would Be 'Flabbergasted' by Gravitational Wave Win," Space.com, October 3, 2017, www.space.com/38350-nobel-prize-physics-gravitational-waves-einstein.html.

3. Kip Thorne, "Warping Spacetime," in *The Future of Theoretical Physics and Cosmology: Celebrating Stephen Hawking's 60th Birthday*, ed. G. W. Gibbons, S. J. Rankin, and E. P. S. Shellard (Cambridge: Cambridge University Press, 2003), 4–5, reprinted by Kip S. Thorne, California Institute of Technology, at www.cco.caltech.edu/~kip/scripts/PubScans/VI-47.pdf.

4. Thorne, "Warping Spacetime," 30.

5. Stephen Hawking, *The Universe in a Nutshell* (New York: Bantam Books, 2001), 14.

6. See video accompanying Ian Sample, "Stephen Hawking Unveils Formulae for England World Cup Success," *The Guardian*, May 28, 2014, www.theguardian.com/science/2014/may/28/stephen-hawking-formulae-england-world-cup-success.

7. "Ridley Turtle Tipped for Oily Exit," Paddy Power press release, May 25, 2010.

8. See Charles Seife, *Proofiness: The Dark Arts of Mathematical Deception* (New York: Viking, 2010), 65–66; Sample, "Stephen Hawking Unveils Formulae."

9. Sample, "Stephen Hawking Unveils Formulae."

10. Sample, "Stephen Hawking Unveils Formulae."

11. Al Zuckerman, personal communication with author.

12. J. H. Taylor and J. M. Weisberg, "A New Test of General Relativity: Gravitational Radiation and the Binary Pulsar PSR1913+16," *Astrophysical Journal* 253 (1982): 908.

13. B. P. Abbott et al. (LIGO Scientific Collaboration and Virgo Collaboration), "Observation of Gravitational Waves from a Binary Black Hole Merger," *Physical Review Letters* 116, 061102 (2016).

14. "The view from GR," presented at KITP Rapid Response Workshop, August 23, 2013, http://online.kitp.ucsb.edu/online/fuzzorfire_m13/hawking/rm/jwvideo.html.

15. Charles Q. Choi, "No Black Holes Exist, Says Stephen Hawking—at Least Not Like We Think," *National Geographic*, January 27, 2014, www.nationalgeographic.com/news/2014/1/140127-black-hole-stephen-hawking-firewall-space-astronomy.

16. S. W. Hawking, "Information Preservation and Weather Forecasting for Black Holes," arXiv:1401.5761v1, January 22, 2014, 3.

17. Gareth Morgan, "Stephen Hawking Says There Is No Such Thing as Black Holes, Einstein Spinning in His Grave," *Express*, January 24, 2014, www.express.co.uk/entertainment/gaming/455880/Stephen-Hawking-says-there-is-no-such-thing-as-black-holes-Einstein-spinning-in-his-grave; Mark Prigg, "Stephen Hawking Stuns Physicists by Declaring 'There Are No Black Holes'—But Says There Are GREY Ones," *Daily Mail*, January 24, 2014, www.dailymail.co.uk/sciencetech/article-2545552/Stephen-Hawking-admits-no-black-holes-GREY-holes.html.

18. Ramin Setoodeh, "Toronto: 'The Theory of Everything' Made Hawking Cry," *Variety*, September 7, 2014.

19. Ramin Setoodeh, "How Eddie Redmayne Became Stephen Hawking in 'The Theory of Everything,'" *Variety*, October 28, 2014.

20. *The Theory of Everything*, directed by James Marsh, November 2014.

21. Catherine Shoard, "Stephen Hawking's First Wife Intensifies Attack on The Theory of Everything," *The Guardian*, October 3, 2018.

22. Ben Travis and Alex Godfrey, "Eddie Redmayne on Meeting Stephen Hawking for The Theory of Everything," *Empire*, March 14, 2018.

23. For Jane Hawking's film rights, see Film and Music Entertainment, Inc., CIK#0001309152, Form 10SB12G/A, 2006-02-06.

24. "The Theory of Everything," Box Office Mojo, www.boxofficemojo.com/movies/?id=theoryofeverything.htm.

25. *Stem Cell Universe with Stephen Hawking*, Discovery Science, February 3, 2014.

26. Errol Morris, personal communication with author.

27. Author's analysis.

28. Morris, personal communication with author.

29. "Stephen Hawking recorded iNNOCENCE + eXPERIENCE monologue," YouTube, posted May 17, 2015, by Bob Mackin, www.youtube.com/watch?v=YZM7uM8OyxA.

30. "Report and Financial Statements for the Period 1 April 2016 to 31 March 2017," Stephen Hawking Foundation, Charity number 1163521.

31. Zuckerman, personal communication with author.

32. BBC Radio 4 Today, March 18, 2014, www.bbc.com/news/av/science-environment -26625791/stephen-hawking-wins-inflation-debate.

33. Dennis Overbye, "Space Ripples Reveal Big Bang's Smoking Gun," *New York Times*, March 18, 2014, A1.

34. Chris Havergal, "Cambridge University's Stephen Hawking Claims Victory in Bet with Neil Turok After Cosmic Wave Discovery," *Cambridge Evening News*, March 18, 2014.

35. George P. Mitchell, "George Mitchell Lays Groundwork for New Texas A&M Science Initiative with $35 Million Gift," press release, Texas A&M University, November 3, 2005; "Hawking Honored with Auditorium at Texas A&M," Associated Press, April 6, 2010.

36. Andy Strominger, personal communication with author; S. W. Hawking, "The Information Paradox for Black Holes," arXiv:1509.01147v1, September 3, 2015.

37. Strominger, personal communication with author.

38. Strominger, personal communication with author. The elision is "information paradox with supertranslation hair." *Supertranslation hair* is one variety of soft hair, which is the subject of the first of three papers by Hawking, Strominger, and Perry; for technical reasons, this variety of soft hair cannot, in fact, solve the paradox.

39. Raphael Bousso and Massimo Porrati, "Soft Hair as a Soft Wig," arXiv:1706.00436v2, September 20, 2017; Marika Taylor personal communication with author.

40. Lawrence M. Krauss (@LKrauss1), Twitter, September 25, 2015, 4:39 p.m., https://twitter .com/lkrauss1/status/647510799678750720.

41. Lawrence M. Krauss (@LKrauss1), Twitter, January 11, 2016, 10:46 a.m., https://twitter .com/lkrauss1/status/686574829542092800; "Thursday: Scientists to Provide Update on Search for Gravitational Waves," LIGO Media Advisory, February 8, 2016.

42. Dr Erin Ryan, (@erinleeryan), Twitter, February 11, 2016, 10:14 a.m., https://twitter.com /erinleeryan/status/697800782175997952.

43. "LIGO Detects Gravitational Waves—Announcement at Press Conference (part 2), February 11, 2016, www.ligo.caltech.edu/WA/video/ligo20160211v12.

44. Stephen Hawking, *A Brief History of Time* (New York: Bantam Books, 1998), 93–94.

45. LIGO Scientific Collaboration, "GWTC-1: A Gravitational-Wave Transient Catalog of Compact Binary Mergers Observed by LIGO and Virgo During the First and Second Observing Runs," arXiv:1811.12907v2, December 16, 2018.

46. Kip Thorne, personal communication with author.

CHAPTER 3: MODELS

1. "Opening Ceremony—London Paralympic Games," YouTube, posted August 29, 2012, by Paralympic Games, www.youtube.com/watch?v=Kd4FgGSY5BY.

2. "Opening Ceremony—London Paralympic Games"; Stephen Hawking, *My Brief History* (London: Bantam Books, 2018), 122.

3. Andy Strominger, personal communication with author.

4. John Boslough, *Stephen Hawking's Universe* (London: HarperCollins, 1995), 93–94.

5. Stephen Hawking, *A Brief History of Time* (New York: Bantam Books, 1998), 13.

6. Edwin M. McMillan, "Current Problems in Particle Physics," *Science* 152, no. 3726 (May 27, 1966): 1212, https://doi.org/10.1126/science.152.3726.1210; Willis Lamb, "Fine Structure of

the Hydrogen Atom," Nobel Prize Lecture, December 12, 1955, www.nobelprize.org/uploads/2018/06/lamb-lecture.pdf (quotation marks omitted).

7. "Stephen Hawking on Higgs: 'Discovery Has Lost Me $100,'" BBC News, July 4, 2012, www.bbc.com/news/av/science-environment-18708626/stephen-hawking-on-higgs-discovery-has-lost-me-100.

8. Peter Guzzardi, personal communication with author; Hawking, *My Brief History*, 122.

9. Amber Goodhand, "Renowned Physicist Stephen Hawking Frequents Sex Clubs," Radar Online, February 24, 2012, https://radaronline.com/exclusives/2012/02/stephen-hawking-sex-clubs-physicist-freedom-acres.

10. "A Brief History of the Time Stephen Hawking Went to a Sex Club: University Says Physicist Visited California Swingers' Club with Friends," *Daily Mail*, February 27, 2012, www.dailymail.co.uk/news/article-2106025/Stephen-Hawking-visits-California-swingers-sex-club.html.

11. "Freedom Acres Resort," Freedom Acres Resort, http://freedomacresresorts.com, accessed May 30, 2019; Russ Thomas, personal communication with author.

12. Errol Morris, personal communication with author.

13. Jane Hawking, *Music to Move the Stars: A Life with Stephen Hawking* (London: Pan, 2000), 328.

14. "NMC Announces Fitness to Practise Hearing Outcome Involving Former Nurse of Professor Stephen Hawking," Nursing and Midwifery Council, press release, March 12, 2019.

15. Andrew Holgate, "The Trouble with Being a Genius: Stephen Hawking Has Had to Overcome Extraordinary Obstacles in His Life. His Memoir Hints at His Travails [Eire Region]," *Sunday Times* (London), September 22, 2013.

16. Chuck Leddy, "'My Brief History' by Stephen Hawking," *Boston Globe*, September 10, 2013.

17. Robert P. Crease, "A Cosmological Life," *Nature* 501, no. 7466 (September 12, 2013): 162.

18. Stephen Hawking, *Black Holes and Baby Universes and Other Essays* (New York: Bantam Books, 1993), 1–39.

19. Martin Rees, "Stephen Hawking—An Appreciation," March 14, 2018, www.ctc.cam.ac.uk/news/Hawking_an_appreciation-Rees.pdf.

20. Stephen Hawking, "This Is the Most Dangerous Time for Our Planet," *The Guardian*, December 1, 2016, www.theguardian.com/commentisfree/2016/dec/01/stephen-hawking-dangerous-time-planet-inequality; Robert Booth, "Stephen Hawking: I Fear I May Not Be Welcome in Donald Trump's US," *The Guardian*, March 20, 2017, www.theguardian.com/science/2017/mar/20/stephen-hawking-trump-good-morning-britain-interview.

21. Stephen Hawking, "The NHS Saved Me. As a Scientist, I Must Help to Save It," *The Guardian*, August 18, 2017, www.theguardian.com/commentisfree/2017/aug/18/nhs-scientist-stephen-hawking.

22. Harriet Sherwood and Matthew Kalman, "Stephen Hawking Joins Academic Boycott of Israel," *The Guardian*, May 7, 2013, www.theguardian.com/world/2013/may/08/stephen-hawking-israel-academic-boycott; Robert Booth and Harriet Sherwood, "Noam Chomsky Helped Lobby Stephen Hawking to Stage Israel Boycott," *The Guardian*, May 10, 2013, www.theguardian.com/world/2013/may/10/noam-chomsky-stephen-hawking-israel-boycott.

23. Iain Thomson, "Israeli Activists Tell Hawking to Yank His Intel Chips over Palestine," *The Register*, May 9, 2013, www.theregister.co.uk/2013/05/09/israel_boycott_stephen_hawking_intel; Averil Parkinson and Jonathan Rosenhead, "Professor Stephen Hawking and the Academic Boycott of Israel," *BRICUP Newsletter* 120, April 2018, 3.

24. Steven Plaut, "Hawking," Zionist Conspiracy, May 8, 2013, http://zioncon.blogspot.com/2013/05/hawking.html.

25. S. W. Hawking, "Virtual Black Holes," *Physical Review D* 53, no. 6 (March 15, 1996): 3099–3107, https://doi.org/10.1103/PhysRevD.53.3099; John Gribbin, "Hawking Throws Higgs into Black Holes," *New Scientist*, December 2, 1995, www.newscientist.com/article/mg14820062-400-hawking-throws-higgs-into-black-holes.

26. Alastair Dalton, "Clash of the Atom-Smashing Academics," *The Scotsman*, September 2, 2002.

27. Dalton, "Clash of the Atom-Smashing Academics."

28. "On the Hunt for the Higgs Boson," BBC, September 9, 2008, http://news.bbc.co.uk /today/hi/today/newsid_7598000/7598686.stm.

29. Mike Wade, "Higgs Launches Stinging Attack Against Nobel Rival," *The Times* (London), September 11, 2008.

CHAPTER 4: GRAND DESIGN

1. *Into the Universe with Stephen Hawking*, Episode 2, "Time Travel," directed by Nathan Williams, April 25, 2010.

2. Stephen Hawking, "Space and Time Warps," available at Internet Archive, Wayback Machine, web.archive.org/web/20170126163529/http://www.hawking.org.uk/space-and-time-warps .html.

3. Personal communication with author.

4. Hélène Mialet, *Hawking Incorporated: Stephen Hawking and the Anthropology of the Knowing Subject* (Chicago: University of Chicago Press, 2012), 152.

5. Mialet, *Hawking Incorporated*, 154.

6. Mialet, *Hawking Incorporated*, 35.

7. Mialet, *Hawking Incorporated*, 36.

8. Dennis Overbye, *Lonely Hearts of the Cosmos: The Story of the Scientific Quest for the Secret of the Universe* (New York: Harper and Row, 1991), 115–116.

9. Al Zuckerman, personal communication with author; Leonard Mlodinow, personal communication with author.

10. Mlodinow, personal communication with author.

11. Mlodinow, personal communication with author.

12. Gilbert Taylor, "The Grand Design," *Booklist* 106, no. 22 (August 2010): 5.

13. "Hawking: God Did Not Create Universe," *Sunday Times* (London), September 2, 2010, www.thetimes.co.uk/article/hawking-god-did-not-create-universe-mrbvqgs50xl; Alastair Jamieson, "Baroness Greenfield Criticises 'Taliban-Like' Stephen Hawking," *The Telegraph*, September 8, 2018, www.telegraph.co.uk/news/science/stephen-hawking/7988785/Baroness -Greenfield-criticises-Taliban-like-Stephen-Hawking.html; Philip Mathias, "A Book About Nothing: Stephen Hawking's 'The Grand Design' Imagines a Universe That Creates Itself 'Spontaneously.' But He Never Quite Explains How Matter and Energy Emerge from a Void," *National Post*, September 27, 2010, at Pressreader, www.pressreader.com/canada/national-post -latest-edition/20100927/281930244321327.

14. "Feynman :: Rules of Chess," YouTube, posted by defjam99b, n.d., www.youtube.com /watch?v=o1dgrvlWML4.

15. "Feynman :: Rules of Chess."

16. Graham Farmelo, "Life, the Universe and M-Theory: Books of the Week. Are We Close to a Unifying Theory of Physics or Are We Nearing the Limits of Knowledge? Graham Farmelo Looks at Two Contrary Viewpoints," *The Times* (London), September 11, 2010, 9.

17. Dwight Garner, "Many Kinds of Universes and None Require God," *New York Times*, September 7, 2010, C1.

18. "Stephen Hawking's Last Speech," You Subtitles, www.yousubtitles.com/Stephen -Hawkings-Last-Speech-id-1978237.

19. Ian Sample, "Stephen Hawking Taken to Hospital After Becoming 'Very Ill,'" *The Guardian*, April 20, 2009, www.theguardian.com/science/2009/apr/21/stephen-hawkings-illness.

20. Kitty Ferguson and S. W Hawking, *Stephen Hawking: An Unfettered Mind* (New York: St. Martin's Press, 2017), 272.

CHAPTER 5: CONCESSIONS

1. Judy Bachrach, "A Beautiful Mind, an Ugly Possibility," *Vanity Fair*, June 2004, www
.vanityfair.com/news/2004/06/hawking200406. See also "Is Stephen Hawking Being Abused
by His Wife?," *People*, updated February 9, 2004, https://people.com/archive/is-stephen
-hawking-being-abused-by-his-wife-vol-61-no-5.

2. Hawking acquaintance, personal communication with author.

3. Various, personal communications with author.

4. Bachrach, "A Beautiful Mind, an Ugly Possibility"; Olga Craig, "Is This the Brief History of a Troubled Marriage?," *The Telegraph*, January 25, 2004, www.telegraph.co.uk/news
/uknews/1452515/Is-this-the-brief-history-of-a-troubled-marriage.html.

5. Assistant, personal communication with author.

6. Natalie Clarke, "Professor Hawking in Assault Probe," *Daily Mail*, January 2004, www
.dailymail.co.uk/news/article-206323/Professor-Hawking-assault-probe.html; "Is Stephen Hawking Being Abused by His Wife?"; Craig, "Is This the Brief History"; Laura Peek, "Hawking Denies
Assault by Wife After Fresh Allegations by Carer," *The Times* (London), January 24, 2004.

7. Leonard Mlodinow, *Stephen Hawking: A Memoir of Friendship and Physics* (New York:
Pantheon Books, 2020), 163.

8. Craig, "Is This the Brief History."

9. David Sapsted, "Hawking Defends His Wife After Assault Claims," *The Telegraph*, January
24, 2004, www.telegraph.co.uk/news/uknews/1452453/Hawking-defends-his-wife-after-assault
-claims.html.

10. Cambridgeshire Constabulary, "Professor Stephen Hawking," press release, March 29,
2004.

11. Nurse, personal communication with author.

12. "Shadowland," *The Age*, April 21, 2004, www.theage.com.au/entertainment/books
/shadowland-20040421-gdxpqh.html.

13. Kitty Ferguson and S. W. Hawking, *Stephen Hawking: An Unfettered Mind* (New York:
St. Martin's Press, 2017), 286; "Stephen Hawking's Grand Design: The Meaning of Life Full
Episode," YouTube, posted September 16, 2015, by S. M. Taif- Ul-Kabir, www.youtube.com
/watch?v=usdqCexPQww.

14. "Shadowland."

15. Max Planck, *The Origin and Development of the Quantum Theory: Being the Nobel Prize
Address Delivered Before the Swedish Academy of Sciences at Stockholm, 2 June 1920*, trans. H. T.
Clarke and L. Silberstein (Oxford: Clarendon Press, 1922).

16. Author's personal recollection of July 21, 2004; Jenny Hogan, "Hawking Cracks the Paradox," *New Scientist* 183, no. 2456 (July 17, 2004), 11.

17. "The Hawking Paradox," *Horizon*, Season 42, Episode 2, directed by William Hicklin,
BBC, aired September 15, 2005.

18. Leonard Susskind, personal communication with author, as described in Charles Seife,
"A General Surrenders the Field, but Black Hole Battle Rages On," *Science* 305, no. 5686 (August
13, 2004): 934, https://doi.org/10.1126/science.305.5686.934.

19. Wager posted at John Preskill's Caltech website, www.theory.caltech.edu/~preskill/info
_bet.html.

20. Charles Seife, "Hawking Slays His Own Paradox, but Colleagues Are Wary," *Science* 305,
no. 5684 (July 30, 2004): 586, https://doi.org/10.1126/science.305.5684.586.

21. Andy Strominger, personal communication with author.

22. S. W. Hawking, "Information Loss in Black Holes," *Physical Review D* 72, no. 8 (October
18, 2005): 084013, https://doi.org/10.1103/PhysRevD.72.084013.

23. Scott Feschuk, "Briefer Madness," *Maclean's* 118, no. 46 (November 14, 2005): 148;
Jim al-Khalili, "Short Cut to Space-Time," *Nature* 438, no. 7065 (2005): 159–159, 161.

24. Matthew Syed, "Only Partially Blinded by Science," *The Times* (London), October 15,
2005.

25. Stephen Hawking, *God Created the Integers: The Mathematical Breakthroughs That Changed History* (Philadelphia: Running Press, 2007), 291.

26. Stephen Gaukroger, *Descartes: An Intellectual Biography* (Oxford: Clarendon Press, 2002), 416.

27. Hawking, *God Created the Integers*, 675.

28. Des MacHale, *George Boole: His Life and Work* (Dublin: Boole Press, 1983), 155–156.

29. Gil King, personal communication with author.

30. "Jeff Koons," *Journal of Contemporary Art*, October 1986, www.jca-online.com/koons.html.

31. Hawking, *God Created the Integers*, 1162.

32. John Moran, "'Theory of Everything' Not for Novices," *Chicago Tribune*, June 13, 2002, 4–5.

33. Hillel Italie, "Hawking Tries to Stop Publication of Lectures," Associated Press, in *Journal Gazette*, May 24, 2002, 4W.

34. Scott Flicker, "Complaint Against New Millennium Entertainment," April 26, 2002, addressed to J. Howard Beales III, director of consumer protection, no. 2130979.

35. Michael Hiltzik, "It's a Safe Bet That He'll Sue in This Town Again," *Los Angeles Times*, June 2, 2003, www.latimes.com/archives/la-xpm-2003-jun-02-fi-golden2-story.html.

36. "Latest News," Stephen Hawking website, Internet Archive, Wayback Machine, June 9, 2003, https://web.archive.org/web/20030609215606/http://hawking.org.uk/home/hindex.html.

37. Stephen Hawking, *My Brief History* (London: Bantam Books, 2018), 89; Court employee, personal communication with author; Elaine Hawking, "Baptism Testimony of Elaine Hawking," Internet Archive, April 2018, https://archive.org/details/chipping-campden-baptist-church -513231/2018-04-01-Baptism-Testimony-Elaine-Hawking-Elaine-Hawking-46453311.mp3.

38. Al Zuckerman, personal communication with author.

39. Hawking, *My Brief History*.

40. "Shadowland."

41. "Stephen Hawking Calls for Mankind to Reach for Stars," *Space Daily*, June 14, 2006, www.spacedaily.com/reports/Stephen_Hawking_Calls_For_Mankind_To_Reach_For_Stars.html.

42. Lucy and Stephen Hawking, with Christophe Galfard, illustrated by Garry Parsons, *George's Secret Key to the Universe* (London: Doubleday, 2007), 4–5.

43. *Edge*, www.edge.org/3rd_culture/krauss06/images/35.jpg; Agency, "Stephen Hawking Pictured on Jeffrey Epstein's 'Island of Sin,'" *The Telegraph*, January 12, 2015, www.telegraph.co.uk /news/science/stephen-hawking/11340494/Stephen-Hawking-pictured-on-Jeffrey-Epsteins -Island-of-Sin.html.

44. "Stephen Hawking Eulogizes George Mitchell as Science Visionary, World-Changer, Friend," Texas A&M College of Science (blog), August 9, 2013, https://science.tamu.edu/news/2013/08 /stephen-hawking-eulogizes-george-mitchell-as-science-visionary-world-changer-friend.

45. Jesse Drucker, "Kremlin Cash Behind Billionaire's Twitter and Facebook Investments," *New York Times*, November 5, 2017, www.nytimes.com/2017/11/05/world/yuri-milner-facebook -twitter-russia.html; Stephen Hawking, "Yuri Milner," *Time*, April 21, 2016, https://time.com /collection-post/4300003/yuri-milner-2016-time-100.

46. Milner met Hawking at a conference in Moscow in 1987. Yuri Milner, "Stephen Hawking: The Universe Does Not Forget, and Neither Will We," *Scientific American* (blog), March 29, 2018, https://blogs.scientificamerican.com/observations/stephen-hawking-the-universe-does-not -forget-and-neither-will-we; Marika Taylor, personal communication with author.

47. Marika Taylor, personal communication with author.

48. "Peter Diamandis: Stephen Hawking Hits Zero G," YouTube, TED Talk, posted July 8, 2008, by TED, www.youtube.com/watch?v=0VJqrlH9cdI; "Reimbursable Space Act Agreement Between National Aeronautics and Space Administration, John F. Kennedy Space Center and Zero Gravity Corporation for Recurring Use of the Shuttle Landing Facility," signed March 20, 2006 (James Kennedy), and March 30, 2006 (Peter Diamandis).

49. Diamandis TED Talk.

50. "Review of NASA's Microgravity Flight Services," NASA Office of Inspector General, Report no. IG-10-015; "Move to New Planet, Says Hawking," BBC, November 30, 2006, http://news.bbc.co.uk/2/hi/6158855.stm.

51. "Professor Stephen Hawking and Virgin Galactic," Virgin, December 17, 2015, www.virgin.com/richard-branson/professor-stephen-hawking-and-virgin-galactic.

52. "Professor Stephen Hawking and Virgin Galactic."

CHAPTER 6: BOUNDARIES

1. Author's personal recollection of APS meeting in Atlanta, March 1999, written down twenty years later.

2. "Hawking Draws Packed House to Atlanta Civic Center," *APS News* 8, no. 5 (May 1999), www.aps.org/publications/apsnews/199905/hawking.cfm.

3. "Remarks by President Clinton and Q&As at Hawking Lecture," White House Millennium Council, March 6, 2000, https://clintonwhitehouse4.archives.gov/Initiatives/Millennium/19980309-22774.html.

4. "Descent," *Star Trek: The Next Generation*, Season 6, Episode 26, directed by Alexander Singer, aired June 19, 1993.

5. *The Simpsons*, Season 10, Episode 22, directed by Pete Michels, aired May 9, 1999; Nicholas Hellen and Steve Farrar, "Hawking Joins the Celestial High Earners," *Sunday Times* (London), March 28, 1999, 3; "Specsavers—Stephen Hawking (1999)," YouTube, posted March 14, 2017, by The Drum/Hosting, www.youtube.com/watch?v=aLMNYxEzXvU.

6. "Remarks by President Clinton and Q&As at Hawking Lecture."

7. Matin Dunani and Peter Rodgers, "Physics: Past, Present, Future," *Physics World* 12, no. 12 (December 1999): 7–14, https://doi.org/10.1088/2058-7058/12/12/2.

8. Errol Morris, personal communication with author.

9. Stephen Hawking, *The Universe in a Nutshell* (New York: Bantam Books, 2001), vii.

10. Kitty Ferguson and S. W. Hawking, *Stephen Hawking: An Unfettered Mind* (New York: St. Martin's Press, 2017), 199.

11. Neel Shearer, personal communication with author.

12. Jon Turney, "Review: The Universe in a Nutshell by Stephen Hawking," *The Guardian*, November 10, 2001, www.theguardian.com/books/2001/nov/10/scienceandnature.highereducation.

13. "Rave Review for Hawking's New Book," *Physics World*, November 8, 2001, https://physicsworld.com/a/rave-review-for-hawkings-new-book.

14. John Gribbin, "The Universe in a Nutshell by Stephen Hawking," *The Independent*, November 3, 2001, www.independent.co.uk/arts-entertainment/books/reviews/the-univese-in-a-nutshell-by-sephen-hawking-9153195.html.

15. Hawking, *Universe in a Nutshell*, 15.

16. "Hawking Rewrites History," *The Australian*, October 15, 2001, 9.

17. Jane Hawking, *Music to Move the Stars: A Life with Stephen* (London: Pan, 2000), 573.

18. Jane Hawking, *Music to Move the Stars*, 580, 586.

19. "Stephen and Me: Libby Brooks Talks to Jane Hawking," *The Guardian*, August 4, 1999, 1; Jane Hawking, *Music to Move the Stars*, 555.

20. Tim Adams, "Jane Hawking: Brief History of a First Wife," *The Guardian*, April 3, 2004; Jane Hawking, *Music to Move the Stars*, 23–24, 46, 572.

21. Veronica Lee, "The Playwright, the Scientist, His Wife, and Her Lover," *Globe and Mail*, August 15, 2000.

22. Robin Hawdon, *God and Stephen Hawking*, Josef Weinberger Plays, 2000.

23. Martin Anderson, "God and Stephen Hawking: The Playhouse, Derby, *The Independent*," September 5, 2000, www.independent.co.uk/arts-entertainment/theatre-dance/features

/god-and-stephen-hawking-the-playhouse-derby-701939.html; Matin Dunani, "Hawking Slams 'Stupid, Worthless' Play," *Physics World*, August 2000, 8.

24. "All My Shootings Be Drivebys," MC Hawking website, www.mchawking.com/all -my-shootings-be-drivebys---lyrics.

25. Susan Carpenter, "Check It! MC Hawking Raps," *Los Angeles Times*, November 2, 2000.

26. Roger Highfield, "I Thought My Time Was Up at the Age of 59.97, Says Hawking," *The Telegraph*, January 12, 2002, www.telegraph.co.uk/news/uknews/1381187/I-thought-my-time -was-up-at-the-age-of-59.97-says-Hawking.html.

27. Hélène Mialet, *Hawking Incorporated: Stephen Hawking and the Anthropology of the Knowing Subject* (Chicago: University of Chicago Press, 2012), 2.

28. Neil Turok, personal communication with author.

29. Peter Stringfellow, personal communication with author, July 2003.

30. "Stephen Hawking at 70: Exclusive Interview," *New Scientist*, January 4, 2012, www.new scientist.com/article/mg21328460-500-stephen-hawking-at-70-exclusive-interview.

31. Hawking, *Brief History*, 136.

32. Don N. Page, "The Hartle-Hawking Proposal for the Quantum State of the Universe," in *The Creation of Ideas in Physics: Studies for a Methodology of Theory Construction*, ed. Jarrett Leplin, University of Western Ontario Series in Philosophy of Science (Dordrecht: Springer Netherlands, 1995), 184.

33. S. W. Hawking, "The Cosmological Constant Is Probably Zero," *Physics Letters B* 134, no. 6 (January 26, 1984): 403.

34. Neil Turok, personal communication with author.

35. Kip S. Thorne, *Black Holes and Time Warps: Einstein's Outrageous Legacy* (New York: Norton, 1994), 419; Andy Strominger, personal communication with author.

36. "The Hawking Paradox, *Horizon*, Season 42, Episode 2, directed by William Hicklin, BBC, aired September 15, 2005.

37. Leonard Susskind, *The Black Hole War: My Battle with Stephen Hawking to Make the World Safe for Quantum Mechanics* (Boston: Little, Brown, 2009), 419.

38. Peter D'Eath, personal communication with author (via email) for this quotation and others that immediately follow.

39. Andrew Farley, personal communication with author (via email).

40. Christophe Galfard, personal communication with author; "The Hawking Paradox"; Mialet, *Hawking Incorporated*, 93.

41. "The Hawking Paradox," Season 42, Episode 2.

42. Pallab Ghosh, "The Day I Thought We'd Unplugged Stephen Hawking," BBC, March 14, 2018, www.bbc.com/news/science-environment-43400021.

43. Neil Turok, personal communication with author.

CHAPTER 7: INFORMATION

1. "A Brief History of Physicist Hawking's Plans to Remarry," *Chicago Tribune*, July 6, 1995, 2.

2. Jane Hawking, *Music to Move the Stars: A Life with Stephen* (London: Pan, 2000), 562, 580; Ray Laflamme, personal communication with author.

3. Jane Hawking, *Music to Move the Stars*, 585; Matrimonial Causes Act of 1973, 28(3), rev. February 1, 1991.

4. Matrimonial Causes Act of 1973, 1(2), rev. February 1, 1991.

5. "Hawking to Marry His Former Nurse," *The Times* (London), July 6, 1995; Jane Hawking, *Music to Move the Stars*, 585; "Hawking Gets Hitched," *The Gazette*, September 16, 1995. Leonard Mlodinow, in his *Stephen Hawking: A Memoir of Friendship and Physics* (New York: Pantheon Books, 2020), 214, states that Robert, the eldest child, was, in fact, in attendance, but that the other two were not.

6. Personal communications with author.

7. For more about information and information theory, see my *Decoding the Universe: How the New Science of Information Is Explaining Everything in the Cosmos, from Our Brains to Black Holes* (New York: Viking, 2006).

8. Leonard Susskind, *The Black Hole War: My Battle with Stephen Hawking to Make the World Safe for Quantum Mechanics* (Boston: Little, Brown, 2009), 184.

9. "1990s," Royal Albert Hall, www.royalalberthall.com/about-the-hall/our-history/explore-our-history/time-machine/1990s.

10. "Show Detail: Aspen Center for Physics Presents: Does God Throw Dice in Black Holes with Stephen Hawking, July 24, 1996," Video On-Demand, Cablecast, http://mc.grassrootstv.org/Cablecast/public/Show.aspx?ChannelID=1&ShowID=947. There might be slight differences between this version of the lecture and the one given several months earlier at Albert Hall.

11. "A Matter of Divine Right: Does God Play Dice with the Universe? Einstein Said No, but Stephen Hawking Says Yes. Peter Millar Reports," *South China Morning Post*, December 4, 1995, 20.

12. Stephen Hawking, *Black Holes and Baby Universes and Other Essays* (New York: Bantam Books, 1993), 49 ff.

13. Hawking, *Black Holes and Baby Universes*, 67.

14. Daniel Z. Freedman, personal communication with author.

15. Sara Lippincott, "Interview with John H. Schwarz," 2002, Caltech Oral Histories, Caltech Institute Archives, https://resolver.caltech.edu/CaltechOH:OH_Schwarz_J, 44.

16. Stephen Hawking, typescript [photocopy] of a draft of *A Brief History of Time* with typed letter signed from Hawking to Mr. S. [Stephen] Zatman, March 24, 1986, Washington University, St. Louis, ID no. MSS/VMF/134, p. 94/101; Stephen Hawking, *A Brief History of Time* (New York: Bantam Books, 1988), 165.

17. Marika Taylor, personal communication with author.

18. Laflamme, personal communication with author.

19. Susskind, *Black Hole War*, 391.

20. Andy Strominger, personal communication with author.

21. Susskind, *Black Hole War*, 394.

22. Taylor, personal communication with author for this quotation and others that immediately follow.

23. Valerie Grove, "The Million-Dollar Cinderella," *The Times* (London), February 7, 1997, 16.

24. Daya Alberge, "Tolkien Wins Title Lord of the Books by Popular Acclaim," *The Times* (London), January 20, 1997.

25. Robert Crampton, "Intelligence Test," *The Times* (London), April 8, 1995.

26. Al Zuckerman, personal communication with author. The elision excises Zuckerman's mistaken reference to *Universe in a Nutshell* (also packaged by Philip Dunn) rather than *The Illustrated Brief History of Time*, which was the subject being discussed.

27. "The Big Bang," *Stephen Hawking's Universe*, Episode 6, aired on PBS in 1997.

28. "US Robotics X2 Commercial (Stephen Hawking)," YouTube, posted May 18, 2012, by vector108, www.youtube.com/watch?v=-HAUiXRuH_I; "Physicist Hawking Getting a Big Bang Out of Internet," *Chicago Tribune* (Associated Press), March 23, 1997, 19.

29. Paul Rodgers, "$6m Hawking Professorship Is 'Too Generous,' Say Critics," *Forbes*, February 5, 2014, www.forbes.com/sites/paulrodgers/2014/02/05/us-philanthropists-6m-for-stephen-hawking-professorship-is-too-generous-say-cambridge-critics; Neil Turok, personal communication with author.

30. Michael D. Lemonick, "Hawking: Is He All He's Cracked Up to Be?," *Time*, February 3, 2014, https://time.com/3531/hawking-myth-or-legend.

31. Peter Kingston, "Curtain Up on the Cosmos," *The Guardian*, November 13, 1997; *Falling Through a Hole in the Air*, written by Judith Goldhaber and Carlton Reese Pennypacker, directed by David Parr, 2007, recording provided by Carlton Reese Pennypacker.

32. Hélène Mialet, *Hawking Incorporated: Stephen Hawking and the Anthropology of the Knowing Subject* (Chicago: University of Chicago Press, 2012), 58.

33. Mialet, *Hawking Incorporated*, 61.

34. Wager posted at Preskill's Caltech website. See www.theory.caltech.edu/people/preskill/old_naked_bet.html.

35. John Preskill, personal communication with author.

36. Kip Thorne, "Warping Spacetime," in *The Future of Theoretical Physics and Cosmology: Celebrating Stephen Hawking's 60th Birthday*, ed. G. W. Gibbons, S. J. Rankin, and E. P. S. Shellard (Cambridge: Cambridge University Press, 2003), 24.

37. Preskill, personal communication with author.

38. Wager posted at Preskill's Caltech website. See www.theory.caltech.edu/people/preskill/info_bet.html.

39. Susskind, *Black Hole War*, 419.

40. Strominger, personal communication with author.

41. Joe Friesen, "Hawking Proves a Good Sport When It Comes to Settling Bets," *Globe and Mail*, July 22, 2004.

42. Marika Taylor, personal communication with author.

CHAPTER 8: IMAGES

1. Emma Wilkins, "Wheelchair Physicist in Love Tangle with Nurse," *Daily Mail*, August 1, 1990.

2. "People," United Press International, July 29, 1990, BC cycle; Wilkins, "Wheelchair Physicist."

3. Jane Hawking, *Travelling to Infinity: My Life with Stephen Hawking* (Richmond, UK: Alma Books, 2014), 473.

4. Wilkins, "Wheelchair Physicist."

5. Peter Guzzardi, personal communication with author.

6. Jane Hawking, *Music to Move the Stars: A Life with Stephen* (London: Pan, 2000), 444.

7. Hawking, *Music to Move the Stars*.

8. Bob Sipchen, "Simply Human: Profile. Wheelchair-Bound Physicist Stephen Hawking Resists Efforts to Deify His Life or His Disabilities," *Los Angeles Times*, June 6, 1990, E1.

9. Stephen Hawking, *Black Holes and Baby Universes and Other Essays* (New York: Bantam Books, 1993), 32; Stephen Hawking, *My Brief History* (London: Bantam Books, 2018), 122; Hélène Mialet, *Hawking Incorporated: Stephen Hawking and the Anthropology of the Knowing Subject* (Chicago: University of Chicago Press, 2012), 71.

10. Leslie Hanscom, "I Have No Enemies," *Sydney Morning Herald*, July 2, 1988.

11. Kip S. Thorne, *Black Holes and Time Warps: Einstein's Outrageous Legacy* (New York: Norton, 1994), 413.

12. Michael S. Morris and Kip S. Thorne, "Wormholes in Spacetime and Their Use in Interstellar Travel: A Tool for Teaching General Relativity," *American Journal of Physics* 56, no. 5 (May 1988): 397.

13. "Citizen Science," 365 Days of Astronomy, December 27, 2012, https://cosmoquest.org/x/365daysofastronomy.

14. Morris and Thorne, "Wormholes in Spacetime and Their Use," 407.

15. S. W. Hawking, "Chronology Protection Conjecture," *Physical Review D* 46, no. 2 (July 15, 1987): 603.

16. Hawking, "Chronology Protection Conjecture.".

17. Hawking, "Chronology Protection Conjecture."

18. Hawking, "Chronology Protection Conjecture."

19. Thorne, *Black Holes and Time Warps*, 521.

20. Sue Lawley, "Desert Island Discs: Stephen Hawking," BBC Radio 4, December 27, 1992, www.bbc.co.uk/programmes/p0093xb2.

21. John Preskill, personal communication with author.

22. Kitty Ferguson and S. W. Hawking, *Stephen Hawking: An Unfettered Mind* (New York: St. Martin's Press, 2017), 168; Judy Bachrach, "A Beautiful Mind, an Ugly Possibility," *Vanity Fair*, January 25, 2004, www.vanityfair.com/news/2004/06/hawking200406.

23. Morgan Strong, "Playboy Interview: Stephen Hawking," *Playboy*, April 1990, 74.

24. Strong, "Playboy Interview," 68, 64.

25. Bryan Appleyard, "A Master of the Universe," *Sunday Times* (London), June 19, 1988, 30; Jane Hawking, *Music to Move the Stars*, 592.

26. Lawley, "Desert Island Discs."

27. "1993 British Telecom—'Hawking' TV Commercial," YouTube, posted March 14, 2018, by History of Advertising Trust, www.youtube.com/watch?v=GH5Q54eIaPk.

28. Sean Michaels, "Stephen Hawking Sampled on Pink Floyd's The Endless River," *The Guardian*, October 8, 2014, www.theguardian.com/music/2014/oct/08/stephen-hawking-sampled-pink -floyd-the-endless-river; "Radio Interview on Opening Night of 1994 Tour: The Division Bell," Pink Floyd & Co., https://pfco.neptunepinkfloyd.co.uk/band/interviews/grp/grpredbeard.html.

29. Don N. Page and Stephen W. Hawking, "Gamma Rays from Primordial Black Holes," *Astrophysical Journal* 206 (May 15, 1976): 2.

30. Most sources give the date of this quotation as Hawking's 2016 Reith Lecture, but it dates back at least several years before that. S. W. Hawking, "Into a Black Hole," lecture ca. 2008, Internet Archive, Wayback Machine, https://web.archive.org/web/20120112012649/http://www .hawking.org.uk/into-a-black-hole.html.

31. S. W. Hawking and J. M. Stewart, "Naked and Thunderbolt Singularities in Black Hole Evaporation," arxiv:hep-th/9207105v1, July 30, 1992.

32. Raphael Bousso, personal communication with author.

33. Mialet, *Hawking Incorporated*, 55; Roger Penrose, "'Mind over Matter': Stephen Hawking—Obituary by Roger Penrose," *The Guardian*, March 14, 2018, www.theguardian.com /science/2018/mar/14/stephen-hawking-obituary.

34. R. Bousso and S. W. Hawking, "Black Holes in Inflation," *Nuclear Physics B* 57 (1997): 204.

35. Bousso, personal communication with author.

36. Hawking, "Into a Black Hole."

37. Leonard Susskind, "Lecture 8: Black Hole Formation, Penrose Diagrams and Wormholes. CosmoLearning Physics," CosmoLearning, posted January 11, 2015, https://cosmolearning.org/ video-lectures/black-hole-formation-penrose-diagrams-wormholes; Mialet, *Hawking Incorporated*, 68–70.

38. Mialet, *Hawking Incorporated*, 68–70.

39. Roger Penrose, personal communication with author.

40. G. H. Hardy, *A Mathematician's Apology* (Cambridge: Cambridge University Press, 1940), 6–7.

41. Andreas Albrecht, personal communication with author.

42. Robert Crampton, "Intelligence Test," *The Times* (London), April 8, 1995.

43. Graham Farmelo, *The Strangest Man: The Hidden Life of Paul Dirac, Mystic of the Atom* (New York: Basic Books, 2009), 288.

44. Hanscom, "I Have No Enemies."

45. Hanscom, "I Have No Enemies."

46. Hawking, *A Brief History of Time*, 116.

47. Hawking, *A Brief History of Time*, 174.

48. William Craig, "'What Place, Then, for a Creator?': Hawking on God and Creation," *British Journal for the Philosophy of Science* 41, no. 4 (December 1990): 483.

49. Hawking, *Black Holes and Baby Universes*, 41–42.

50. Jane Hawking, *Music to Move the Stars*, 536. The other journalist was Pauline Hunt, who wrote for the local Cambridge paper.

51. "Master of a Narrow Universe: Stephen Hawking Is on a Voyage to Stardom but Unable to Navigate in the Human Realm," *The Independent*, October 13, 1993, www.independent.co.uk

/voices/master-of-a-narrow-universe-stephen-hawking-is-on-a-voyage-to-stardom-but-unable
-to-navigate-in-the-1510340.html.

52. Robert Crampton, "Intelligence Test," *The Times* (London), April 8, 1995.

53. Martyn Harris, "A Brief History of Hawking," Spectator Archive, June 27, 1992, http://
archive.spectator.co.uk/article/27th-june-1992/18/a-brief-history-of-hawking.

CHAPTER 9: FLASH

1. Ray Laflamme, personal communication with author.

2. Peter Guzzardi, personal communication with author.

3. Stephen Hawking, typescript [photocopy] of a draft of *A Brief History of Time* with typed
letter signed from Hawking to Mr. S. [Stephen] Zatman, March 24, 1986, Washington University,
St. Louis, ID no. MSS/VMF/134 (also available at the University of Cambridge Library, Add. MS
9222, 95).

4. Stephen Hawking, "Is the End in Sight for Theoretical Physics?," in *Black Holes and Baby
Universes and Other Essays* (New York: Bantam Books, 1993), 65.

5. Al Zuckerman, personal communication with author; John Gribbin, personal communication with author.

6. Guzzardi, personal communication with author for this quotation and others that immediately follow.

7. Archana Masih, "The Stephen Hawking I Knew," *Rediff*, March 22, 2018, www.rediff.com
/news/special/the-stephen-hawking-i-knew/20180322.htm.

8. Hawking, *Brief History* typescript, 3.

9. Stephen Hawking, "Drafts of Stephen Hawking's A Brief History of Time," ca. 1983–ca.
1987, University of Cambridge Library, Add. MS. 9222, 171.

10. Stephen Hawking, *A Brief History of Time* (New York: Bantam Books, 1998), 1.

11. Hawking, *Brief History* typescript, 95.

12. Hawking, *A Brief History of Time*, 175.

13. Peter Guzzardi, personal communication with author; Stephen Hawking, "A Brief History
of A Brief History," *Popular Science*, August 1989, 70 ff.

14. Masih, "The Stephen Hawking I Knew."

15. Guzzardi, personal communication with author.

16. Don Page, "Hawking's Timely Story," *Nature* 332 (April 21, 1988): 742–743; Hawking,
A Brief History of Time, 73, 95.

17. Hawking, "A Brief History of a Brief History"; Elizabeth Mehren, "Reagan Administration Book Beat Picks Up," *Los Angeles Times*, April 3, 1988, www.latimes.com/archives/la-xpm
-1988-04-03-bk-1130-story.html; Michael White and John Gribbin, *Stephen Hawking: A Life in
Science* (New York: Dutton, 1992), 286.

18. Jane Hawking, *Music to Move the Stars: A Life with Stephen* (London: Pan, 2000), 486.

19. Andy Albrecht, personal communication with author.

20. White and Gribbin, *Stephen Hawking*, 326, 327.

21. Hawking, "Drafts of Stephen Hawking's A Brief History of Time," ca. 1983–ca. 1987, 282;
Stephen Hawking, *A Brief History of Time*, 131.

22. White and Gribbin, *Stephen Hawking*, 327.

23. Albrecht, personal communication with author; White and Gribbin, *Stephen Hawking*,
328; Jerry Adler, "Reading God's Mind," *Newsweek*, June 13, 1988.

24. Dennis Overbye, *Lonely Hearts of the Cosmos* (New York: Harper and Row, 1991), 253.

25. Stephen Hawking, "Inflation Reputation Reparation," *Physics Today* 42, no. 2 (February
1989): 16.

26. Albrecht, personal communication with author.

27. Stephen Hawking, "Why Does the Universe Inflate?," in *Quantum Mechanics of Fundamental Systems: The Quest for Beauty and Simplicity*, ed. Marc Henneaux and Jorge Zanelli (New

York: Springer Science + Business Media, 2009), 147; Daniel Z. Freedman, personal communication with author.

28. Adler, "Reading God's Mind."

29. "Suiting Science to a T (Shirt), Two Chicago Bar Owners Set Up a Stephen Hawking Fan Club," *People*, September 11, 1989, https://people.com/archive/suiting-science-to-a-t-shirt-two-chicago-bar-owners-set-up-a-stephen-hawking-fan-club-vol-32-no-11; Steven Pratt, "His Fan Club Suits Hawking to a T," *Chicago Tribune*, July 26, 1989, www.chicagotribune.com/news/ct-xpm-1989-07-26-8902200569-story.html.

30. "Suiting Science to a T."

31. Laflamme, personal communication with author.

32. Nigel Farndale, "A Brief History of the Future," *The Independent*, January 8, 2000, www.independent.ie/irish-news/a-brief-history-of-the-future-26125776.html.

33. Laflamme, personal communication with author; Jane Hawking, *Music to Move the Stars*, 501.

34. Charles Oulton, "Cosmic Writer Shames Book World," *Sunday Times* (London), August 28, 1988.

35. Hawking, "A Brief History of a Brief History," 70 ff.

36. David Blum, "The Tome Machine," *New York Magazine*, October 24, 1988, 40.

37. Hawking, "A Brief History of a Brief History."

38. Guzzardi, personal communication with author.

39. Simon Mitton, personal communication with author.

40. Morgan Strong, "Playboy Interview: Stephen Hawking," *Playboy*, April 1990, 64.

41. Hawking, *Black Holes and Baby Universes*, 1.

42. Hawking, *My Brief History*, 6.

43. Errol Morris, personal communication with author.

44. Morris, personal communication with author.

45. Morris, personal communication with author.

46. Morris, personal communication with author.

47. Morris, personal communication with author.

48. Raphael Bousso, personal communication with author.

49. Jane Hawking, *Music to Move the Stars*, 504, 505.

50. Hawking, *My Brief History*, 85; Jane Hawking, *Music to Move the Stars*, 339.

51. Jane Hawking, *Music to Move the Stars*, 455, 460.

52. Jane Hawking, *Music to Move the Stars*, 487, 538.

53. Jane Hawking, *Music to Move the Stars*, 540.

54. Jane Hawking, *Music to Move the Stars*, 548.

55. Jane Hawking, *Music to Move the Stars*, 546.

56. Jane Hawking, *Music to Move the Stars*, 553.

57. Richard Jerome, "Of a Mind to Marry," *People*, August 7, 1995, https://people.com/archive/of-a-mind-to-marry-vol-44-no-6.

58. Jane Hawking, *Music to Move the Stars*, 557.

59. Bryan Appleyard, "Stephen Hawking Changed My Life," *Sunday Times* (London), December 14, 2014.

60. Pliny, *Natural History*, Book 33, Chapter 4; Tertullian, *Apology*, 33.3.

61. Bryan Appleyard, "A Master of the Universe," *Sunday Times* (London), June 19, 1988, 30.

CHAPTER 10: IGNITION

1. Stephen Hawking, *A Brief History of Time* (New York: Bantam Books, 1998), 116.

2. S. W. Hawking, "The Boundary Conditions of the Universe," in *Astrophysical Cosmology: Proceedings of the Study Week on Cosmology and Fundamental Physics, September 28–October*

2, 1981, vol. 48, Pontificiae Academiae Scientiarum Scripta Varia, ed. H. A. Brück, George V. Coyne, and Malcolm S. Longair (Vatican City: Pontificia Academia Scientiarum, 1982), 564.

3. Neil Turok, personal communication with author.

4. James Gleick, *Genius: The Life and Science of Richard Feynman* (New York: Pantheon, 1992), 245.

5. Gleick, *Genius*, 250.

6. Turok, personal communication with author.

7. James Hartle, personal communication with author.

8. James Hartle, personal communication with author.

9. Stephen Hawking, *Black Holes and Baby Universes and Other Essays* (New York: Bantam Books, 1993), 81.

10. Neil Turok, personal communication with author.

11. Hawking, *Black Holes and Baby Universes*, 94.

12. Neil Turok, personal communication with author.

13. Stephen Hawking, *My Brief History* (London: Bantam Books, 2018).

14. Malcolm W. Browne, "Does Sickness Have Its Virtues?," *New York Times*, March 10, 1981, www.nytimes.com/1981/03/10/science/does-sickness-have-its-virtues.html.

15. "Michael Harwood, "The Universe and Dr. Hawking," *The New York Times Magazine*, January 23, 1983, 1, 16. "Mild cetacean smile" from Timothy Ferris, "Mind over Matter," *Vanity Fair*, June 1984, 58.

16. "Professor Hawking's Universe," *Horizon*, Season 20, Episode 4, directed by Fisher Dilke, BBC, aired October 17, 1983.

17. Stephen Hawking, "A Brief History of A Brief History," *Popular Science*, August 1989, 70.

18. Hawking, "A Brief History of A Brief History."

19. Stephen Hawking, *Black Holes and Baby Universes*, 50.

20. Simon Mitton, personal communication with author.

21. As many of Hawking's draft manuscripts of *Brief History* (including his earliest one) don't bear dates and the memories of the key players in the story don't always line up exactly, there is some ambiguity in the early timeline of events surrounding the book. It is unclear, for example, precisely when Hawking formally engaged Zuckerman, and whether it was before or after Hawking approached Mitton with the typescript. I consider the former to be more likely, which would mean that Hawking approached Mitton as a courtesy (with the side benefit of receiving some free feedback) after getting the draft manuscript together (likely spurred on by Zuckerman) by the fall of 1983. The timeline firms up a bit by the sale of the book to Bantam in early 1984 and then Guzzardi's first face-to-face meeting with Hawking in Chicago in mid-1984.

22. Mitton, personal communication with author.

23. Leon Jaroff, "Stephen Hawking: Roaming the Cosmos," *Time*, February 8, 1988, http://content.time.com/time/subscriber/article/0,33009,966650-1,00.html; Mitton, personal communication with author.

24. Al Zuckerman, personal communication with author.

25. Peter Guzzardi, personal communication with author.

26. "History of A Brief History of Time," *The Bookseller*, October 21, 1988, 1625–1627.

27. Guzzardi, personal communication with author. It's worth noting that Mitton distinctly remembers the "airport bookstores" idea coming up in his early conversation with Hawking about *Brief History*; however, that conversation would probably have taken place well before Guzzardi mentioned it to Hawking in the letter. In this instance, I tend to think that Guzzardi is the most likely person to have first planted the airport bookstore idea in Hawking's head. If so, barring a fairly striking coincidence, that would suggest Mitton might be incorporating the airport bookstore trope from a later discussion with Hawking into his memory of their initial conversation. Without external documents or events to pin these memories to, it's all but impossible to tell for certain.

28. Michael White and John Gribbin, *Stephen Hawking: A Life in Science* (New York: Dutton, 1992), 284; Mitton, personal communication with author.

29. Jane Hawking, *Music to Move the Stars: A Life with Stephen* (London: Pan, 2000), 403–404.

30. John Boslough, *Stephen Hawking's Universe* (London: HarperCollins, 1995), 128.

31. Ray Laflamme, personal communication with author.

32. Ray Laflamme, personal communication with author; Raphael Bousso, personal communication with author.

33. Ray Laflamme, personal communication with author.

34. Laflamme, personal communication with author.

35. Jane Hawking, *Music to Move the Stars*, 417.

36. Stephen Hawking, typescript [photocopy] of a draft of *A Brief History of Time* with typed letter signed from Hawking to Mr. S. [Stephen] Zatman, March 24, 1986, Washington University, St. Louis, ID no. MSS/VMF/134 (typescript also available at the University of Cambridge Library, Add. MS 9222, 81).

37. Laflamme, personal communication with author.

38. Jane Hawking, *Music to Move the Stars*, 426.

39. Laflamme, personal communication with author.

40. Jane Hawking, *Music to Move the Stars*, 429.

41. Jane Hawking, *Music to Move the Stars*, 435.

42. Fergus Walsh, "Hawking: 'I Support Assisted Dying,'" BBC News, July 16, 2014, www.bbc.com/news/health-28337443.

43. Jane Hawking, *Music to Move the Stars*, 443.

44. *Larry King Live*, December 25, 1999; Jane Hawking, *Music to Move the Stars*, 453.

45. "Walt Woltosz on the Passing of World-Famous Astrophysicist, Professor Sir Stephen Hawking," Simulations Plus, May 16, 2018, www.simulations-plus.com/resource/walt-woltosz-passing-world-famous-astrophysicist-professor-sir-stephen-hawking; Joao Medeiros, "How Intel Gave Stephen Hawking a Voice," *Wired*, January 13, 2015, www.wired.com/2015/01/intel-gave-stephen-hawking-voice.

46. Emma Wilkins, "Wheelchair Physicist in Love Tangle with Nurse," *Daily Mail*, August 1, 1990, 4.

47. Elaine Hawking, "Baptism Testimony of Elaine Hawking," Internet Archive, April 2018, https://archive.org/details/chipping-campden-baptist-church-513231/2018-04-01-Baptism-Testimony-Elaine-Hawking-Elaine-Hawking-46453311.mp3.

48. *Dara O Briain Meets Stephen Hawking*, BBC One, June 16, 2015.

49. *Dara O Briain Meets Stephen Hawking*.

50. Laflamme, personal communication with author.

51. Laflamme, personal communication with author.

52. Hawking, *A Brief History of Time*, 150.

53. Stephen Hawking, "Disability Advice," March 10, 2009, Internet Archive, Wayback Machine, https://web.archive.org/web/20090310162145/hawking.org.uk/index.php/disability/disabilityadvice; Errol Morris, personal communication with author; Kip Thorne, personal communication with author.

54. Jane Hawking, *Music to Move the Stars*, 549.

55. *Dara O Briain Meets Stephen Hawking*.

56. Hawking, "A Brief History of A Brief History," 70 ff.

CHAPTER 11: INFLATION

1. Jane Hawking, *Music to Move the Stars: A Life with Stephen* (London: Pan, 2000), 331–332.

2. Jane Hawking, *Music to Move the Stars*, 332.

3. Jane Hawking, *Music to Move the Stars*, 326, 332; *Dara O Briain Meets Stephen Hawking*, BBC One, June 16, 2015.

4. Jane Hawking, *Music to Move the Stars*, 338.

5. Jane Hawking, *Music to Move the Stars*, 338–339.

6. *The Key to the Universe*, directed by John Gorman, presented by Nigel Calder, BBC Two, aired January 30, 1977.

7. Nigel Calder, "What Did Hawking Discover?," Calder's Updates, April 6, 2010, https://calderup.wordpress.com/tag/the-key-to-the-universe.

8. *Key to the Universe*.

9. S. W. Hawking, "The Quantum Mechanics of Black Holes," *Scientific American* 236, no. 1 (January 1977): 41.

10. S. W. Hawking, "Breakdown of Predictability in Gravitational Collapse," *Physical Review D* 14, no. 10 (November 15, 1976): 2464, https://doi.org/10.1103/PhysRevD.14.2460.

11. Stephen Hawking, "Does God Play Dice," 1999, Stephen Hawking website, www.hawking.org.uk/in-words/lectures/does-god-play-dice.

12. Andrei Linde, personal communication with author.

13. Linde, personal communication with author.

14. Kip S. Thorne, *Black Holes and Time Warps: Einstein's Outrageous Legacy* (New York: Norton, 1994), 278.

15. Linde, personal communication with author.

16. Linde, personal communication with author.

17. Jane Hawking, *Music to Move the Stars*, 349.

18. Jane Hawking, *Music to Move the Stars*, 349.

19. Jane Hawking, *Music to Move the Stars*, 361.

20. Jane Hawking, *Music to Move the Stars*, 109, 141, 110.

21. Original as quoted in Jane Hawking, "The Dawn—A Study of the Traditional Love Lyric of Medieval Spain and Portugal" (PhD diss., Westfield College, University of London, 1979), 63, translation by the author, based upon various translations available at Jarchas, www.jarchas.net/jarcha-23.html.

22. Jane Hawking, *Music to Move the Stars*, 122.

23. Jane Hawking, *Music to Move the Stars*, 291, 412.

24. Letter from Stephen and Jane Hawking, October 1981, John Archibald Wheeler Papers, American Philosophical Society Library, Philadelphia; Jane Hawking, *Music to Move the Stars*, 402, 480.

25. Stephen Hawking, *A Brief History of Time* (New York: Bantam Books, 1998), 131.

26. John Preskill, personal communication with author.

27. Dennis Overbye, *Lonely Hearts of the Cosmo* (New York: Harper and Row, 1991), 254; Alan H. Guth, *The Inflationary Universe: The Quest for a New Theory of Cosmic Origins* (Reading, MA: Addison-Wesley, 1997), 230.

28. John D. Barrow and Michael S. Turner, "The Inflationary Universe—Birth, Death and Transfiguration," *Nature* 298, no. 5877 (August 1982): 802.

29. Michael White and John Gribbin, *Stephen Hawking: A Life in Science* (New York: Dutton, 1992), 326.

30. Stephen Hawking, *Black Holes and Baby Universes and Other Essays* (New York: Bantam Books, 1993), 50.

31. *A Brief History of Time*, directed by Errol Morris, 1991.

32. Timothy Ferris, *The Whole Shebang: A State of the Universe Report* (New York: Touchstone), 93.

CHAPTER 12: BLACK SWAN

1. Jane Hawking, *Music to Move the Stars: A Life with Stephen* (London: Pan, 2000), 250.

2. Jane Hawking, *Music to Move the Stars*, 262.

3. Murray Gell-Mann, "Dick Feynman—The Guy in the Office Down the Hall," *Physics Today* 42, no. 2 (February 1989): 54, https://doi.org/10.1063/1.881192.

4. Jack Birkinshaw, "Bottomless Helps Nobel Physicist with Figures," *Los Angeles Times*, November 8, 1969, sec. 2, 9.

5. "Richard P. Feynman: Al Seckel (Some Amazing Stories from a Close Friend). Personal Observations on the Reliability of the Shuttle, by R. P. Feynman. The Mysterious Number 137," Internet Archive, Wayback Machine, https://web.archive.org/web/20070219131111/http://www .brew-wood.co.uk/physics/feynman.htm.

6. George Johnson, "The Jaguar and the Fox," *The Atlantic*, July 2000, www.theatlantic.com /magazine/archive/2000/07/the-jaguar-and-the-fox/378264.

7. Letter from Bob Geroch to John Archibald Wheeler, John Archibald Wheeler Papers, American Philosophical Society Library, Philadelphia, Box 10.

8. "Richard Feynman: Al Seckel."

9. Michael White and John Gribbin, *Stephen Hawking: A Life in Science* (New York: Dutton, 1992), 147.

10. Neil Turok, personal communication with author.

11. Jane Hawking, *Music to Move the Stars*, 249, 265.

12. Bernard Carr, personal communication with author (email); Bernard Carr, "Stephen Hawking: Recollections of a Singular Friend," *Paradigm Explorer*, no. 127 (2018): 11.

13. Jane Hawking, *Music to Move the Stars*, 254.

14. Jane Hawking, *Music to Move the Stars*, 253.

15. Jane Hawking, *Music to Move the Stars*, 265, 265–266.

16. Jane Hawking, *Travelling to Infinity: My Life with Stephen Hawking* (Richmond, UK: Alma Books, 2014), 234.

17. Stephen Hawking, *A Brief History of Time* (New York: Bantam Books, 1998), 93–94.

18. Kip Thorne, personal communication with author.

19. An image of the handwritten wager appears at the University of Southampton personal webpage of Dr. Ian Jones, www.personal.soton.ac.uk/dij/GR-Explorer/bh/Hawking_Thorne _wager.jpg; Thorne quote from personal communication with author.

20. Kip S. Thorne, *Black Holes and Time Warps: Einstein's Outrageous Legacy* (New York: Norton, 1994), 315.

21. Jane Hawking, *Music to Move the Stars*, 281.

22. A. A. Jackson and Michael P. Ryan, "Was the Tungus Event Due to a Black Hole?," *Nature* 245, no. 5420 (September 1973): 88–89, https://doi.org/10.1038/245088a0; Walter Sullivan, "Curiouser and Curiouser," *New York Times*, July 14, 1974.

23. Malcolm W. Browne, "Clues to Origin of Universe Expected," *New York Times*, March 12, 1978, 1.

24. Letter from S. Coleman to R. Feynman, July 26, 1976, Richard Feynman Papers, Caltech Archives, Pasadena, California.

25. Leonard Susskind, personal communication with author.

26. Susskind, personal communication with author.

27. Robert Penrose, "The Grand Design," *Financial Times*, September 3, 2010.

28. Jane Hawking, *Music to Move the Stars*, 298, 299; Roger Dobson, "An Exceptional Man," *British Medical Journal* 324, no. 7352 (June 22, 2002): 1478.

29. Jane Hawking, *Music to Move the Stars*, 376, 306.

30. Jane Hawking, *Music to Move the Stars*, 309; "Jane Hawking: Cry to Dream Again," *The Ryan Tubridy Show*, aired July 10, 2018, RTE Radio One, available at www.rte.ie/radio1/ryan -tubridy/programmes/2018/0710/977619-the-ryan-tubridy-show-tuesday-10-july-2018 (the audio is at www.rte.ie/radio/radioplayer/html5/#/radio1/21396421).

31. Jane Hawking, *Music to Move the Stars*, 299, 373.

32. Jane Hawking, *Music to Move the Stars*, 377.

33. Jane Hawking, *Music to Move the Stars*, 298.

34. Stephen Hawking, *Black Holes and Baby Universes and Other Essays* (New York: Bantam Books, 1993), 121.

35. James Gleick, *Genius: The Life and Science of Richard Feynman* (New York: Pantheon, 1992), 296.

36. Letter from Lewis H. Strauss to John Archibald Wheeler, August 2, 1977, Wheeler Papers.

37. Letter from John Archibald Wheeler to Lewis H. Strauss, August 1977, Wheeler Papers; letter from Richard Feynman to Lewis Strauss, August 9, 1977, Feynman Papers.

38. "Einstein Award," press release, January 29, 1978, Wheeler Papers.

39. Kevin C. Knox and Richard Noakes, *From Newton to Hawking: A History of Cambridge University's Lucasian Professors of Mathematics* (New York: Cambridge University Press, 2006), 468.

40. Hélène Mialet, *Hawking Incorporated: Stephen Hawking and the Anthropology of the Knowing Subject* (Chicago: University of Chicago Press, 2012), 124.

CHAPTER 13: BLACK BODY

1. Stuart Clark, "A Brief History of Stephen Hawking: A Legacy of Paradox," *New Scientist*, March 14, 2018, www.newscientist.com/article/2053929-a-brief-history-of-stephen-hawking-a -legacy-of-paradox; Stephen Hawking, *A Brief History of Time* (New York: Bantam Books, 1998), 108.

2. Hawking, *A Brief History of Time*, 99–100.

3. Hawking, *A Brief History of Time*, 102.

4. Roger Penrose, personal communication with author for this quotation and others that immediately follow.

5. Interview with John Wheeler, Part 86, "Entropy of a Black Hole: Bekenstein, Stephen Hawking," available at Web of Stories, www.webofstories.com/people/john.wheeler/86.

6. Wheeler, "Entropy of a Black Hole."

7. Jane Hawking, *Music to Move the Stars: A Life with Stephen* (London: Pan, 2000), 175.

8. *A Brief History of Time*, directed by Errol Morris, 1991; Penrose, personal communication with author.

9. Robert Babson, "Gravity Research Foundation," Gravity Research Foundation, www .gravityresearchfoundation.org/historic; Michael White and John Gribbin, *Stephen Hawking: A Life in Science* (New York: Dutton, 1992), 177.

10. J. D. Bekenstein, "Black Holes and the Second Law," *Lettere al Nuovo Cimento* 4, no. 15 (August 12, 1972): 737–740.

11. Wheeler, "Entropy of a Black Hole."

12. Kip S. Thorne, *Black Holes and Time Warps: Einstein's Outrageous Legacy* (New York: Norton, 1994), 425.

13. Jacob D. Bekenstein, "Black-Hole Thermodynamics," *Physics Today* 33, no. 1 (January 1980): 28, https://doi.org/10.1063/1.2913906; Jacob D. Bekenstein, "Black Holes and Entropy," *Physical Review D* 7, no. 8 (April 15, 1973): 2333–2346.

14. Letter from Jacob Bekenstein to John Archibald Wheeler, October 13, 1972, John Archibald Wheeler Papers, American Philosophical Society Library, Philadelphia; letter from Bekenstein to Wheeler, January 23 1973, Wheeler Papers; letter from Wheeler to Bekenstein, October 2, 1972, Wheeler Papers.

15. J. M. Bardeen, B. Carter, and S. W. Hawking, "The Four Laws of Black Hole Mechanics," *Communications in Mathematical Physics* 31 (1973): 168.

16. Thorne, *Black Holes and Time Warps*, 434; Jane Hawking, *Music to Move the Stars*, 209, 207–208.

17. Kip Thorne, personal communication with author; Thorne, *Black Holes and Time Warps*, 433; Timothy Ferris, "Mind over Matter," *Vanity Fair*, June 1984, 59.

18. Jane Hawking, *Music to Move the Stars*, 219, 217.

19. Hawking, *A Brief History of Time*, 105.

20. Hawking, *A Brief History of Time*, 105.

21. Hawking, *A Brief History of Time*, 105.

22. *A Brief History of Time*, directed by Errol Morris, 1991.

23. Letter from Dennis Sciama to John Archibald Wheeler, January 8, 1974, Wheeler Papers.

24. *A Brief History of Time*, directed by Errol Morris, 1991.

25. Dennis Overbye, *Lonely Hearts of the Cosmos* (New York: Harper and Row, 1991), 112.

26. Ferris, "Mind over Matter," 102.

27. Penrose, personal communication with author; *A Brief History of Time*, directed by Errol Morris, 1991.

28. John Boslough, *Stephen Hawking's Universe* (London: HarperCollins, 1995), 70. A slightly different account can be found in Jane Hawking, *Music to Move the Stars*, 238–239, in which Taylor reportedly says: "Well, this is quite preposterous! I have never heard anything like it. I have no alternative but to bring this session to an immediate close!" I find Boslough's account somewhat more credible.

29. Overbye, *Lonely Hearts*, 113.

30. Kitty Ferguson and S. W Hawking, *Stephen Hawking: An Unfettered Mind* (New York: St. Martin's Press, 2017), 73–74.

31. Jacob D. Bekenstein, "Black-Hole Thermodynamics," *Physics Today* 33, no. 1 (January 1980): 28, https://doi.org/10.1063/1.2913906; JTA, "Israeli Academic Inspired One of Stephen Hawking's Biggest Discoveries," *Times of Israel*, March 14, 2018, www.timesofisrael.com/israeli-academic-inspired-one-of-stephen-hawkings-biggest-discoveries; letter from John Archibald Wheeler to Moshe Carmeli, October 1, 1975, Wheeler Papers, Box 4.

32. Wheeler to Carmeli, October 1, 1975.

33. Jane Hawking, *Music to Move the Stars*, 240.

34. Penrose, personal communication with author; letter from Roger Penrose to John Archibald Wheeler, November 14, 1973, Wheeler Papers, Box 11; letter from John Archibald Wheeler to Thomas Cowling, January 11, 1974, Wheeler Papers, Box 11.

35. Penrose, personal communication with author.

36. "City Man Youngest Ever RS Fellow," *Cambridge Evening News*, March 22, 1974.

CHAPTER 14: BLACK HOLE

1. Stephen Hawking, *My Brief History* (London: Bantam Books, 2018), 51.

2. Hawking, *My Brief History*, 51; Jane Hawking, *Music to Move the Stars: A Life with Stephen* (London: Pan, 2000), 76–77.

3. Roger Penrose, personal communication with author.

4. Hawking, *My Brief History*, 63; Penrose, personal communication with author; *A Brief History of Time*, directed by Errol Morris, 1991. There are accounts that put Hawking at the lecture and describe the moment of his realization on the train back, such as Michael White and John Gribbin, *Stephen Hawking : A Life in Science* (New York: Dutton, 1992), 84. However, Hawking's first-person account, as well as Penrose's—he remembers the incident distinctly—are much more likely to be true.

5. *A Brief History of Time*, directed by Errol Morris, 1991.

6. Penrose, personal communication with author.

7. Hawking, *My Brief History*, 65.

8. F. J. Tipler, C. J. S. Clarke, and G. F. R. Ellis, "Singularities and Horizons—A Review Article," in *General Relativity and Gravitation*, ed. A. Held (New York: Plenum Press, 1980), 134; Penrose, personal communication with author.

9. White and Gribbin, *Stephen Hawking*, 85.

10. Jane Hawking, *Music to Move the Stars*, 86–87.

11. Jane Hawking, *Music to Move the Stars*, 80.

12. S. Hawkins [*sic*] and G. F. R. Ellis, "Singularities in Homogeneous World Models," *Physics Letters* 17, no. 3 (July 15, 1965): 246–247, https://doi.org/10.1016/0031-9163(65)90510-X.

13. Stephen Hawking, *Black Holes and Baby Universes and Other Essays* (New York: Bantam Books, 1993), 10.

14. Letter from John Archibald Wheeler to Walker Bleakney, January 24, 1967, John Archibald Wheeler Papers, American Philosophical Society Library, Philadelphia, Box 4.

15. Letter from Robert Geroch to John Archibald Wheeler, February 8, 1967, Wheeler Papers, Box 4.

16. Letter from Stephen Hawking to John Archibald Wheeler, June 22, 1966, Wheeler Papers, Box 4.

17. Jane Hawking, *Music to Move the Stars*, 114.

18. Geroch to Wheeler, February 8, 1967.

19. Letter from Roger Penrose to John Archibald Wheeler, February 26, 1967, Wheeler Papers, Box 11.

20. Penrose, personal communication with author.

21. Letter from Roger Penrose to John Archibald Wheeler, March 28, 1967, Wheeler Papers, Box 11.

22. Penrose, personal communication with author; letter from John Archibald Wheeler to R. H. Dalitz, March 19, 1970, Wheeler Papers, Box 25.

23. Letter from John Archibald Wheeler to G. K. Batchelor, December 10, 1970, Wheeler Papers, Box 6. Though this was written in the context of a recommendation for Carter, it is consistent with what Wheeler wrote in other contexts at the time.

24. Jane Hawking, *Music to Move the Stars*, 111.

25. Jane Hawking, *Music to Move the Stars*, 111, 479.

26. Hawking, *My Brief History*, 54; Jane Hawking, *Music to Move the Stars*, 140.

27. Interview with John Wheeler, Part 83, "1967: Naming the Black Hole," available at Web of Stories, www.webofstories.com/people/john.wheeler/83. There's an earlier claim—a journalist used the term in 1964. But most scientists, Hawking included, give Wheeler the credit for coining the term in 1967.

28. Hawking, *Black Holes and Baby Universes*, 116–117.

29. Jane Hawking, *Music to Move the Stars*, 121, 167; Bernard Carr, "Stephen Hawking: Recollections of a Singular Friend," *Paradigm Explorer*, no. 127 (2018): 10.

30. Unsigned letter to Edmund Leach, June 5, 1972, St. John's College Library (the letter is possibly from Fred Hoyle); Simon Mitton, personal communication with author.

31. Unsigned Letter to Leach, June 5, 1972.

32. Hawking, *My Brief History*, 56–57.

33. Hawking, *My Brief History*, 58.

34. Jane Hawking, *Music to Move the Stars*, 162.

35. Jane Hawking, *Music to Move the Stars*, 164, 162–165.

36. Jane Hawking, *Music to Move the Stars*, 165.

CHAPTER 15: SINGULARITY

1. John Boslough, *Stephen Hawking's Universe* (London: HarperCollins, 1995), 22, 23.

2. Stephen Hawking, *Black Holes and Baby Universes and Other Essays* (New York: Bantam Books, 1993), 14.

3. *A Brief History of Time*, directed by Errol Morris, 1991.

4. Boslough, *Stephen Hawking's Universe*, 23.

5. *A Brief History of Time*, directed by Errol Morris, 1991.

6. Simon Mitton, *Fred Hoyle: A Life in Science* (Cambridge: Cambridge University Press, 2011), xi.

7. *A Brief History of Time*, directed by Errol Morris, 1991.

8. *A Brief History of Time*, directed by Errol Morris, 1991.

9. Jane Hawking, *Music to Move the Stars: A Life with Stephen* (London: Pan, 2000), 18, 25.

10. Jane Hawking, *Music to Move the Stars*, 22.

11. Jane Hawking, *Music to Move the Stars*, 23; Stephen Hawking, *My Brief History* (London: Bantam Books, 2018), 47.

12. Stephen Hawking, "The Theory of the NHS," *Journal of the Royal Society of Medicine* 110, no. 12 (December 1, 2017): 469, https://doi.org/10.1177/0141076817745764; *A Brief History of Time*, directed by Errol Morris, 1991.

13. Dennis Overbye, "Wizard of Space and Time," *Omni*, February 1979, 104; Boslough, *Stephen Hawking's Universe*, 25; Morgan Strong, "Playboy Interview: Stephen Hawking," *Playboy*, April 1990, 68.

14. Michael Harwood, "The Universe and Dr. Hawking," *The New York Times Magazine*, January 23, 1983.

15. Boslough, *Stephen Hawking's Universe*, 25; Michael White and John Gribbin, *Stephen Hawking: A Life in Science* (New York: Dutton, 1992), 74.

16. Jane Hawking, *Music to Move the Stars*, 33, 29.

17. Jane Hawking, *Music to Move the Stars*, 48, 49.

18. Adrian Melott, "Dennis Sciama and *The Theory of Everything*," *APS News*, February 2015, www.aps.org/publications/apsnews/201502/letters.cfm.

19. Harwood, "The Universe and Dr. Hawking"; Hawking, *Black Holes and Baby Universes*, 15.

20. Hawking, *My Brief History*, 46.

21. White and Gribbin, *Stephen Hawking*, 79.

22. Hawking, *My Brief History*, 46.

23. *Hawking*, directed by Stephen Finnigan, 2013; Harwood, "The Universe and Dr. Hawking."

24. Jane Hawking, *Music to Move the Stars*, 51.

25. Celia Walden, "Jane Hawking: 'Living with Stephen Made Me Suicidal but I Still Love Him,'" *The Telegraph*, May 16, 2015, www.telegraph.co.uk/news/features/11609068/Jane-Hawking-Living-with-Stephen-made-me-suicidal-but-I-still-love-him.html.

26. Jane Hawking, *Music to Move the Stars*, 58, 59.

27. Jane Hawking, *Music to Move the Stars*, 65; Hawking, *My Brief History*, 50.

28. Hawking, *My Brief History*, 50.

CHAPTER 16: YLEM

1. Stephen Hawking, *My Brief History* (London: Bantam Books, 2018), 34–35 (photo).

2. Hawking, *My Brief History*, 32; *A Brief History of Time*, directed by Errol Morris, 1991.

3. Michael White and John Gribbin, *Stephen Hawking: A Life in Science* (New York: Dutton, 1992), 61–62.

4. Michael Church, "Games with the Cosmos," *The Independent*, June 6, 1988; *A Brief History of Time*, directed by Errol Morris, 1991.

5. Andrew Grant, "Stephen Hawking (1942–2018)," *Physics Today*, March 14, 2018, https://doi.org/10.1063/PT.6.4.20180314a; *A Brief History of Time*, directed by Errol Morris, 1991; Stephen Hawking and Gene Stone, *Stephen Hawking's A Brief History of Time: A Reader's Companion* (New York: Bantam Books, 1992), 42–43.

6. "Photo 3 of 3," in "Photos of Frank Hawking," uploaded April 1, 2008, Geni, www.geni.com/photo/view/5213819130880079421?album_type=photos_of_me&end=&photo_id=5593127504360016751&start=&tagged_profiles=.

7. Hawking, *My Brief History*, 27 (photo).

8. Hawking, *My Brief History*, 24.

9. White and Gribbin, *Stephen Hawking*, 49.

10. Hawking, *My Brief History*, 28 (photo).

11. White and Gribbin, *Stephen Hawking*, 22–23.

12. "Stephen Hawking on the Teacher That Changed His Life | #TeachersMatter," YouTube, posted March 8, 2016, by Talking Education, www.youtube.com/watch?v=2srGloZ673A&feature=youtu.be; Hawking, *My Brief History*, 27–28.

13. Hawking, *My Brief History*, 30–31.

14. "Filariasis," *Scientific American* 199, no. 1 (1958): 94–101.

15. Hawking, *My Brief History*, 26.

16. Justin D. Arnold and Saurabh Singh, "How Frank Hawking, DM, and Table Salt, Helped Eliminate Lymphatic Filariasis from China," *JAMA Dermatology* 153, no. 8 (2017): 780, https://doi.org/10.1001/jamadermatol.2017.2494.

17. Hawking, *My Brief History*, 15 (photo).

18. Hawking and Stone, *Stephen Hawking's A Brief History of Time: A Reader's Companion*, 10.

19. Jane Hawking, *Music to Move the Stars: A Life with Stephen* (London: Pan, 2000), 20; Hawking, *My Brief History*, 7.

20. *A Brief History of Time*, directed by Errol Morris, 1991.

21. Hawking, *My Brief History*, 18.

22. Hawking, *My Brief History*, 19.

23. "Billy Graham Starts Crusade (1954)," YouTube, posted April 13, 2014, by British Pathé, www.youtube.com/watch?v=Fx_Tsc9IheM.

24. White and Gribbin, *Stephen Hawking*, 16–17.

25. Hawking and Stone, *Stephen Hawking's A Brief History of Time: A Reader's Companion*, 23.

26. *A Brief History of Time*, directed by Errol Morris, 1991.

27. John Boslough, *Stephen Hawking's Universe* (London: HarperCollins, 1995), 22.

28. *A Brief History of Time*, directed by Errol Morris, 1991.

29. *A Brief History of Time*, directed by Errol Morris, 1991.

30. Hawking, *My Brief History*, 24.

31. Hawking, *My Brief History*, 24.

32. Hawking and Stone, *Stephen Hawking's A Brief History of Time: A Reader's Companion*, 21.

33. Michael Church, "Games with the Cosmos," *The Independent*, June 6, 1988; Hawking, *My Brief History*, 13; *A Brief History of Time*, directed by Errol Morris, 1991.

34. *A Brief History of Time*, directed by Errol Morris, 1991.

35. Diario de Ibiza, "Stephen Hawking: Un Pequeño Turista en Ibiza," *Diario de Ibiza*, March 14, 2018, www.diariodeibiza.es/pitiuses-balears/2018/03/14/stephen-hawking-pequeno-turista-ibiza/975762.html.

36. Hawking and Stone, *Stephen Hawking's A Brief History of Time: A Reader's Companion*, 21.

37. Hawking, *My Brief History*, 20.

38. Hawking, *My Brief History*, 20.

39. Charles Marlow, "Stephen Hawking in Deia: A Very Brief History," Charles Marlow (blog), March 18, 2018, www.charlesmarlow.com/blog/2018/03/18/stephen-hawking-deia-brief-history.

40. Robert Graves, *Goodbye to All That* (London: Penguin, 2000), 220.

41. Marlow, "Stephen Hawking in Deia"; Hawking and Stone, *Stephen Hawking's A Brief History of Time: A Reader's Companion*, 22.

42. Hawking, *My Brief History*, 12 (photo).

43. Hawking, *My Brief History*, 12; Hawking and Stone, *Stephen Hawking's A Brief History of Time: A Reader's Companion*, 6.

44. Hawking, *My Brief History*, 14–16.

45. Hawking, *My Brief History*, 4.

CHAPTER 17: ON THE SHOULDERS OF GIANTS

1. Charlie Chaplin, *A Comedian Sees the World* (Columbia: University of Missouri Press, 2014), 60.

2. "Lights All Askew in the Heavens," *New York Times*, November 10, 1919, 17; Marshall Missner, "Why Einstein Became Famous in America," *Social Studies of Science* 15, no. 2 (1985): 276.

3. Philipp Frank, *Einstein: His Life and Times* (Cambridge, MA: Da Capo Press, 2002), 178–179.

4. Missner, "Why Einstein Became Famous in America," 282.

5. Frank, *Einstein*, 179.

6. Abraham Pais, *Einstein Lived Here* (Oxford: Oxford University Press, 1984), 138; Jürgen Neffe, *Einstein: A Biography* (New York: Farrar, Straus and Giroux, 2007), 373.

7. Kip Thorne, personal communication with author.

8. "Stephen Hawking's Wheelchair Sells for €340,000 at Auction of His Personal Items," TheJournal.ie, November 10, 2018, www.thejournal.ie/stephen-hawkings-wheelchair-and -thesis-fetch-more-than-expected-at-auction-4333052-Nov2018; Yvonne Nobis, Gordon Moore Library, personal communication with author (via email); Mark Chandler, "Stephen Hawking Official Biography Goes to John Murray," *The Bookseller*, September 27, 2019, www.thebook seller.com/news/stephen-hawking-official-biography-signed-john-murray-1088216.

9. Isaac Butler, "Errol Morris on His Movie—and Long Friendship—with Stephen Hawking," *Slate*, March 16, 2018, https://slate.com/culture/2018/03/errol-morris-on-stephen-hawking -and-his-movie-a-brief-history-of-time.html.

INDEX

SIGRID ESTRADA

CHARLES SEIFE is a professor of journalism at New York University's Arthur L. Carter Journalism Institute and has been writing about science and mathematics for nearly three decades. The author of numerous books, including the best-selling *Zero: The Biography of a Dangerous Idea*, he lives in New York.